STRATEGIC
GEOGRAPHY
AND THE
CHANGING
MIDDLE EAST

STRATEGIC GEOGRAPHY AND THE CHANGING MIDDLE EAST

GEOFFREY KEMP AND
ROBERT E. HARKAVY

CARNEGIE ENDOWMENT FOR INTERNATIONAL PEACE
IN COOPERATION WITH
BROOKINGS INSTITUTION PRESS

Strategic Geography and the Changing Middle East may be ordered from:
Brookings Institution Press
1775 Massachusetts Avenue N.W.
Washington, D.C. 20036
Tel: 1-800-275-1447
(202) 797-6258
Fax: (202) 797-6004

Library of Congress Cataloging-in-Publication data:
Kemp, Geoffrey.
Strategic geography and the changing Middle East / Geoffrey Kemp and Robert E. Harkavy.
p.   cm.
Includes bibliographical references and index.
ISBN 0-87003-022-1 (cloth: alk. paper)—ISBN 0-87003-023-X (pbk.: alk. paper)
1. Geopolitics—Middle East.   2. Military geography—Middle East.
I. Harkavy, Robert E.   II. Title.
DS63.1.K44   1997
355′.033056—dc21   97-12997
                                            CIP

9 8 7 6 5 4 3 2
The paper used in this publication meets the minimum requirements of the American National
Standard for Information Sciences—Permanence of Paper for Printed Library Materials: ANSI
Z39.48-1984

Typeset in Times Roman

Composition by Princeton Editorial Associates
Scottsdale, Arizona, and Roosevelt, New Jersey

Printed by R. R. Donnelley and Sons Co.
Harrisonburg, Virginia

Contents

Preface and Acknowledgments

THE CONCEPTUAL GENESIS of this book derives from the long-standing interests of both authors in the role of geography in international politics. With support from grants provided between 1991 and 1995 by the Ford Foundation, the United States Institute for Peace, and especially the Carnegie Corporation of New York, Geoffrey Kemp's Middle East Arms Control Project at the Carnegie Endowment was able to undertake a series of studies on the Middle East arms race, U.S. relations with Iran, and the Arab-Israeli peace process. Owing to the significant geopolitical changes brought about by the end of the cold war it was decided that a substantive study on the role of strategic geography in the Middle East would be a welcome addition to the analysis of regional conflict and discussion of the prospects for conflict resolution. Throughout this book's long production and preparation the Carnegie Endowment's president, Morton Abramowitz, has provided strong intellectual and personal support, as have several vice presidents of the Endowment, especially Larry Fabian, Michael O'Hare, and Paul Balaran.

The manuscript was read in its entirety by Eliot A. Cohen and W. Seth Carus. Both provided extremely detailed and constructive criticism, which has been incorporated into the final version. James Placke reviewed the section on energy and Michael Eisenstadt the chapters that deal primarily with military affairs. Shaul Cohen made useful suggestions on the early chapters dealing with geopolitics and geography and history in the Middle East. Many colleagues made valuable, substantive contributions to the study over the years, including Harold Bernsen, Patrick Clawson, Anthony Cordesman, Ahmed Fakhr, Shai Feldman, Stephen Jones, Andrew Marshall, William Martin, Julia Nanay, Ze'ev Schiff, Jasjit Singh, Janice Stein, K. Subrahmanyam, Abdoullah Toukan, and

vii

Michael Vlahos. Geoffrey Kemp also acknowledges the support and inspiration he received in the late 1970s when he first undertook work on strategic access questions with Harlan Ullman, Robert McFarlane, and Robert Gard at the National Defense University. At that time he was ably helped by John Maurer, then a graduate student at the Fletcher School of Law and Diplomacy. In addition he acknowledges the special contribution of his former Carnegie colleague Jeremy Pressman, who worked with him for five years on the Middle East Arms Control Project, as well as a number of Carnegie junior fellows, including Farah Godrej, Paula Hacopian, Assia Ivantcheva, Hala Tayyarah, and Elaine Weiss. Chris Bicknell contributed to early work on the project while at Carnegie. Maria Sherzad undertook the huge and demanding task of typing and endlessly revising the manuscript as it underwent review, as well as coping with two busy authors operating in different cities and workplaces. The Carnegie librarians, Jennifer Little, Kathleen Daly, and Chris Henley, provided unique support that only those who know how the Carnegie library works can truly appreciate.

Although the first draft of this manuscript was completed many months before publication, numerous updates and additions were completed after Geoffrey Kemp left the Carnegie Endowment to become Director of Regional Strategic Programs at the Nixon Center for Peace and Freedom. He thanks his colleagues at the Nixon Center, especially Dimitri Simes and Paul Saunders, for their understanding and support during this period, and he is grateful for the help provided by Betsy Schmid.

Additional support for Robert Harkavy was provided by a grant from the Office of Net Assessment, Department of Defense (Andrew Marshall, director), and a portion of the project was developed in the course of work sponsored by the Strategic Capabilities Assessment Center of the National Defense University (Robert L. Butterworth, then director). Other support was provided by the Institute for Policy Research and Evaluation (IPRE) and the Political Science Department of the Pennsylvania State University. Secretarial and administrative assistance was provided by IPRE's secretaries, Tammi Aumiller, Michelle Aungst, and Angela Narehood. Robert Harkavy conducted some of the research for this project while funded by an Alexander von Humboldt Stiftung grant during a sabbatical leave at the Institute for Political Science, Christian-Albrechts University, Kiel, Germany, directed by Professor Werner Kaltefleiter. He had previously conducted research on basing access under the auspices of the Stockholm International Peace Research Institute, and that research formed the basis for sections of this study. A number of Penn State undergraduates provided research assistance at various stages of the project: Thomas Beresnyak, Heather Chapin, Rudy Glocker, Nicole Kadingo, Kristin Keiser, Evan Welch, and Charles Wilfong. If others have been neglected, the authors regret the omission.

The final volume would not have been produced without the ingenuity and skills of Carnegie's director of publications, Valeriana Kallab, and her close cooperation with Robert Faherty, director of the Brookings Institution Press. The editing, composition, and production management were undertaken in the most professional way by Princeton Editorial Associates, and the cover was designed by Paddy McLaughlin. David Merrill has provided sterling service in designing many of the maps for the project.

Introduction

TOWARD THE END of the 1980s the Gorbachev revolution in the Soviet Union precipitated the end of the cold war and the demise of communism as a global challenge. In the United States a new preoccupation with the economic dominance of Japan emerged and there was discussion of the replacement of "geopolitics" with "geoeconomics." This was paralleled with a debate about "the end of history"—the end of armed conflict among the major industrial powers. Then on August 2, 1990, unexpectedly and with brutal swiftness, Saddam Hussein's army marched into Kuwait and was positioned to threaten the huge oil fields of Saudi Arabia. The stage was set for the Gulf crisis and the subsequent war. The invasion of Kuwait was a vivid reminder that the most elemental components of geopolitical power are still very much a part of the international system, especially in the Middle East. As one colleague put it at the time, "It's back to basics."

Historically the Middle East has been a crossroad linking empires, dynasties, cultures, and armies in both peace and war, and those who controlled access to its vital land and water trade routes wielded great power and frequently amassed huge fortunes. Today the region's role as a crossroad among the continents has changed and in some ways has been diminished, though perhaps not permanently. While it still has considerable importance in the context of world transportation, new technologies and trading patterns have opened up alternate routes. Europe's vast trade with Asia is no longer dependent upon land or sea routes through the Middle East. Huge container ships from Rotterdam to Tokyo can make the journey around the Cape of Good Hope at little or no extra cost as compared with using the Suez Canal, and many of today's oil supertankers are too big to use the canal in any case. Long-range aircraft such as the Boeing 747 SP can fly nonstop from London to Southeast Asia without refueling and can fly over Eastern Europe and Russia rather than the Middle East.

However, these trends could change again. A political upheaval in Moscow could result in the loss of the new air routes over Russia. If a well-developed road network between Europe and the Middle East were built, land traffic through the region could expand geometrically. It is worth noting that Turkish road links with Central Asia have become a key element of the economic changes under way in the Caucasus and Central Asia. Since 1996 Turkish trucks carrying European products have been traveling regularly by road through Georgia and Azerbaijan and then cross the Caspian Sea by ferry. Turkish, rather than Russian, trucks are the most notable feature in this new network.

The region's strategic geography has changed in another way. During the cold war both the traditional Middle East and South Asia (including Afghanistan) were major arenas for superpower competition. The southern borders of the Soviet Union constituted one of the front lines in the East-West struggle. The Soviet occupation of northern Iran in 1946 triggered the first post–World War II crisis with the West. For much of the next forty-five years, Soviet and American client states in the region competed with each other for military aid and sought the services of their respective superpower as a guarantor of their security. This had a profound impact on intraregional politics. Most of the major confrontations in the Middle East since 1945, though often local in origin, became part of the East-West conflict, and there was always the danger that the superpowers would become directly involved in the fighting. This bipolar world ended with the breakup of the Soviet Union, which some would argue was accelerated by the Soviet defeat and withdrawal from Afghanistan. New political divisions have emerged and the definition and parameters of the region have changed to take account of the new geopolitical reality. Today the political fault lines in the region are more likely to be demarcated by whether or not a state is "radical" or "traditional," autocratic or democratic, rather than pro-Washington or pro-Moscow.

However, we should underscore the fact that current alignments may be temporary. A return to communist or extreme nationalist leadership in Russia could "redraw" the post-Soviet map and if this were accompanied by a more assertive policy toward its southern neighbors, elements of the old fault lines might reemerge. There could also be radically new alignments with equally dangerous implications. A Sino–Russian alliance embracing Iran is the sort of nightmare few in the United States wish to think about but there is considerable speculation about this possibility in Russian writings, especially if NATO expansion continues east to the Russian border.[1]

Despite these changes the Middle East remains a vital region of the world because of the continuing need for access to its resources and, as a corollary, to secure access for the deployment of military forces to protect—or threaten—those resources. If there are future wars in the region that involve the external powers, they will most likely be triggered by broad geostrategic fears about the control of energy supplies.

Indeed, what we term the greater Middle East and its energy resources may now be *the* strategic fulcrum and prize in the emerging arena of world politics. Approximately 70 percent of the world's proven oil reserves and over 40 percent of its natural gas reserves lie within an egg-shaped catchment area from southern Russia and Kazakstan to Saudi Arabia and the United Arab Emirates. What is of special relevance are the growing energy needs of Asia, including China, India, and Southeast Asia, and the fact that they will all have to compete with Europe and North America for greater Middle East energy supplies. This will lead to significant changes in the patterns of diplomacy and security relationships that have evolved since the energy crisis of the 1970s and the 1990–91 Gulf War. At least for the next fifteen years or so, growing dependence on Middle East energy sources will have a major impact on political and economic relationships, both within the region itself and between the countries of the region and the rest of the world. It is clear that the greater Middle East will retain its strategic value, though just where it will fit in the changing international system remains a question for speculation.

No matter what international system evolves, there is no doubt that the greater Middle East will continue to be a source of anxiety because of both energy concerns and the issue of weapons proliferation. With the exception of North Korea virtually all of the world's concern about weapons proliferation focuses on countries in the greater Middle East, where rivalries are intense, distances short, and enemies contiguous. This is the region where there is most expectation that weapons of mass destruction will be used in war (chemical weapons have already been used extensively during the Iran-Iraq war). It is difficult to foresee any breakthrough in the short term to change this disturbing trend. India, Pakistan, and Israel are effectively established nuclear powers. Iran and Iraq are clearly still interested in pursuing programs to develop weapons of mass destruction, including nuclear weapons. Proliferation dangers are compounded because of the unstable situation in Central Asia and the Caucasus, the distinct possibility of the leakage of nuclear and other deadly materials from the former Soviet Union, unresolved primal hatreds that go back hundreds of years, and the reality that the region is an enormous prize because of its oil and natural gas.

The energy and weapons proliferation problems must be viewed against a backdrop of changing power relationships throughout the region. In the event that there is a reemergence of a more confrontational nationalist Russia, many of the cold war concerns will be resurrected, though the dimension of the military problem would clearly be less demanding on the West. The West itself is unlikely to remain a united bloc in the decade ahead. The sharp difference of opinion between the United States and its European allies, especially France and Germany, over the way to deal with Iraq and Iran is a case in point. A major debate between the United States and the Europeans over NATO's future is inevitable. The growing importance of Persian Gulf and Caspian oil and gas for Japan, China, and India parallels the emergence of these three countries as powerful players on the world scene. Certainly all three are going to have an important say in the diplomacy of Middle East energy, and, in the case of China and India, we may even see security roles developing. The possibility of the deployment of a blue water Chinese navy into the Indian Ocean in the next century is a matter of concern in India and elsewhere. Yet India itself is believed by its neighbors to be on a quest for regional hegemony.

The political futures of Turkey, Saudi Arabia, and Egypt remain critical variables in the geopolitical equation. Turkey's centrality to the region has been enhanced by the end of the cold war. So long as Turkey remains a relatively open and pro-Western society there will be great support in both Europe and the United States for its growing regional role. But were a fundamentalist Islamic regime to gain control, overturning Turkey's long history of secularism, the power equation could change radically. There would be a similar concern if either Egypt or Saudi Arabia were to move out of the Western orbit into a more neutral or anti-Western mode. These changes—should they occur—would have a profound effect, not only on the politics and economics of the region but on the ability of the United States to project power and to defend its interests there.

Concern about these possible changes in regional alignments comes at a time when the United States, while the dominant world power, is facing severe pressures on its own defense budget and therefore the quality and size of its armed forces. The ability of the United States to intervene around the world is being questioned not only by neoisolationist politicians but also by the U.S. military itself, which wonders how much longer global dominance can be sustained on reduced outlays. A key question here is whether the current vast American superiority in the high-tech weaponry that has generated the revolution in military affairs can be sustained in the face of the inevitable spread of such technologies to the regional players themselves. Will the United States be able to exploit the weather and terrain and win a quick victory again in a repeat of Desert Storm? Ten years from now, if the regional powers are equipped with the sorts of conventional technology the United States had in 1991, a Middle East conflict would be devastating. While it may be difficult for the countries of the region to obtain access to advanced reconnaissance systems, such as satellites, there is no reason why Iraq or Iran could not eventually acquire much of the technology that was used so efficiently during Desert Storm. The new technology has great advantages in the open desert terrain in which most Middle East wars have been fought, but there are other military theaters in the region where it may be much less relevant, for example, in the mountains of the Caucasus or Afghanistan. More likely, and from a U.S. standpoint equally dangerous, is the prospect that hostile Middle East leaders—like Saddam Hussein or Muammar Quaddafi—will get hold of weapons of mass destruction and use them *either* for megaterrorism *or* to deter possible U.S. military intervention in the region.

We do not believe war and conflict are inevitable but we caution against undue optimism for a comprehensive peace. However, should there be resolution of the traditional enmities in the region, including the Arab-Israel, Gulf, and India-Pakistan conflicts, new political maps will emerge, which might have far-reaching economic, political, and social consequences. This could lead to a major expansion in economic activity, including joint development of regional infrastructure, especially roads, airports, canals, water and energy projects, and communications systems. If this happens, all parties will have a much higher stake in avoiding war.

This book begins with an overview of the greater Middle East within the context of an on-going debate about the relevance and substance of modern geopolitics. This is followed by a definition of "strategic geography" and a brief review of the region, including its physical geography and its history as it relates to geographical considerations. We then proceed to an analysis of contemporary Middle East conflicts and the energy issues, particularly the demand and supply of oil, that are likely to emerge over the next ten to fifteen years. We then analyze the changing military environment, including the impact of weapons of mass destruction, the meaning of the revolution in military affairs for conventional warfare, and the prospects for power projection capabilities for the next century. In the concluding section, we review the role of geography in influencing new patterns of Middle East wars and the prospects for economic cooperation and conflict resolution.

Part One

BACKGROUND:
GEOGRAPHY AND HISTORY

Chapter 1 provides an overview of new concepts in international relations that have emerged since the end of the cold war and their impact on a "new" or a "changing" Middle East. It defines the framework for our discussion of strategic geography and the greater Middle East and illustrates the enduring relevance of the subject in the region. It concludes with a physical description of the greater Middle East, including its key peripheral and internal geographic features. Chapter 2 examines the lessons of history in the context of Middle East resources, changing technology, and the quest for strategic access.

ONE

Strategic Geography and the Changing Middle East: Concepts, Definitions, and Parameters

THE END of the cold war has led to new configurations of global power relations and has changed the definition and the geostrategic parameters of the traditional Middle East. In this chapter these issues are examined to provide background for the subsequent analysis.

The Middle East and the Emerging International System: Conceptual Issues

During the cold war the international system was characterized by a bipolar structure and an ideologically driven basis of enmity and rivalry between the two major power blocs. By the mid 1990s a number of competing images had emerged to describe the new configuration in international relations. These images are not necessarily discrete; indeed, each may capture one aspect of the evolving new set of realities. But one way or another the nature of the new international system will have a major impact on the greater Middle East. Each of these competing images is rooted in geography and can be related to the basic issues of traditional geopolitical theory.

Some analysts see a competition developing among three major economic blocs: a U.S.-led America bloc based on the North American Free Trade Agreement (NAFTA), a Germany-led European bloc centered on the European Union, and a Japan-led Asia bloc.[1] Such bloc competition is predicted to be primarily economic in character, that is, without a security dimension (no arms races and potential warfare). In a broader sense, some analysts see geopolitics being superseded as the main focus of international relations by geoeconomics.

This geoeconomics model is related to another image—that of the "zones of peace–zones of turmoil" theme propounded by Aaron Wildavsky and Max Singer.[2] According to this thesis and the related theme of the "end of history," there will likely be a permanent peace among the industrialized democracies of Europe, North America, Asia, and Oceania juxtaposed against increasing chaos, bloodshed, and ethnic-racial fragmentation within what used to be called the third world. This model accepts the idea central to the three-bloc configuration of evolving, peaceful economic competition among the major power regions.

Others have a more traditional outlook. They regard the current period of cooperation among the major powers represented in the Group of Seven (G-7)—United States, Canada, United Kingdom, Germany, France, Italy, and Japan—as a temporary interregnum reminiscent of the periods after the Congress of Vienna (1815) and after the Paris Peace Conference (1919), and anticipate the eventual resumption of a multipolar, balance of power global competition, including a strong national security dimension.[3] These "realists" see the United States, Russia, the European Union, China, Japan, and perhaps India forming the poles of such a system, with various possibilities for shifting alliances among the major powers.[4]

Still others, however, see an emerging unipolar system rooted in American hegemony. They reject the declinist perspective on the United States popularized by Paul Kennedy in his book, *The Rise and Fall of the Great Powers,* and see the United States as the only superpower at the first rank of both military and economic power.[5] American technological and economic power is now seen to be on the rise vis-à-vis Japan and Europe as "Rising Sam" regains the lead in many high-technology endeavors after a period of seeming stagnation. Further, the United States is seen as having a marked and growing lead in the new military technologies associated with the Revolution in Military Affairs (RMA). In short, the United States is seen as still well established in a "long cycle" of commercial and maritime dominance, the rightful and perpetual heir of the British Empire, maybe only decades later to be challenged by China, Japan, the European Union, or a resurgent Russia or USSR.[6]

Still another relevant image of the emerging international system, at least one worth commenting upon as it applies to the evolving role of the Middle East in that system, is that of the heartland–rimland face-off familiar in traditional geopolitical theory. A brief sketch of geopolitical theory in this century would go as follows. U.S. admiral Alfred Thayer Mahan and British geographer Halford Mackinder advanced what appeared to be contrary views on the relative importance of seapower and landpower for global dominance. Both focused on the permanence and centrality of a global struggle for power between Eurasian-based landpower and rimland-based seapower in the context of global maritime dominance. However, Mackinder thought that landpower was destined to prevail because of emerging technological developments, such as motorized transport and road and rail networks, which would simplify land logistics from the Eurasian core out to the rimland periphery; indeed, these might also allow for the heartland power to achieve maritime superiority as well. Mahan read the opposite into emerging technological trends, seeing the advent of growing possibilities for dominance by a modern maritime nation more easily able to project power all around the rimland.

One recent writer has asserted that "this geopolitical tradition had some consistent concerns, like the geopolitical correlates of power in world politics, the identification of international core areas, and the relationships between naval and terrestrial capabilities."[7] In other words, it is said to involve "the endemic antagonism between the British-American seapower and Russian landpower; the inherent dangers of the German 'Drang nach Osten'; the strategic importance of different geographical areas; the reshuffle of geostrategic relationships by technological innovations in warfare and transport . . . or the debate between the Blue Water school of strategists and the advocates of vast continental areas as the strategic key to world power."[8]

Nicholas Spykman further developed the "rimland thesis" in contrast to Mackinder's heartland doctrine. Both believed that at given times certain geographical regions become pivotal in relation to global power. Mackinder saw the Russia-Eastern Europe area as pivotal. Spykman contended that considerations such as population, size, resource availability, and economic development all combined to make the rimland—peninsular Europe and the coastal Far East—the most significant geopolitical zone, which if dominated by one power would translate into global hegemony.[9] American interests then dictated the prevention of the unification of the European or the Far Eastern coastland by any hostile coalition.

In a related conceptualization Saul Cohen used the term shatterbelts as roughly equivalent to the concept of the rimland. Cohen defines a shatterbelt as "a large, strategically located region that is occupied by a number of conflicting states and is caught between the conflicting interests of adjoining Great Powers."[10] He refers to similar earlier use of the term crush zones by James Fairgrieve.[11] Cohen saw the Middle East and Southeast Asia as the primary shatterbelt regions, and he says that "the shatterbelt appears to be incapable of attaining political and/or economic unity of action," and that whereas some parts of the shatterbelt may be committed to neutrality, others are enmeshed in external ties. Hence, too, referring to the Middle East, he says that "it is because internal differences are so marked, and because they are found in a region that is crushed between outside interests, that we have defined the Middle East as a shatterbelt."[11] That was, of course, during the cold war.

In the recent past still another, contrary, predictive image or model of the emerging world order has captured the attention of students of international affairs, that is, the "clash of civilizations" model formulated by Samuel Huntington in an article in *Foreign Affairs*.[13] The gist of Huntington's arguments is captured in the following excerpt:

> It is my hypothesis that the fundamental source of conflict in this new world will not be primarily ideological or primarily economic. The great divisions among humankind and the dominating source of conflict will be cultural. Nation states will remain the most powerful actors in world affairs, but the principal conflicts of global politics will occur between nations and groups of different civilizations. The clash of civilizations will dominate global politics. The fault line between civilizations will be the battle lines of the future.[14]

Actually Huntington sees this as the latest phase in the evolution of conflict over the last couple of centuries. He contends that during the century and a half following the Peace of Westphalia the primary conflicts in the Western World "were largely among princes—emperors, absolute monarchs, and constitutional monarchs attempting to expand their bureaucracies, their armies, their mercantile economic strength, and, most important, the territory they ruled." In the next phase, ushered in by the French Revolution, the principal lines of conflict were between nation-states rather than princes, and this phase—so often characterized by others with the terminology of balance of power or multipolarity—is said to have lasted until after World War I. Then, according to Huntington, "as a result of the Russian Revolution and the reaction against it, the conflict of nations yielded to the conflict of ideologies, first among communism, fascism-Nazism, and liberal democracy, and then between communism and liberal democracy. . . . During the cold war, this latter conflict became embodied in the struggle between the two superpowers, neither of which was a nation state in the classical European sense and each of which defined its identity in terms of ideology." In all of the above phases, it is pointed out that the conflicts were within Western civilizations, in short, were "Western civil wars." Now he sees a conflict of civilizations.[15]

Stating that "fault lines between civilizations are replacing the political and ideological boundaries of the cold war as the flash points for crisis and bloodshed," Huntington focuses particularly on the cultural lines of demarcation between Western Christianity and Orthodox Christianity in Europe and between the latter and Islam. The most significant dividing line in Europe, he says, may be the old eastern boundary of Western Christianity in the year 1500. He also points out that conflict along the fault line between Western and Islamic civilizations has been going on for 1,300 years, with the Arabs and Moors having reached as far as Tours in 732 and the gates of Vienna in the seventeenth century, after which there was the reversal that led in the twentieth century to European domination over the Middle East and North Africa.[16] Like many other contemporary analysts, both in the West and Islam, Huntington sees these two civilizations as potentially pitted against each other as a defining feature of an evolving world order. Hence:

The centuries-old military interaction between the West and Islam is unlikely to decline. It could become more virulent. The Gulf War left some Arabs feeling proud that Saddam Hussein had attacked Israel and stood up to the West. It also left many feeling humiliated and resentful of the West's military presence in the Persian Gulf, the West's overwhelming military dominance, and their apparent inability to shape their own destiny.... Some openings in Arab political systems have already occurred. The principal beneficiaries of these openings have been Islamic movements. In the Arab world, in short, Western democracy strengthens anti-Western political forces. This may be a passing phenomenon, but it surely complicates relations between Islamic countries and the West.[17]

Huntington's thesis is clearly underscored by the several current conflicts along this old fault line in Bosnia, Kosovo, the Turkish-Bulgarian frontier, Armenia versus Azerbaijan, Chechnya, and the areas in Central Asia (Kazakstan, Tajikistan), where large Russian ethnic populations have been left amid Muslim peoples in the wake of the receding power of the former USSR. Indeed, this fault line seems to have replaced the old frontier demarcating the USSR and its ideological client states from the Middle East.

There are some other fault lines between civilizations that may also contain some potential for large-scale future conflict. One is that between the Arab Islamic civilization and the pagan, animist, or Christian cultures to the south; another between Muslim and Hindu civilizations in South Asia, which has long been in evidence in the antagonism between India and Pakistan. With the recent conflicts in Bosnia, the Gulf, and the Caucasus in mind, Huntington predicts that "the next world war, if there is one, will be a war between civilizations."[18]

Of course not everyone agrees with Huntington's now widely discussed thesis. Fouad Ajami, for instance, finds Huntington wrong in underestimating the tenacity of modernity and secularism in the Middle East and elsewhere in the third world, and downplays the threat to the West represented by traditionalist movements in Egypt, Algeria, Iran, Turkey, and India, among others. He sees Western culture and values as having been totally and irretrievably internalized in these places, and further sees Huntington as having underestimated the continuing power of the nation-state.[19] John Esposito, among others, in opposing this new form of "orientalism," also offers a countering view, insisting that most Islamic movements are not necessarily anti-Western, anti-American, or anti-democratic, and that it is a mistake by Westerners to interpret Islam as a monolith, rather than a complex and diverse realm.[20] But, nonetheless, Huntington insists on the validity of the "clash of civilizations paradigm as the appropriate successor to the cold war paradigm," and to the extent that this turns out to be accurately predictive to one degree or another, it will be a defining feature of a new strategic map.[21]

For the remainder of this volume, we will be emphasizing again and again the point that the Middle East (including the Caspian Basin region) has now assumed the role of the strategic high ground, a key strategic prize in the emerging global system at the juncture between the twentieth and twenty-first centuries. This will undoubtedly be the case whether this new system most closely approximates the three-bloc geoeconomics model, a new multipolar balance of power model featuring the United States, the European Union, Russia, China, and Japan as poles, a "clash of civilizations" focus on the West versus Islam divide, unipolar American hegemony (which would probably eventually impel a countervailing coalition), or a bipolar confrontation between the United States and either Japan or China or both, or a resurgent Russia.

What Is Strategic Geography?

Strategic geography refers to the control of, or access to, spatial areas (land, water, and air, including outer space) that has an impact—either positive or negative—on the security and economic prosperity of nations. It embraces all dimensions of geography, which includes both physical and human geography. This is a more focused definition than the classical concepts of geopolitics succinctly defined by Saul Cohen as "the relation of international political power to the geographical setting."[22] It has a more specific meaning in that it is more directed at the tactical elements of geography that contribute to grand strategy.

The *physical geography* of a region generally changes very slowly, though some features change at different rates than others. Over centuries topographical features such as mountains, lakes, rivers, and shore lines can be altered significantly, with far-reaching consequences, and the climate of a region can change, often more rapidly than its topography. Continents that today are separated by oceans were once joined; rivers once flowed across what is now the Sahara; during the last ice age much of the northern hemisphere was uninhabited; and the great forests that covered India, Europe, and parts of the Middle East are no more. Some physical changes can occur within decades, such as the depletion of natural resources, and some can happen over ten to fifteen years, which is the approximate time range covered in this study. There could be significant physical changes in the Middle East if major water projects are developed that will literally change the physical landscape.

The *human geography* of a region—which can and does change very rapidly, depending upon a number of factors—can be broken down into dozens of subcategories. For our purposes we will focus on three of them. *Political geography* describes the control and organization of territory, including people and assets. *Economic geography* refers to the infrastructure and industrial and rural facilities that contribute to the economy of a region, including roads, ports, airports, pipelines, energy utilities, factories, farms, and patterns of trade.

Military geography concerns the deployment and power projection of military assets as they relate to space, time, and distance and the impact that physical constraints have upon both offensive and defensive military operations.

Political geography can be radically changed overnight as a result of war, revolution, or other political upheavals. The breakup of the Soviet Union with its impact on the political borders of Eurasia is the most dramatic and important instance in recent years but it is not the only example. The 1947 partition of India and the 1971 creation of Bangladesh radically changed the political configuration of the subcontinent. In 1967 Israel's military victory over the Arabs changed the political map of the Middle East for thirty years. Richard Nixon's historic visit to China in February 1972 changed the political balance with the Soviet Union overnight. An influx of refugees or mass migration of people can alter the political map of a region very rapidly. A second exodus of Palestinians to Lebanon in the 1970s was one of the precursors of a fifteen-year civil war.

The economic geography of a region or country or city can change radically in positive and negative ways as a result of innovative technologies and new market conditions. Venetian traders must have been appalled the day it was announced that the Portuguese had pioneered a route to the Spice Islands via Africa. The impact of the Suez Canal on international commerce was enormous. Yet in the 1970s the closure of the canal owing to war and the appearance around the same time of supertankers too large to use it showed how alternative commercial routes can frequently be developed if the economics are right. The mining of phosphates by Israel and Jordan along the Dead Sea coupled with a reduction in the flow of water from the north created a land mass that divides the Dead Sea. A proposed canal from the Red Sea to the Dead Sea along the Jordan-Israel border could restore the Dead Sea to its traditional shape and radically change the commerce of what is now a desert region. In Central Asia and the Caucasus decisions about the laying of oil and natural gas pipelines will have a profound impact on regional economies in the decades ahead. Some countries will benefit while others will not, and this will have political consequences.

Military geography has been strongly influenced over the centuries by developments in technology. Ports and routes that were essential for navies during the age of sail ceased to be important once steamships dominated fleets, and instead coal bunker ports and access to coal supplies became critical. But in less than sixty years oil had replaced coal as the best fuel for ship propulsion and a new set of logistical priorities to provide oil for ships became important. Britain's interest in the Persian Gulf evolved rapidly in the early twentieth century because of its desire to control Mesopotamian oil and not be dependent on the United States.[23] In the 1950s the development of nuclear propulsion for ships virtually eliminated the need for overseas base access for U.S. nuclear-powered submarines and aircraft carriers.

In more recent years, aircraft, missiles, and new types of ground vehicles have radically altered the importance of traditional geographical constraints upon military power projection. Most dramatically, the use of outer space to manage battlefield communications has made former barriers to military operations less significant. On the other hand, equipped with Stinger surface-to-air missiles, Afghani resistance fighters were able to defeat the Soviet army by exploiting terrain and the vulnerability of modern armed forces to wars of attrition.

The Relevance of Strategic Geography in the Middle East

Applying this concept of strategic geography to the greater Middle East, we can identify a great many contemporary examples of its continuing importance.

In terms of great power competition:

—The United States continues to have vital interests in the Middle East, including the survival of allies, especially Israel, and the denial of control of Persian Gulf energy resources to hostile powers. To secure these interests it must be able, in the last resort, to project military power over great distances and ensure access to forward bases and facilities in the region. In this regard Egypt, Israel, Saudi Arabia, and Turkey are especially important. The challenges to American power projection capabilities will grow if the proliferation of advanced weapons to regional adversaries accelerates.

—Russian control or dominance of the Caspian Sea and Basin will ensure Moscow's control of the key oil and gas distribution systems from the region to the outside world and give it great leverage over supply. Continued instability in the North Caucasus and Transcaucasus, Turkey, and Afghanistan poses serious potential threats to the various pipelines that have been proposed for transporting oil and gas to the international markets. Any deployment of significant Russian forces in Iran, Turkmenistan, or Afghanistan could resurrect concerns about Russian ambitions in regard to the Persian Gulf and the Indian Ocean. It will be recalled that the Soviet invasion of Afghanistan was viewed by the United States as a strategic threat to the Persian Gulf since it brought Soviet air power 600 miles closer to the Strait of Hormuz. It was the trigger for the enunciation of the Carter Doctrine and the development of the Rapid Deployment Joint Task Force.

—China's growing need for energy may be met by increased oil supplies from the Persian Gulf. The deployment of Chinese blue water naval forces into the Indian Ocean in the next century to protect its seaborne supply lines would clearly represent a new challenge to the existing maritime powers and to India and the Gulf countries.

—Increased military ties between China and Pakistan would further intensify Indian efforts to develop very-long-range missiles which, in turn, would have strategic implications beyond South Asia. India already uses the presence of Saudi Arabia's Chinese-supplied missiles as one justification for its own programs even though it has very friendly relations with the Saudi regime. However, the most likely catalyst of another India-Pakistan war is the unresolved conflict over Kashmir.

—The deployment of weapons of mass destruction (WMD) by an Islamic regime in either Algeria or Libya would have widespread repercussions in southern Europe and could trigger a major debate as to whether NATO's areas of responsibility should extend further south as well as further east.

Conflict over access to fresh water supplies remains a casus belli in the Middle East:

—Attempts by either Ethiopia or Sudan to divert the Nile River and reduce or cut off water supplies to Egypt would threaten its very existence and trigger an immediate crisis and probably war.

—Significant and protracted interference by Turkey with the flow of water from the Tigris and Euphrates to Syria and Iraq could lead to military confrontation.

—Any unilateral or bilateral cut-off of Israel's water supplies by Syria and Jordan would, as in the past, cause a grave crisis.

In the Persian Gulf region territory remains a potential casus belli:

—Iraq's potential threat to Kuwait will remain a major crisis flash point for the foreseeable future. The occupation of Kuwait by either Iraq or Iran would pose unacceptable threats to Saudi Arabia, especially the oil fields and terminals in its Eastern Province.

—Iraq's desire for greater maritime access to the Persian Gulf will remain a major obstacle to a regional security regime no matter who controls Baghdad.

—The occupation of key islands in the Persian Gulf and Red Sea by Iran and Eritrea, respectively, could lead to an armed conflict in the first instance between Iran and the United Arab Emirates and in the second between Eritrea and Yemen, primarily because of the location of the islands with respect to potential off-shore oil and gas deposits.

In the Arab-Israel case territorial issues remain the key to the conflict and its resolution:

—The relevance of topography for military operations is probably most clear in the context of Arab-Israeli peace negotiations, especially concerning Israel's prospective withdrawal from the West Bank of the Jordan River and the Golan Heights. Some believe that if Israel relinquishes control of the West Bank it will be threatened by a Palestinian state or federation. For this reason no Israeli government will give up the approaches or permit the Palestinians to control all of the Jordan Valley. Thus any eventual Israeli agreement with the Palestinians would have to include iron-clad security arrangements and would never involve handing back the entire West Bank to a Palestinian authority. There are similar arguments concerning the Golan Heights. Those in favor of an Israeli withdrawal argue that it can never happen unless tough conditions are negotiated that preclude Syria ever being able to reintroduce ground forces onto the Heights. On the other hand, Syria feels threatened by Israeli forces on the Golan and insists on their being withdrawn. Thus topography remains a critical ingredient in Arab-Israeli security calculations, and even in the missile and information age control of high ground remains a key military principle.

—The introduction of foreign armies into Jordan would be treated by Israel as a major military threat and could lead to a massive preemptive attack on that force if it proceeded toward the Jordan River. Israel has insisted that Jordan be a *cordon sanitaire* and that no military forces from Iraq, Syria, or Saudi Arabia be deployed in strength anywhere in the country. This fear reflects the reality that in conflict regions such as the Arab-Israel theater distances are so short that modern armies with heavy armor can make rapid progress along modern roads and can change the balance of power in a particular region in a matter of hours. For this reason Israel will continue to rely on nuclear weapons and conventional high technology to ensure its ability to deter and if necessary fight a hostile neighbor.

The Dynamics of Geographic Factors

While the list above confirms the enduring importance of Middle East strategic geography, it provides little guidance as to the dynamics of geographical factors. How have key variables, including changes in technology and the demand for resources, affected the relative importance of specific geographical features of the region at different times? Today it is the Middle East's energy resources that command attention but in the past it was its grain and timber and the routes from Asia that carried silks and spices that were at issue. As technologies of transportation have changed, so has the strategic value of specific trade and military access routes. With the coming of steam- and nuclear-powered ships, airpower, and the missile and space age, the geographical dynamics of the military balance has changed once more.

Throughout this text these two factors—technology and resources—are considered prime variables. However, to fully explain the magnitude and direction of the changes currently taking place in the geopolitical configuration of the region two additional, very significant, factors have to be taken into account—changing political ideologies and population dynamics. In the past fifty years the relative influence of nationalism, communism, and religious radicalism has changed with dramatic consequences for regional political alignments and military alliances. Similarly, the migration of people and the increase in the population of the region have had a profound impact on economic growth and expectations.

Defining the Middle East

There is no single, agreed definition of the political and geographic boundaries of the Middle East. Geographers, historians, journalists, and bureaucrats all use the term, yet frequently have different definitions of what they mean. In parts of Asia it is fashionable to refer to the region as West Asia but this then excludes Egypt, Sudan, and the Magreb, which are in Africa yet are generally thought of as Middle East countries. In the nineteenth century the major European powers regarded the East or Orient as the region of Eurasia (excluding Russia) that began where Western civilization ended, which is to say with the African continent and the Ottoman Empire. The great strategic competition among Britain, France, Russia, and Germany for access to and control of this area came to be known as the Eastern Question. With the expansion of Western influence further into Asia, however, it became necessary to distinguish between the Near East and Far East. According to Bernard Lewis, the term Middle East was first used by Mahan to refer to the area between Arabia and India that had particular relevance for naval strategy—that is, the Persian Gulf. During World War I the command for the British forces in the region was designated the Middle East Command. Since that time the term has been used, sometimes synonymously with the term Near East, to mean the area from North Africa up to but not including the Indian subcontinent.[24] The U.S. Department of State refers to the region as the Near East and includes within that designation North Africa, the Levant, and the Gulf countries but not Turkey since the latter is a member of NATO.[25] In contrast, the U.S. Department of Defense divides the region in yet another way. U.S. Central Command (CENTCOM) has responsibility for military operations in a region that includes Egypt, Sudan, Ethiopia, Djibouti, Kenya, Somalia, Jordan, Saudi Arabia, Iraq, Iran, the states of the Gulf Cooperation Council (GCC), Afghanistan, and Pakistan. Excluded are Turkey, Israel, Syria, and India—the first three remaining the responsibility of the European Command (EUCOM) and India falling under the Pacific Command (PACOM).

The breakup of the Soviet Union and the establishment of the newly independent republics of the Caucasus (Georgia, Azerbaijan, and Armenia) and Central Asia (Turkmenistan, Uzbekistan, Kyrgyzstan, Tajikistan, and Kazakstan) has once more raised questions about exactly where the Middle East begins and where it ends and whether it can be comprehensively, consistently defined. Given the strategic thrust of this study we believe the definition of the region must include those countries directly involved in four main conflicts in the area—Arab-Israeli, Persian Gulf, Caspian Basin, and South Asia.[26]

How then do we define the Middle East? One option would be to use the phrase "Greater Middle East," which has gained some currency, to cover the areas we think are most significant to our basic thesis.[27] Yet such a formal designation implies a degree of precision that we do not believe is presently

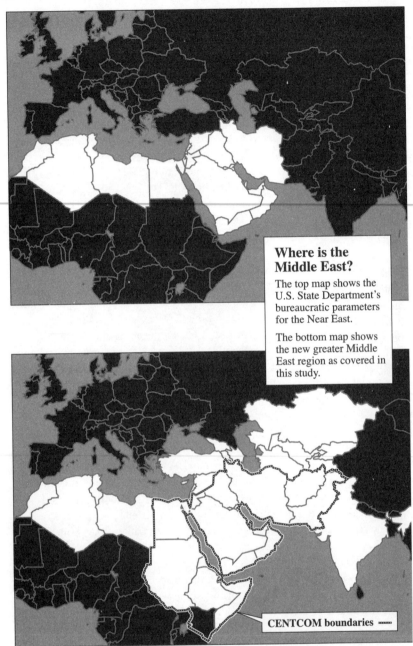

Where is the Middle East?

The top map shows the U.S. State Department's bureaucratic parameters for the Near East.

The bottom map shows the new greater Middle East region as covered in this study.

CENTCOM boundaries ----

Map 1

justified and embraces more countries than we are examining in this study. It assumes there is a consensus concerning which countries to include and which to exclude (as in the case of defining continents, for example, Asia or Africa). Yet selection is bound to be arbitrary because the rationale for including one country and excluding another is based on judgments about which are the determinant variables. Since we are primarily interested in strategic geography rather than religion or political alliances our selection of countries is necessarily different from those who would wish to analyze, say, the Muslim world or the East–West cold war confrontation states. As will become clear our focus is on the strategic importance of energy resources, water scarcity, and weapons proliferation, all of which have critical geographical components.

Which then should be included in our grouping? First the traditional U.S. State Department list for the Near East must be included (Morocco, Algeria, Tunisia, Libya, Egypt, Israel, Syria, Lebanon, Jordan, Iran, Iraq, the GCC states, and Yemen). We believe Sudan, Ethiopia, and Somalia have to be discussed, albeit briefly, because of their strategic and political importance, though we refer to them mainly in the context of military geography. Given the strategic developments in the eastern Mediterranean, the Caucasus and Central Asia, Greece, Russia, and China are all important players but to include them within a definition of the Middle East is inappropriate. However, Afghanistan, India, Pakistan, Sudan, and Turkey must be included as should the energy-producing countries of the Caspian Basin (Azerbaijan, Kazakstan, and Turkmenistan). Given their location Georgia and Armenia must also be included but we have not gone into any detailed discussion of the other Central Asian states (Uzbekistan, Tajikistan, and Kyrgyzstan) even though they are shown on our map of the new Middle East and we occasionally refer to them. Our inclusion of India and Pakistan raises a question about the definition of South Asia. We have not included Bangladesh, Nepal, and Sri Lanka in the analysis and some South Asians would argue that Burma (Myanmar) and even Tibet should be included in a comprehensive definition of South Asia.

We believe the most accurate way to describe the region covered by this study would be the Middle East (including North Africa, Turkey, Sudan, and the Horn of Africa), the Transcaucasus, west Central Asia, and South Asia. Yet this is too clumsy. We have therefore decided to include all the above countries and groupings under the phrase greater Middle East region. We realize this will not sit well with some analysts but short of convening a quorum of geographers to iron out an agreed definition this is our best alternative, and we hope our usage is acceptable for the purposes of this study (see map 1).

Geographic Parameters and Access Routes

One way to think of the greater Middle East region is as a vast quadrilateral, a geographic centerpoint and crossroad where Europe, Russia, Asia, and Africa

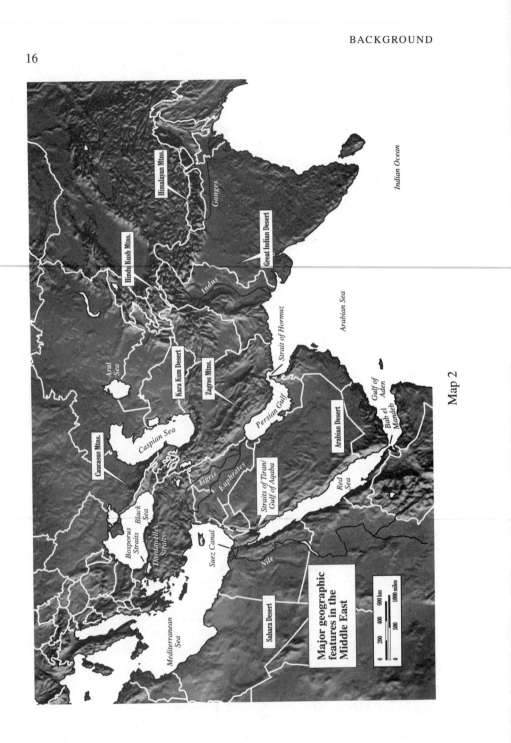

Map 2

Major geographic
features in the
Middle East

Indian Ocean

Himalayan Mtns.

Ganges

Hindu Kush Mtns.

Great Indian Desert

Indus

Strait of Hormuz

Arabian Sea

Aral
Sea

Kara kum Desert

Zagros Mtns.

Persian Gulf

Gulf of
Aden

Caucasus Mtns.

Caspian Sea

Arabian Desert

Bab el
Mandeb

Tigris

Euphrates

Straits of Tiran/
Gulf of Aqaba

Red
Sea

Bosporus
Straits

Black
Sea

Dardanelles
Straits

Suez Canal

Nile

0   200   400   600 km

0   500   1000 miles

Mediterranean
Sea

Sahara Desert

intersect. It is bounded to the west by the Sahara Desert and the Mediterranean Sea, to the north by the Black Sea, the Caucasus Mountains, the Elburz Mountains, and the Hindu Kush, to the east by the plains of India and to the south by the Indian Ocean. Within this region lie many intraregional geographic barriers, numerous ethnic and religious groups, and sovereign states as well as an abundance of important natural resources, especially oil and natural gas. The key physical features of this region are shown in map 2.

Traders, travelers, warriors, and kings and all those projecting power and seeking access and wealth in the Middle East have had to find ways to overcome the unique geographical constraints of the region. Middle East topography remains a key factor influencing access and power projection, even in the current age of missiles and weapons of mass destruction.

Peripheral Barriers

The boundaries of this region we call the greater Middle East contain formidable land and maritime barriers. In the south and west land access is restricted by the Sahara Desert, movement from Russia in the north by the Caucasus Mountains, and access in the northeast by the imposing Elburz and Hindu Kush mountain ranges. These terrain barriers have had a profound impact on the history of the region. As Mahnaz Z. Ispahani explains, natural creations and human decisions regulate access to any given region. Mountains, deserts, legal boundaries, and tariffs function as "routes," increasing movement and centralization, or as "antiroutes," restricting access and distribution of resources. Ispahani delineates five different dimensions of routes: geographical, political, economic, military, and ideational (flow of ideas).[28]

The Sahara, one of the world's largest deserts, is over 3.5 million square miles, stretching from the Atlantic Ocean to the Red Sea and separating northern Africa from Black Africa. In that sense the desert limits the reach of the Middle East. Since physical access between the north African states and the rest of Africa (usually labeled sub-Saharan Africa) is so difficult, the former are often included in a broadly defined Middle East.

As we move north past the Mediterranean, the Turkish straits, and the Black Sea, the next key feature is the Caucasus Mountains. Serving as a 700-mile wall between the Black Sea and the Caspian Sea, the Caucasus range has frustrated invaders and counterinvaders for centuries. With few strategic passes these mountains make the passage from Russia to the Middle East, or vice versa, very difficult. Several of the peaks top 15,000 feet, with the highest, Mount Elbrus, reaching 18,481 feet. As recently as World War II the area was a major battleground for the German and Russian armies; at that time the Germans invaded and were not driven out by the Soviet forces until January 1943.

Among the Caucasus range's few passes, the best known are the Mamison and the Daryal. Other passes include the Klukhor and the Pass of the Cross. This last defile may have acquired the name in the fifth or sixth century A.D., when the Persians controlled the area, and it is one of the few places where ancient troops and modern tanks have been able to traverse the Caucasus range.[29]

The Caspian Gates Pass lies south of the Caspian Sea near the Elburz Mountains. With desert to the south and mountains to the north, control of the Gates was vital for armies seeking to traverse ancient Persia (Iran). As historian George Rawlinson wrote: "The latter [pass], now known as the *Girduni Sudurrah* Pass, constitutes the famous '*Pylae Caspiæ.*' Through this pass alone can armies proceed from Armenia, Media, and Persia eastward, or from Turkestan, Khorasan, and Afghanistan into the more western parts of Asia."[30]

It is the imposing nature of the Elburz range that made the Caspian Gates so important. With many peaks over 10,000 feet, the Elburz Mountains separate the southern shores of the Caspian Sea from central Iran; the highest peak, Damavand, rises to 18,376 feet. The lowlands near the Caspian are never more than twenty-five miles wide, so the Elburz restrict north-south travel and together with the Iranian Great Salt Desert restrict east-west travel as well.[31] In ancient times regional powers focused on control of the Gates and, when necessary, established military garrisons to achieve it.

Completing the topographical barriers that surround the land mass of the Middle East are the two in the Pakistani-Afghani region: the Hindu Kush Mountains and the Indus River. The Hindu Kush have served as a Central Asian roadblock for centuries. With connecting ranges such as the Koh-i-Baba near Kabul and the Karakoram, the Hindu Kush are the dominant feature of the area. Many peaks top 20,000 feet, and the Tirich Mir reaches 25,260 feet. From the nineteenth-century Afghan wars to the modern Kashmir dispute, the mountains have been factors in the political and military conflicts. Similar to the Caucasus Mountains the Hindu Kush contain several major passes. The Khyber Pass, thirty-three miles long, is the most famous; others include the Tochi, Kurram, Gomal, and Bolan.[32] Each has been used by invaders and almost all are guarded by fortresses or local tribes.

Originating in the Kailas Mountains of Tibet, the Indus River flows for 1,800 miles to the Arabian Sea (Indian Ocean). The river is a border in some provinces and a source of water for agriculture, and its valley has been the site of several military conflicts.

Maritime access is restricted by narrow straits. Gibraltar controls access from the Atlantic Ocean to the western Mediterranean. There is only one transit route for ships coming to the eastern Mediterranean from Bulgaria, Rumania, Ukraine, and Russia: through the Bosporus and the Dardanelles, which connect the Black Sea with the Mediterranean and the Suez Canal.[33] Both the Dardanelles and the Bosporus are narrow but permit deep-draft shipping. The Dardanelles are thirty-eight miles long but only three-quarters of a mile to four miles wide. Only nineteen miles long, the Bosporus is even narrower, being only half a mile to nearly three miles wide. Today with six countries bordering the Black Sea and the large Russian naval fleet calling the area home, the gates of the Black Sea retain their strategic importance.

The Suez Canal and the entire Sinai peninsula constitute a second maritime choke point on the eastern Mediterranean. The canal is only a part of the strategic water network around the Sinai peninsula. The extensions of the Red Sea—the Gulf of Suez and the Gulf of Aqaba—also have strategic value and have seen significant battles. Israel and Jordan depend on unfettered use of the Gulf of Aqaba for the transport of cargo to Eilat and Aqaba, respectively. The Suez Canal has been the focal point of much fighting, including the Anglo-French invasion in 1956 following its nationalization by Egyptian president Gamal Abdel Nasser. On May 22, 1967, Nasser closed the Straits of Tiran, the entrance to the 100-mile-long Gulf of Aqaba, to Israeli shipping. Nasser's action, combined with Egyptian and Syrian belligerencies, led first to Israeli air strikes on June 5 and then to full-fledged war.

Bab el Mandeb is another important maritime choke point. At the junction of the Red Sea and the Gulf of Aden, the strait, which at its narrowest is fifteen miles wide, separates the southern portion of the Arabian peninsula from the African horn. The strategic location of the strait heightened the importance of Yemen, Ethiopia, and Somalia, especially during the cold war, when the Americans and Soviets competed for the support of the local regimes. With naval control of the strait, any power can close off the Red Sea and block access to the Suez Canal, which would pose a direct threat to Saudi Arabia, Egypt, Ethiopia, Eritrea, Sudan, Israel, and Jordan. It can be said that if the Suez Canal serves as the main gateway to both Europe and the Indian subcontinent, then Bab el Mandeb is the second portal.

At the world's oil hub the Strait of Hormuz, which separates Iran from the Arabian peninsula and links the Persian Gulf and the Gulf of Oman, is another vital choke point. A longer strait, it also contains three major islands: Qeshm, Hormuz, and Hanjam. During the 1980–88 Iran-Iraq war and the 1991 Gulf War, the strait assumed even greater strategic importance as the gateway for much of the world's oil supply. World powers, including the United States, relied on the strait for transport of oil tankers and warships. Free access to the strait allowed the United States to bring in its navy to help contain Iraqi forces on the eastern shores of Kuwait and Iraq in 1990–91, whereas restricted access would have prevented this deployment, depriving the Americans of a significant portion of their firepower.

Internal and Local Barriers

Internal and local barriers within the Middle East region are both numerous and noteworthy. In some cases they affect mainly one country. In all cases they

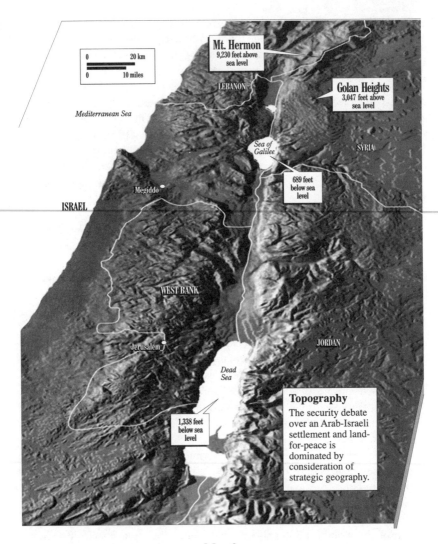

Map 3

are factors in determining access routes and thereby facilitate or impede commerce, military conflict, and cultural interaction.

For Egypt the Nile is vital both economically and politically. The fertile Nile Valley provides the lifeblood for much of the Egyptian population, and the river is a critical factor in the country's stability. If the Nile dried up or was diverted in the south, Egypt would, quite literally, cease to exist.

The Nile River is of major importance to several states. The longest river in the world, it stretches, by some calculations, nearly 4,200 miles. It flows through Sudan and Egypt, but the headwaters also drain northern Tanzania, southwestern Kenya, and parts of Uganda. The Blue Nile, which originates in Ethiopia, joins the White Nile at Khartoum and the Sobat River joins the Nile in southern Sudan. Nearly all of the Nile is navigable, especially during the rainy season. The river supplies much of the region's water needs, so any projected plan to alter the flow or path would in all probability call down heated opposition. The Nile's long and winding path also means that it ties the fates of several African states together. This common and overriding importance of the Nile helps to explain why tension in any one country—such as the ongoing battles in Sudan—is cause for serious concern in neighboring capitals.

No part of the Middle East better demonstrates the historic and enduring role of geography in influencing national policies concerning peace and war than the tiny piece of land on the eastern shore of the Mediterranean known historically as Palestine or Eretz Israel (see map 3).[34]

The famous Via Maris stretched from Egypt along the Sinai, through Gaza to what is now Hadera, north of Tel Aviv. There the road turned inland through the Musmus Pass in the Wadi Arah to Megiddo. Megiddo was a crossroad. One route proceeded to the northwest along the Mediterranean coast and the other northeast to Hazor and on to Damascus. Numerous armies fought their way both north and south through the Musmus Pass. Some of the greatest battles involving Egyptians, Babylonians, Israelites, Canaanites, and others—including some of the famous battles described in the Bible—were fought in and around Megiddo. Whoever controlled Megiddo could deploy mobile forces, including chariots, to block the passes or intercept armies approaching from the north or east. Megiddo's importance in antiquity has carried through to modern times. In September 1918, General Edmund Allenby defeated the Turkish armies in a decisive battle at Megiddo in his drive north from Egypt. He then was able to liberate all of Palestine and march on Damascus.

The Judean and Samarian mountain ranges, much of which are today known as the West Bank, provided sanctuary for the ancient Hebrews who occupied the highlands. Zerah the Ethiopian and Sennacherib the Assyrian tried to conquer the highlands and were defeated by the natural barriers; Egypt avoided attacking them altogether.[35] Roman emperors Vespasian, Titus, and Hadrian succeeded in conquering the highlands, but Vespasian and Titus were careful to occupy all of the surrounding plains before undertaking the conquest, while

Hadrian's victory was one of the bloodiest in Roman history. After he had taken Gaza Napoleon Bonaparte refused even to attempt to conquer Jerusalem, which straddles the highlands.

As the Judean mountains and Samarian ridge of the West Bank, the Golan Heights overlook a flat, narrow section of Israel—the Hula Valley—that contains the northern settlements. Between 1965 and 1967, before Israel captured the Golan Heights in the 1967 war, Syrian forces shelled Israeli border settlements from the Heights, some 1,970 feet above Israel's northeast frontier.[36] Under Israeli control the Golan Heights now provide a vantage point from which to survey Syria. Damascus is only thirty-seven miles from the current Israeli front line, and there are no natural barriers between the frontier and the Syrian capital. Thus from a Syrian perspective, Israel's continued military occupation of the Golan poses an unacceptable danger.

The Gaza Strip covers an area of 140 square miles in southwest Palestine, stretching twenty-five miles along the coast south of Tel Aviv, and contains no natural fortifications.[37] Because of its location on the Via Maris, this strip of land has fallen to numerous invaders. It was under Israeli military occupation between 1967 and 1994. The strategic importance of this area is due to its access to the sea and its border with Egypt along the Sinai. However, the population of Gaza, primarily Palestinian refugees, is extremely hostile to Israel and for years Gaza has been a center for rebellion and terrorism. The Palestinian uprising began as a riot in Gaza in early December 1987, and quickly spread to the West Bank and Jerusalem.

Saudi Arabia is mostly desert, with only 2 percent naturally arable land. The Rub al-Khali, or Great Sandy Desert, which is the largest of the Saudi deserts, subsuming several smaller ones, covers 250,000 square miles. As do mountains and waterways, the desert limits the advances of societies and armies. Those few who have known how to live and travel in the desert have always had a distinct advantage over outsiders. In a larger sense the desert splits the traditional Middle East region, with Yemen, Oman, and parts of Saudi Arabia on one side and Jordan, Israel, and Syria on the other, though the two parts are accessible to each other via the Red Sea.

Iraq's geography has strongly influenced both its domestic ethnic politics and its military strategy. The northern region is dominated by mountains and fertile valleys and remains home to much of Iraq's Kurdish population. South of Baghdad the marshes of the Tigris and Euphrates rivers have been home for hundreds of years to Shia Marsh Arabs. Saddam Hussein's recent decision to drain the marshes was precipitated by the need to deny opposition forces sanctuary and permit his army access to a region within which it had been difficult to search and destroy.

The Zagros Mountains in Iran are another feature in the Middle East that dramatically affects its strategic geography, on two levels. First, the mountains, many of which exceed 9,000 feet (the highest peak, Zardeh Kuh, reaches

14,921 feet), sit in a contentious spot, even for the Middle East. Beginning in Azerbaijan the Zagros ranges straddle the Iran-Iraq border, and several parallel ranges of the system cover southern and southwestern Iran. The United States always considered the mountains a natural barrier to a Soviet invasion of the Gulf. Second, the Zagros's Khuzestan foothills are the source of much of Iran's oil supply, always a strategic consideration.

As most arenas of modern-day tension, the Indian subcontinent is deeply affected by geography and topography. Political problems have been laid over a rich tapestry of high mountains, long rivers, and broad plains that defies an easy division of land and resources.

While today the term India is usually understood to refer to the political entity, in a geographic sense India has traditionally included the entire subcontinent, ignoring the boundaries that set Pakistan and Bangladesh apart. If we use this expansive definition, India includes three major regions: the Extra-Peninsular Mountain Wall or Himalayas in the north, the Indo-Gangetic Plain, and the old Peninsular Block or the tablelands in the south.

The mountain region includes perhaps the most breathtaking ranges in the entire world. Many areas cannot be traversed, save for a few high passes; seasonal variations also determine the viability of many mountain routes. The Himalayas and Hindu Kush form the backbone of the region, with peaks rising as high as Mount Everest (29,028 feet), K2 (28,250 feet), and Kanchenjunga (28,168 feet). The Himalayas themselves are divided into three parallel ranges, the Greater, Lesser, and Outer Himalayas. As many of the subcontinent's rivers originate there, the mountains are a key element in the regional water system and also constitute a containing wall for the monsoons.

Throughout history the mountains of India have served a variety of purposes, being, on the one hand, walls and defensive barriers that have kept areas like Tibet relatively isolated and, on the other, determining trade routes and dictating the paths of empires. But they have also served as a clear topographical marker for the subcontinent. As a result India is separated from the rest of Asia by mountains and desert.

The Indo-Gangetic Plain stretches from the border with Afghanistan and the Arabian Sea in the west to Assam and the Bay of Bengal in the east. The plain, which is a product of the alluvial deposits of the major river systems, covers close to 297,000 square miles. South of the plain lie the Thar Desert, the salt flats of the Rann of Kutch, and the hills of central India (such as the Satpura Range). These are all part of the border with the last of the subcontinent's regions, the peninsula.

The peninsula includes coastland, hills, and a large plateau. Hills, known as the Western and Eastern Ghats, run parallel to both coasts. In the center of the peninsula lies the Deccan plateau, rising 980 to 2,460 feet above sea level. India's coastline runs close to 4,660 miles.

With so much of Indian society still employed in the agricultural sector, rivers and irrigation systems play a central role throughout the subcontinent. There are far too many large rivers to list and discuss here, but it is worth noting a handful of the major systems. Both the 1,800-mile Brahmaputra and the 1,560-mile Ganges empty into the Bay of Bengal. Originating in the Kailas range of Tibet, the Indus River flows for 1,800 miles to the Arabian Sea (Indian Ocean), and serves as a border in some provinces as well as a major source of water for agriculture.

Reviewing the natural layout of the subcontinent serves as a reminder that the geographic parameters differ from the political boundaries. The units of the modern day nation-state are at odds with the features created by millions of years of natural development. In some cases border disputes occur where a political line needs to be drawn over a naturally unified area. When bilateral relations are strong and peaceful, such distinctions may not be relevant. But when ties between countries are already frayed, ambiguous borderlands spell trouble. In addition, human settlement patterns are often dictated by geographic and climatic considerations that collide with political necessities. Thus populations following natural dividers can create an extra burden during times of turmoil, as has been demonstrated by the countless examples of refugees throughout the Middle East and South Asia.

Summary

What now follows is an examination of the elements that will most influence the strategic geography of the greater Middle East region for at least the next fifteen years. We chose this time frame in part because it is used by most energy analysts in estimating future demand for oil and natural gas and in part because it permits some leeway for speculation about political, economic, and military developments without being too futuristic. What emerges, we hope, is a realistic appraisal of the impact on a critical region of the world of key elements of Middle East strategic geography—wealth, weapons, and unresolved political conflict.

TWO

Strategic Access and Middle East Resources: Lessons from History

SINCE EARLIEST TIMES, the Middle East has been important for both its economic resources and the fact that it is a strategic crossroad linking Europe, Asia, and Africa.[1] However, over the years the region has assumed greater or lesser importance as a function of three key factors: technology, resources, and political ideology. The lessons of history are not only intrinsically interesting, but have great relevance today and will have for the foreseeable future. Access to Middle East grain supplies influenced the strategy of the Athenian, Roman, and Byzantine empires, and the cedars of Lebanon provided timber for their palaces and warships. In the Middle Ages the impact of the Crusades coupled with the economic revival that followed upon Europe's recovery from the plague led to European demand for silks and spices from the East. That demand was the impetus for the establishment of highly lucrative land and sea trade routes across Asia and through the Middle East to the markets of the Mediterranean littoral as well as to the development of the new naval technologies that enabled European maritime powers to bypass the Middle East and use alternative routes to Asia via the Cape. In more recent times access to Middle East oil has been the prize and that factor has profoundly influenced strategic planning by the major powers. Over the next fifteen years or more, oil and natural gas from the Persian Gulf and Caspian Sea will be key components of a new geopolitical equation with major consequences for the West, the former Soviet Union, and Asia, as well as for the countries of the greater Middle East. Similarly, advances in military technology have had a most significant impact upon strategic access problems.

Early Resource and Access Issues

During the heyday of the Roman Empire, trade via both land and sea routes—of which the Silk Road was perhaps the most famous—flourished with the East, and Roman, Arab, Persian, Indian, and Chinese merchants realized huge profits. Following the fall of Rome in the fifth century, the exploitation of the sea for access remained restricted, as in the past, to coastal trade. The eastern Mediterranean was soon to fall under the control of the forces of Islam, which spread from Mecca in all directions. Over a period of 100 years a new Muslim Europe was established from northern Spain to India. While this resulted in renewed contact between the Arabs and the East, Europe remained more isolated. However, with the Mongol conquest of Asia in the twelfth century, significant changes emerged in the patterns of European commerce. Trade links were reestablished between China and Europe, and by the end of the thirteenth century two routes were in use: the first went from the Crimea through Central Asia and Mongolia to Peking; the second crossed the Black Sea and continued through Persia, Afghanistan, and Sinkiang. The Mongol conquests also opened up a sea channel, known as the Spice Route, which complemented the existing land-based trade routes that paralleled the old Silk Road. This new route crossed the Black Sea and went overland through Persia to Hormuz, at the mouth of the Persian Gulf (sailing ships could not always make the journey to the north of the Gulf because of the prevailing northerly winds), and on to India and the Far East by sea.

The roads were kept in good repair and were safe for the first time since the conquests of Islam severed the links between China and Europe in the seventh century. This new access lasted until 1370, when the Chinese attacked the Mongols, heralding the decline of the Mongol Empire. The Chinese broke off all contact and trade with Europe and shut down the land routes across Asia. However, for a period in the first half of the fifteenth century, China did expand its maritime presence in Asia and, under the famed "eunuch admiral" Zheng He, extended it as far as East Africa, preceding the Europeans on that continent by 100 years.[2]

In Asia Minor the rise of the Ottoman Empire led to conflict with the outposts of Christendom, culminating in the fall of Constantinople in 1453. In the wake of the fall, European trade with China was severely curtailed precisely at a time when the demand for exotic Eastern products was rising and Arab and Ottoman middlemen were positioned to control the Middle East land and sea routes to India and the Spice Islands. Goods destined for Europe were "taxed," stolen, or otherwise appropriated as they passed through the various stages of the long journey.

In the late Middle Ages spices came from India, Ceylon, and Indonesia via Arab ships and were distributed throughout Europe by the Venetians. The spice routes were complicated: in the Far East Chinese junks carried nutmeg and cloves from the Spice Islands to Malacca; the spices then proceeded by sea from Malacca to India, carried in Arab, Malaysian, or Indian vessels. The Indian coast of Malabar housed many spice ports that sold the Far Eastern products, along with cinnamon from Ceylon and pepper from India. From Malabar the Arabs carried the spices to ports in Persia, Arabia, and East Africa. At the end of the fifteenth century there were two different routes to the Mediterranean, by way of Hormuz or Aden.[3] The route from Aden via the Red Sea required that the spices be transferred from the large *baggalas* to smaller coastal vessels and then to caravans to reach Alexandria and, beyond it, Europe. From Hormuz the coastal craft carried the spices up the Gulf into the Shatt al-Arab, where they were transferred to caravans. They were then sent overland through Asia Minor to Constantinople or across Iraq via Baghdad to the port of Tripoli, the outlet for the great bazaar at Aleppo. From Alexandria and Tripoli Venetian ships carried them to Venice, from where they were transported by land and sea to various destinations in Europe.

The importance of spices to the Europeans should not be underestimated. During the late fourteenth and fifteenth centuries, the demand was so high that some prices quintupled at the source over a five-year period.[4] Since the European middle classes were prepared to pay well for spices, numerous entrepreneurs emerged to meet the demand and reaped very high profits.

This, then, was the situation in the mid-fifteenth century: Europe was relatively prosperous and there was sufficient surplus wealth to spend on "luxury" goods from the East. Trade with the East, however, was controlled by Arab and Ottoman middlemen, and only the city-states of Venice and Genoa (primarily the former, as the power of Genoa had declined after its defeat in the fourth war with Venice, which ended in 1380[5]) were able to profit from the situation.

The Age of Discovery

The attempts by spice-consuming countries to break the Arab, Ottoman, and Venetian stranglehold on the spice trade were partially responsible for the major technological changes that led to the Age of Discovery. Another impetus for the search for new routes might also have been the new tax the Ottomans imposed on the Venetians and Genovese in 1460.[6] The latter had enjoyed both political power and a privileged economic position, which were lost when the Ottomans rescinded their immunity from customs duties. Duties were levied at the rate of 5 percent for non-Muslims (4 percent for non-Ottoman Muslims) which further inflated European prices.

To credit the upsurge of Western discovery of Asia that began in the mid-fifteenth century entirely to economic incentives would be simplistic, for the evangelical ambitions of Christian Europe were also very important. Yet there can be no doubt that the ultimate goal of economic reward was never far from the minds of those who supported and undertook the search for new routes and new lands. The economic incentive that made the Portuguese so eager to find an Atlantic sea route to Asia was Arab, Venetian, and Genovese domination of the Mediterranean. Later, however, Spain and Britain both sought alternative routes in the face of Portuguese control of those around the Cape of Good Hope; Magellan's famous voyage, for instance, was an attempt to find a southwest route to the Indies for Spain.

However, no matter how strong their economic and ideological ambitions, the early maritime pioneers could not have succeeded without significant improvements in naval technology, specifically the design of sailing vessels, more accurate navigation techniques, including better charts and the use of astronomy for pinpointing positions at sea, and increased knowledge of the prevailing winds in the Atlantic and Indian Oceans. Much of the credit for these developments goes to Prince Henry of Portugal, who was the leading inspiration behind these advances.[7]

Indeed the first country to fully exploit the new routes to the East was Portugal, but in spite of its pioneering efforts that established the Cape route and opened up the coast of Africa, the Portuguese dominance of the spice trade was short-lived. Although they had no problem securing spices, their efforts to control the trade led to endless skirmishes with competing Arabs and taxed their limited resources. The Portuguese were never able to conquer Aden, which would have given them control of the Red Sea route. There is also evidence that Portuguese ships fared badly during the rough passage from India and that some consignments of spices were soaked with saltwater.

It was the Dutch, rather than the Portuguese, who eventually succeeded in eliminating the Arab traders from the spice markets by establishing a permanent presence in the Indian Ocean and dealing shrewdly with the local powers. However, in the long run it was the British that were most effective at controlling the sea routes and trade to the Indies. As one of the northernmost European countries, Britain paid the highest prices for spices and therefore had major incentives to control the trade. The British not only established workable political relationships with the local rulers, but were more successful at managing the finances of international trade than the Dutch.

The Transportation Revolution and the Middle East: Canals, Coal, Railroads, and Oil

During the nineteenth century the technology of transportation had a radical impact on the geopolitical importance of the Middle East and intensified conflict among the great powers. It altered the patterns of sea and land travel, changing both military and commercial concepts of time and space. By the end of the nineteenth century steamships and railroads enabled the rapid movement of powerful concentrations of naval and military forces, thereby changing the nature of warfare, and by the early twentieth century Middle East oil had become a strategic asset.

The development of the steamship, which replaced the sailing ship as the principal vessel for bulk transportation radically changed the key commercial and military maritime routes. The great sailing ships were restricted to routes that made the best use of prevailing seasonal winds and these rarely followed a straight line, nor did they closely follow coastlines. Thus nominal nautical distances between ports rarely reflected the time it would actually take to sail from one point to another. The prevailing winds across the Atlantic blow from west to east and, especially in summer, all of South America below the Brazilian Cape was closer to Liverpool than to New York in terms of sailing days. With steam propulsion prevailing winds could be disregarded and the routes were nearly equidistant (see maps 4 and 5).

Steamships could traverse the oceans in more direct lines and cut traveling times. However, while the only constraint on the independence of sailing ships was the need for access to fresh water and food, steamships also required coal. Thus throughout the nineteenth century an elaborate system of coal bunkering was developed for both commercial and military use, and this network had profound strategic consequences for the balance of naval power among the major maritime adversaries.

By the mid-nineteenth century trade and military access between Europe and the East had become so important that there was tremendous incentive to develop the Mediterranean and Red Sea route, which could reduce the five-month trip via the Cape from Britain to India to two months. However, this required building either a railway or a canal across Egypt from Port Said to Suez. Plans for a Suez Canal first surfaced in 1841. Britain was initially opposed to the idea, believing that it would undercut its maritime monopoly on the route to India and the Far East, and instead built a railroad across Egypt, which was completed in 1854. That same year, over the strong objections of the English, Egypt's new leader, Mohammed Said Pasha, granted his old friend the Frenchman Ferdinand Mari de Lesseps a concession to build a canal. Despite de Lesseps's involvement the French government remained uncommitted. While they supported the idea of a canal, they did not want to antagonize the British, who were their allies in the Crimean War. As a result they regarded de Lesseps's effort as a private affair. Austria also favored a canal but did not want to side with a Frenchman over England and it, too, adopted a position of benevolent neutrality.

After a great deal of diplomatic maneuvering, de Lesseps finally sold shares in the Compagnie Universelle de Canal Maritime de Suez in 1858, with approximately 52 percent being bought by Frenchmen and work began on the

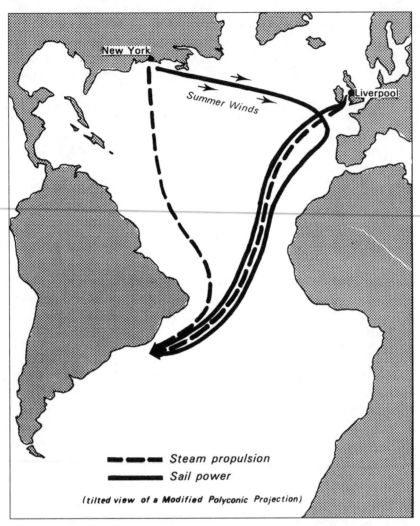

Map 4. Relative distances under sail and steam propulsion. Under sail power all of South America below the Brazilian Cape was much closer to Liverpool than to New York, while under steam propulsion it was nearly equidistant to both.

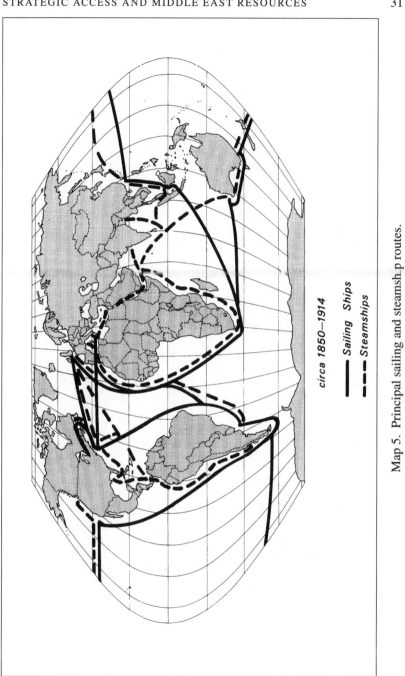

circa 1850—1914

— Sailing Ships

- - - Steamships

Map 5. Principal sailing and steamsh p routes.

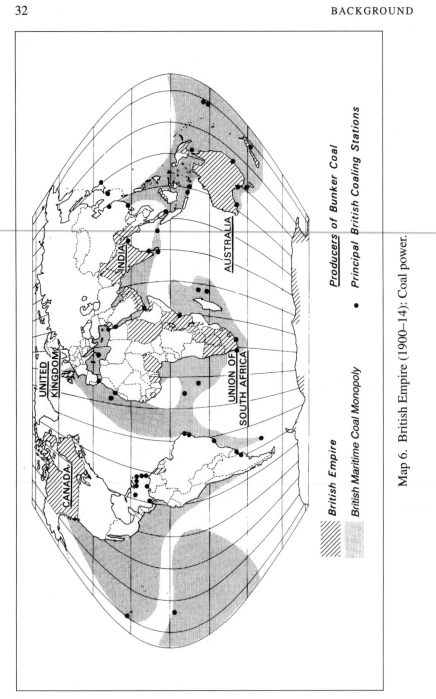

Map 6. British Empire (1900–14): Coal power.

canal in April of the following year. Britain continued to protest and was now joined by Austria, which had become an opponent of France and Sardinia in the 1859 war for Italian liberation. In the fall of 1859 the French government abandoned its neutral position and threw its weight behind the canal project. When the Italian war ended Austria reversed its position and supported the canal. Spain and Russia also offered support. The canal opened on November 17, 1869.

Ironically—but to be expected—Britain derived the greatest economic benefit from the canal. For many years from 60 to 80 percent of the ships using the canal were British. In 1875 Britain purchased over 175,000 Egyptian-owned shares in the Compagnie and became the largest single shareholder (over 40 percent). With the establishment of a British protectorate in Egypt in 1882, the canal effectively came under British control, which meant that the Royal Navy could steam directly to India from the Mediterranean making use of coaling bunkers in Egypt and at the entrance to the Red Sea in Aden.

The need for coal stations around the world was very much to Britain's advantage during the last century. Until about 1880 Britain produced and exported more coal than the rest of the world combined. Even after the industrial revolutions in Germany and the United States and those countries' subsequent rapid increases in coal production, Britain dominated the world's maritime coal trade thanks to the commanding position of its merchant marine and its possession of a chain of strategic coaling stations throughout the Empire. Furthermore, coal from Newcastle, South Wales, the Clyde, and the Mersey was of unsurpassed quality and was preferred by all navies.[8]

Because of its control over the world's seaborne coal trade, Britain maintained a system of coaling stations around the globe unmatched by the other great powers. Its maritime coal dominance was possible because of the worldwide spread of its empire and the fact that few countries of the empire, aside from Britain itself, produced good coal (Australia, Canada, India, and South Africa). As map 6 shows, the areas in which the monopoly existed coincided with the most important maritime trade routes. As a consequence Britain could aid or greatly hinder the movements of any other naval power.[9]

Britain possessed another great advantage over her rivals during the nineteenth century—the domination of the world's telegraph cable communication links outside the continent of Europe. Strategic and commercial interests in London readily appreciated the significance of this development: in time of crisis or war warning and instructions could be swiftly transmitted throughout the far-flung empire. The Treasury received startling proof of the efficiency of telegraphic communication in 1857, when the rapid transmission of news about the defeat of the great mutiny halted the movement of troops in Canada destined for India and thereby saved over £50,000.[10] Cable links could make imperial defense a more coordinated and less expensive undertaking.

Cable, however, presented military planners with a new set of problems vis-à-vis its influence on strategy. One of the most basic questions that had to be answered was how to protect cable communications from interruptions. A cable running on land through the territory of another country could well be cut at those moments when security communications would be deemed most vital. The solution to this problem was the development of a network of submarine cables that passed only through the territory of the British Empire. This "red line" cable system gave Britain an important advantage over the other great powers in diplomatic disputes and military operations outside Europe.[11]

British Competition with Russia and Germany: The Great Game and the Role of Railways

Though Britain was able to control the maritime approaches to its far-flung empire through the power and presence of the Royal Navy, land threats were a different matter. No part of the empire was more vulnerable to land invasion than its prize possession, India. Britain's most serious strategic challenge to India in the nineteenth century was Russia's relentless expansion east and south from its European base, which included the building of a modern railway system to provide easy access for the army to control the newly acquired Russian Empire. British-Russian rivalry over Central Asia, especially Afghanistan, was termed the "Great Game" by famed author Rudyard Kipling. In the early twentieth century Britain was also concerned about Germany's encroachment into the Middle East, especially the proposed Berlin-to-Baghdad railway via Constantinople. This threat was never perceived to be as dangerous as the one posed by Russia, though had the railroad been completed it could have strengthened Turkey's strategic position in World War I and increased the threat to Egypt and the Suez Canal (see map 7). To ensure against the potential threat to its empire, Britain shored up its defenses in the Persian Gulf.[12] It is noteworthy that with the breakup of the Soviet Union and the growing conflicts among the new independent republics of the Caucasus and Central Asia, the term "Great Game" has reemerged in Western discussions about the geopolitical importance of the region.

Russo-English competition in South Asia dates back to the eighteenth century and the reign of Czar Peter the Great. While the British were busy consolidating their position in India, the Russian Empire expanded in every direction under Peter, who began acquiring parts of Persia with an eye to the fabulous wealth of India, which lay further afield. Peter sent a force of 3,500 troops to conquer Khiva and find a road to India, and although the mission failed miserably, Peter's Asiatic ambitions did not go unnoticed in London.[13]

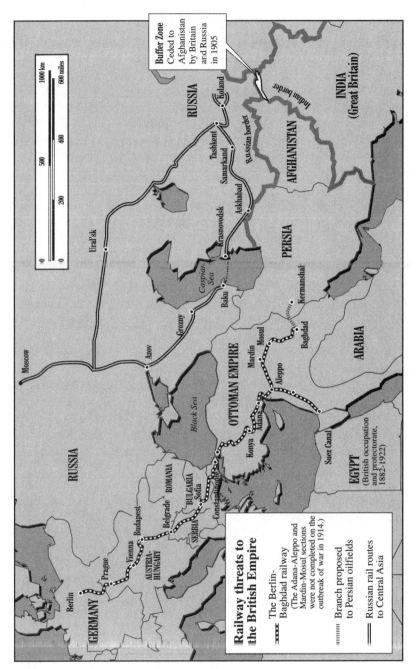

Buffer Zone Ceded to Afghanistan by Britain and Russia in 1905

Railway threats to the British Empire

- The Berlin-Baghdad railway (The Adana-Aleppo and Mardin-Mosul sections were not completed on the outbreak of war in 1914.)
- Branch proposed to Persian oilfields
- Russian rail routes to Central Asia

Map 7

By the early nineteenth century the unexplored capitals of Persia, Afghanistan, and Central Asia began occupying the public imagination in Europe. London and Paris were abuzz with stories of the legendary wealth and splendor of Samarkand, Bokhara, Herat, and Khiva. As Russian troops fought their way southward through the Caucasus, subsequently subduing the Kazakhs, Uzbeks, and Turkomans, the frontier between the British and Russian empires began closing rapidly, further disturbing the British. Over the course of that century, "many officers and explorers, both British and Russian . . . took part in the Great Game. . . . The ultimate prize, or so it was feared in London and Calcutta, and fervently hoped for by ambitious Russian officers serving in Asia, was British India."[14]

In 1807 British intelligence learned that Napoleon Bonaparte planned to march 50,000 French troops across Persia and Afghanistan, where they would join forces with the Cossack regiments of Czar Alexander I for the "final thrust across the Indus into India." This threat, too, eventually dissipated as Napoleon and Alexander fell out, but British fears for the safety of India intensified as the Russians, having subdued the Caucasus, took a renewed interest in the lands further south, advancing toward the ancient Muslim khanates of the former Silk Road. Tashkent, Samarkand, Bokhara, and Khiva all fell to the Russians by the late nineteenth century and the gap between the two empires—some 2,000 miles at the beginning of the century—had closed to a few hundred miles and at one point—the Pamir region—to less than twenty.[15]

Of greatest concern to Britain was the development of the Trans-Caspian railway, which the Russians had begun building in 1880 and continued to push on with as neighboring lands were conquered. Baku on the Caspian Sea was served by two railroads, one from Grozny in the North Caucasus and a west-east line across Transcaucasia from the Black Sea. From Baku the Caspian Sea ferry carried goods and services to the railhead at Krasnovodsk. By 1888 the line had reached Bokhara and Samarkand and was on its way to Tashkent. With its rapidly increasing capacity for transporting troops and artillery, the Trans-Caspian railway exposed the inadequacy of India's frontier links, particularly its roads and railways, leaving the border of the British Empire barely protected from a potential Russian advance to the south. Key British army leaders began calling for a countermove to Russia's railway encirclement of northern India and Afghanistan, arguing that India's defense budget would be "better spent on enabling commanders to rush troops to a threatened sector of the frontier, than on building forts and entrenchments which might never have to be defended." Some even advocated forming an alliance with Abdur Rahman of Afghanistan, then extending the railway into that country and stationing British troops there, on the grounds that the Russians were certain to try and occupy the whole of Afghanistan when Abdur Rahmand died, taking advantage of any ensuing power struggle.[16]

Despite these pressures, however, there were not more than fifty miles of British railways in India's frontier region, and it was not until the appointment of Lord George Nathaniel Curzon as Viceroy of India that railways were recognized as a vital component of India's defense. As a young aspiring statesman, Curzon had traveled the length of the Trans-Caspian railway during the summer of 1888, curious to gauge for himself the extent of this great Russian threat. Upon his return he wrote that the existence of the Trans-Caspian line had "dramatically altered the strategic balance in the region."[17] When the Russians completed the final 200-mile stretch linking Samarkand and Tashkent, they were able to move as many as 100,000 troops from as far away as the Caucasus and Siberia to the Persian and Afghan frontiers as well as artillery and other heavy equipment, once a nightmare to transport across vast distances and rough terrain. The expansion of the Russian rail network was one of the key themes addressed by Halford Mackinder in his famous 1904 essay, "The Geographical Pivot of History," which touted the growing importance of the major land powers of Eurasia.

Along with many other historians, Curzon also theorized that the Russians were not necessarily interested in the conquest of India; rather they wanted to "keep England quiet in Europe by keeping her employed in Asia," for their real objective was not Calcutta but Constantinople. In any event the Trans-Caspian railway, according to Curzon, made the Russians "prodigiously strong" in the Central Asian game, though Britain's policy of denying Russia rail routes in Afghanistan resulted in there being an important buffer between the rail line and the borders of the British Empire.[18]

Coal versus Oil

In the nineteenth century Britain possessed a tremendous strategic advantage over any naval rival because of her control of the world's seaborne coal trade, but it began to lose that edge in the twentieth century because of radical changes in the engine design of capital ships. The reason can be summed up in one word—oil. Because of the rapidity of the shift from coal to oil to fuel ships, this period of the early twentieth century provides a particularly good example of how the power of an empire can be eroded by lack of access to a vital resource. Britain became increasingly dependent upon one of its major industrial rivals, the United States. Equally important it demonstrates how quickly a hitherto backward region such as Mesopotamia can assume great strategic importance in a relatively short time. The lessons for today are very clear if one considers the new interest in the Caspian Basin (see chapter 4).

Despite its dominance of the world's coal industry and control of a world-wide logistic infrastructure based on coal, new modes of propelling warships at greater speed forced Britain to acquire bases and sources of oil that could support an oil-fueled fleet. Winston Churchill, who was appointed first sea lord in 1911, was an early supporter of oil-fueled ships. He set up a Royal Commission on Fuel and Engines chaired by a former first sea lord, Jackie Fisher. Oil clearly possessed many advantages over coal as fuel for warships. It produced more heat, which in turn produced steam under greater pressure and turned the propellers faster. It could also be transferred at sea from tankers and far fewer sailors were needed to refuel.[19] Britain, unlike the United States, did not have a domestic supply of oil. This meant immense expenditures on tankers and maintenance of a large reserve; Churchill estimated that over £10 million had to be spent creating the logistic infrastructure for ships driven by oil.[20]

Most of the world's oil production, and the overwhelming majority of British investments in oil, was divided between Russia and the United States. Britain could not afford to be caught during a time of crisis or war in a position of dependence on the good will of another major power, so the Commission began to search the world for areas that could be closely controlled by the British government and had the potential of providing the Royal Navy with oil. During the course of its hearings, the Commission heard pleas for financial support from the British government from oil companies operating in Newfoundland, Egypt, Nigeria, and even Scotland, where shale oil had been found. They offered to provide for the oil requirements of the Royal Navy in return.[21] The Commission also received an offer from Royal Dutch Shell to supply the Royal Navy with oil in return for British diplomatic support. However, a small company operating in Persia—the Anglo-Persian Oil Company (APOC)— managed to convince the British government that it, rather than Shell, should receive British support. APOC argued that the world's oil industry was falling under the domination of two "monopolistic" companies—Standard Oil of the United States and Shell—and maintained that unless it received support from the British government it would be swallowed up by one of these two giants. Neither of the big firms could be depended upon in a crisis, APOC asserted, because they were controlled by major foreign powers: Standard Oil by the United States and Shell, with 60 percent of the company's shares in the hands of the Dutch government, vulnerable to pressure from Germany on the Netherlands. As a result the British government acquired a controlling share of APOC for £2,200,000. While seeking parliamentary approval for this purchase in the summer of 1914, Churchill described the alleged dangers of the Shell "octopus" and the need for Britain to have a dependable source of oil. In the face of weak opposition Churchill easily received parliamentary approval for the measure, which has been compared to Disraeli's purchase of a controlling interest in the Suez Canal.[22]

Impact of World War I on Oil Supplies

The outbreak of war in August 1914 prevented the implementation of the program recommended by Fisher's Royal Commission, which had called for a reserve equal to four years' worth of wartime consumption. In any case Churchill, mindful of colleagues within the cabinet who opposed huge naval expenditures, felt that four and a half months' wartime supply would be enough.[23] Even had a war not begun, however, the cost of stockpiling such a large fuel reserve would have been prohibitive.

Britain's failure to amass sufficient oil supplies did not greatly affect its strategy or its conduct of operations early in the war, with the exception that it considered it essential to land an army from India at Basra, at the head of the Persian Gulf, beginning in October 1914, to protect its newly acquired oil fields and the pipeline from Turkey. However, in February 1917 the Germans resumed their campaign of unrestricted submarine warfare, which nearly immobilized the Royal Navy for lack of oil as the number of tankers sunk began to rise drastically. At the onset of the campaign there were reserves of 5.1 months' consumption; by May the reserve level had been reduced to 2.9 months, with some bases down to only six days' supply. At that point the Admiralty staff calculated that if the trends in sinkings of tankers continued unabated and consumption rates of the fleet remained the same, by the end of 1917 the Royal Navy would have only six weeks' reserve on hand.[24]

Owing to the dire situation the Admiralty had to send what one commentator has called urgent and humiliating telegrams to the United States saying that the Royal Navy would be immobilized unless more American tonnage became available to carry oil across the Atlantic. Despite its control of APOC the shortage of its own tankers forced Britain to draw about 80 percent of its oil supplies from North America. The Admiralty director of stores noted in September 1917 that, "without the aid of oil fuel from America our modern oil-burning fleet cannot keep the seas."[25] One recent writer has compared this fear on the part of British policymakers of having their oil supply cut off to a "castration complex," one which would haunt them from the submarine-induced crisis of 1917 to the Suez Crisis and until today (see map 8).[26]

Britain managed to overcome the worst of its oil crisis in a novel manner: it used the ballast tanks of liners and cargo steamers as fuel-carrying "double bottoms." The British war cabinet ordered this measure in June 1917, and by November the 443 ships that had been outfitted had carried approximately 243,519 tons of oil from the United States, or the equivalent of the total capacity of fifteen tankers. The introduction during the summer of 1917 of the convoy system for the protection of trade further eased the fuel crisis.

Despite these measures, however, the Allies barely managed to meet their oil needs. The situation was sometimes so critical that the shipping controller suggested to the war cabinet in August 1917 that the Royal Navy should stop

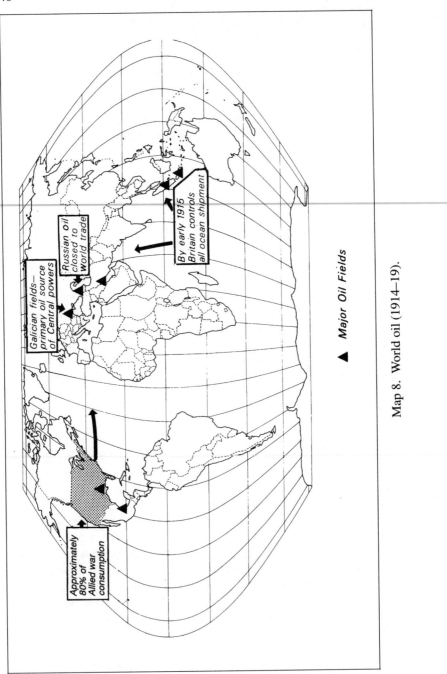

Galician fields—
primary oil source
of Central powers

Russian oil
closed to
World trade

By early 1915
Britain controls
all ocean shipment

Approximately
80% of
Allied war
consumption

▲   *Major Oil Fields*

Map 8. World oil (1914–19).

building ships fueled by oil and revert to coal.[27] Furthermore, the tonnage used to ship oil across the Atlantic resulted in cutbacks in other vital commodities—most notably foodstuffs. By December 1917 approximately 900,000 tons of wheat originally destined for Britain remained at American ports owing to the shortage of vessels for shipping.[28] The widely discussed food shortage that eventually led to rationing in Britain at the beginning of 1918 can be seen as the result of a conscious decision on the part of the British war cabinet to solve the oil crisis by giving oil priority over foodstuffs in shipping.

Britain's Quest for Middle East Oil

In the immediate postwar period Britain set out to make herself independent of the United States in oil supplies in times of crisis. While the war was still being fought the Admiralty began producing position papers concerning the oil needs of the empire. These papers called for exclusive British control of the Persian and Mesopotamian oil fields, a view reflected in the war aims subsequently put forward by Britain after the war.[29] The British decision to achieve independence from American oil added to the Anglo-American naval rivalry over the building of battleships before the Washington Naval Conference in 1922. In early 1920 First Lord of the Admiralty Walter Long made a speech declaring that, "if we secure the supplies of oil now available in the world we can do what we like." Long warned that if Britain did not take this opportunity to gain control over the oil fields of the Persian Gulf, "others will take it, and with it the key to all future success." Tensions in Anglo-American relations eased after the Washington conference and plans to create a worldwide oil reserve comparable to what Britain had once controlled in coal came under constant Treasury attack for economic reasons throughout the 1920s.[30] Thus Britain never amassed sufficient oil reserves, so the logistic infrastructure of the British Empire did not dominate the world's oceans as it had in earlier years when warships were propelled by coal.

World War II and Middle East Oil

Access to oil supplies was even more vital to the combatants during World War II than in World War I. By 1939 the mechanization of armies and the increasing use of airpower generated an unprecedented demand for petroleum fuels.

Most of world's oil supplies came from the United States, the Soviet Union, the Caribbean, South America, and the Middle East (see map 9). The Axis powers (Germany, Italy, and Japan) were desperately short of fuel supplies, and

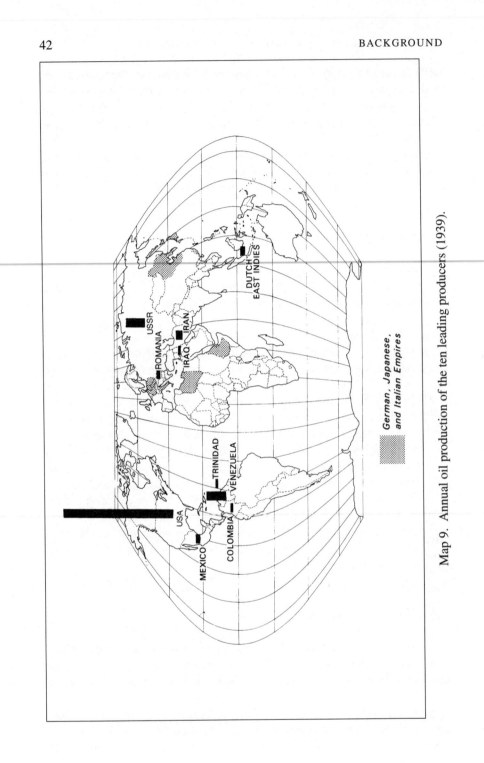

Map 9. Annual oil production of the ten leading producers (1939).

as a consequence German and Japanese grand strategy was greatly influenced by the need for access to secure sources. The Nazi-Soviet Pact of 1939 contained several secret clauses that concerned oil and the Middle East that were to influence strategy in the early days of the war. The Soviet Union was promised a free hand in the Persian Gulf if the West was defeated by Germany. Furthermore, Stalin's decision to supply Germany with oil from the Caucasus was a bitter blow to the Allies. At one point Britain and France contemplated bombing Soviet oil facilities in the Caucasus from French air bases in Syria. However, this would have required overflying Turkey, which was neutral.

The proposal was satirized by A. P. Herbert in a poem entitled "Baku or the Map Game," written in April 1940[31]:

It's jolly to look at the map
And finish the foe in a day.
It's not easy to get at the chap;
These neutrals are so in the way.
But if you say "What would *you* do
To fill the aggressor with gloom?"
Well, we might drop a bomb on Baku,
Or what about bombs on Batum?

However, once Hitler's armies had overrun Western Europe in 1940 and his planned invasion of Britain had failed, he turned his attention once more to the East. The German invasion of the Soviet Union in June 1941 was in part an effort to gain access to the enormous resources of the Ukraine and to protect the oil supplies coming from Romania. At this point in time Germany was cut off from maritime supplies from the rest of the world by the Royal Navy and had no trust that Stalin would continue to deliver vital resources as he had done since the signing of the Nazi-Soviet Pact in 1939.

German strategy in 1941–42—the *Drang nach Osten*—was eventually to drive to Stalingrad and to move simultaneously on the Caucasus, which contained the great three oil fields at Maikop, Grozny, and Baku, as well as the pipelines (see maps 10 and 11). This move was to be paralleled by Rommel's Panzerarmee Afrika eastward drive along the North African coast toward the Nile Delta, the Suez Canal, and Palestine. If Rommel had succeeded and driven Britain out of Egypt and the northern forces had broken through the Caucasus, Germany would have been in a position seriously to threaten the remaining British forces in the Persian Gulf and gain access to Middle East oil.

The German invasion of Russia radically altered the strategic balance and led to a Soviet military agreement with Britain, including joint plans to protect the Persian Gulf. In August 1941 British and Russian units invaded Iran on the pretext of enforcing its neutrality, which was jeopardized by German fifth columnists and the generally pro-German attitude of Shah Mohammad Reza Pahlavi. In reality the attack was designed to do three things: first, to protect the

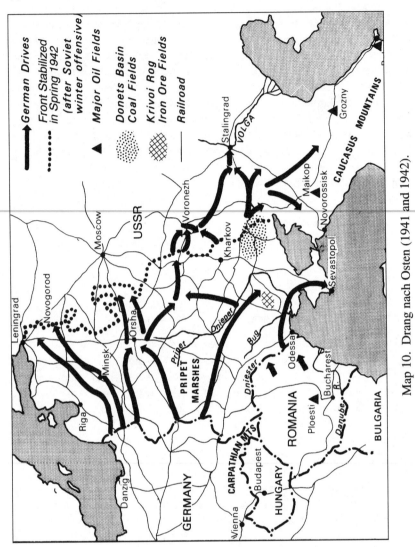

Map 10. Drang nach Osten (1941 and 1942).

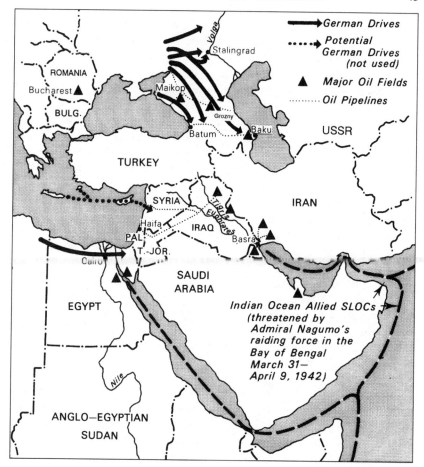

Map 11. German drive for victory (summer 1942).

oil fields in Iraq and Iran; second, to secure a line of communication to protect the Gulf in case the German army broke through the Caucasus; and third, to establish a southern supply route to Russia. The Soviet government's legal pretext for invasion was the peace treaty it had signed with Iran in 1921, which permitted it to place troops temporarily on Iranian territory for the purpose of self-defense.

With Iran secure but with the German armies threatening Allied positions in Egypt and the Caucasus, in May 1942 the commander in chief of the British Middle East Forces, General Sir Claude Auchinleck, made plans in the event of a German attack through Iran from the Caucasus. Auchinleck declared that his objective was to ensure the security of bases, ports, oil supplies, and refineries in Iraq and Iran, all of which were located in the center and south of the region, by stopping the Germans as far north as possible. His belief was that if the Germans succeeded in breaking through Iran's northern provinces it would be much more difficult to prevent them from eventually launching a two-pronged attack on the Gulf through Iran and Iraq.

This German threat to the Middle East must be seen in conjunction with the threat posed to the area by the Japanese navy. The Japanese decision to attack the United States at Pearl Harbor was in part triggered by the effectiveness of the U.S. oil embargo that had been imposed in the summer of 1941 and the need to secure the resources, especially oil, of the Dutch East Indies. With the defeat of the British and Dutch in Southeast Asia in 1942 the way was now clear for Japan to move on India and send its navy into the Indian Ocean toward the Middle East. During the period March 31–April 9, 1942, a major Japanese naval force (far superior to anything the Allies had in the area), consisting of five fleet carriers, four fast battleships, three cruisers, and eight destroyers commanded by Admiral Chuichi Nagumo, raided the Bay of Bengal and the east coast of India. If the Japanese had decided to continue their advance into the Indian Ocean there would have been nothing to stop them, and they could have seriously disrupted the vital Allied sea lines of communication (SLOCs).

By the end of 1942 the Axis threat to the Middle East was over. Rommel's army in North Africa was defeated, the Soviet victory at Stalingrad had turned the tide of the war in the East, and any German threat to the Caucasus and Japan's plan for an invasion of India had been forestalled. For the rest of the war Allied oil supplies were never seriously threatened. In contrast, the Allied bombing campaign against Germany's oil facilities, including the synthetic plants in Germany and the Romanian fields at Ploesti, posed increasing constraints on German mobility and contributed significantly to the eventual Allied victory. Furthermore, with the Middle East secure, the Allies were able to supply the Soviet Union with vital war equipment provided under lend-lease.

The history of American military aid to Russia during World War II highlights the importance of Middle East geography and the search for strategic access. Following the German invasion of Russia in June 1941, there was optimism in the West that the Germans would become bogged down in an eastern campaign but also fear that the Russian army would collapse, leaving the enormous resources of the Soviet Union at Hitler's disposal. The British understood that support for Russia was essential to their own war effort and, although Franklin Roosevelt—who at that time was president of a technically nonbelligerent country—was under pressure not to help Russia (especially by the expressed doubts of his military advisers), the United States brought the Soviet Union quickly into the Lend-Lease Program.

The problems involved in supplying the Soviet Union—aside from the difficulty of getting straight answers from Soviet officials as to their requirements and needs—were logistic. To understand their magnitude, it is well to recall the geography of the Soviet Union and the various lines of supply between the Allies and the Russian armed forces in late 1941 and 1942. The heaviest fighting was taking place in European Russia and the threat at that time was to its north, center, and south. The German offensive in Europe had effectively sealed off the Baltic route and the Mediterranean routes to the Black Sea. That left three supply lines that were, in theory, open: the Arctic route to Murmansk and Archangel, the Atlantic and Indian Ocean route to the Soviet Union via Persia, and the Far East route via Vladivostok.

Each of the three routes posed different problems. The Arctic route was the quickest and most direct, and the Soviet ports and logistical facilities were the best equipped for rapid dispersal of materiel, but the German occupation of Norway presented an enormous hazard. This threat, added to the horrendous weather conditions, made the Arctic convoy routes extremely perilous and they had to be suspended intermittently.

The Pacific route by Vladivostok had both advantages and disadvantages. On the plus side, the Soviet Union and the Japanese Empire had signed a nonaggression pact and were not belligerents for most of the war. Thus steamships flying Soviet flags plied across the Pacific carrying war goods made in the United States, and for a time this source of supply was greater than the combined total of the more famous Persian route and Murmansk run. On the minus side, first, as Russian-Japanese relations deteriorated the ships were vulnerable to Japanese interdiction from bases in China; second, Vladivostok was thousands of miles from the main battlefronts and the Trans-Siberian railway, apart from the fact that it was a tortuous route, was not equipped to carry the quantities of matériel that the Russian forces needed.

The growing difficulties of the Vladivostok run led to increasing interest in the Persian Gulf route since Britain and Russia had jointly occupied Persia in August 1941. Here the problems were the inadequacies of the facilities, as the ports and railroads did not have the capacity to handle the tonnage. However, once matériel had arrived in the Caspian, it could quickly be dispersed to the Soviet forces. The difficulties were eventually resolved when the U.S. Army Corps of Engineers took over management of the Persian corridor; that is to say they became responsible for the running of the ports and the railway, but the British retained overall strategic control of the theater. On a day-to-day basis the U.S. Army ran the lend-lease operation but Britain retained the option of dictating priorities in terms of where materiel went and when. Great quantities of equipment were shipped through the Persian corridor throughout the war.

The Cold War, Europe, and Middle East Oil

World War II put enormous demands for oil on all the combatants, including the United States, which was the primary producer. So great was the need for U.S. oil that there was serious fear of a shortage. This concern led to the need for "conservation," which meant that for the future the United States would have to develop foreign sources of oil in order to conserve its domestic fields. Similar to Britain's concern after World War I, the United States realized the vital importance of the Middle East as a strategic petroleum reserve.[32]

The World War II experience left U.S. military professionals extremely pessimistic about America's ability to "oil" another protracted war. The assumptions were that a war with the Soviet Union could last three to four years, that the few atomic weapons in the U.S. arsenal would not be decisive, and that the United States did not have enough oil reserves to fight a war lasting more than a year or two. War plans thus placed great emphasis upon the need for the West to secure the Middle East oilfields early in the conflict.

On February 10, 1947, a study prepared by the Joint Logistics Committee was forwarded and approved by the Joint Chiefs of Staff. It stated:

> In a future major war of five years' duration, during the period 1947–1951 inclusive, the total United States military and civilian consumption requirements cannot be met after M + 3 years by all the then current production in the United States and United States controlled foreign sources, including that in the Near and Middle East, even with the proposed new drilling, new refinery, and synthetic plant building programs proposed herein as optimistic but realizable.[33]

Due to the contention of the U.S. Army and Army Air Force that the use of atomic weapons would permit the defeat of Russia within six months, thereby eliminating the necessity for long-range planning for oil consumption, the response of the Joint Chiefs of Staff to the study was anemic. Secretary of the Navy James Forrestal pleaded for the early development of the Alaskan oil reserves, and a member of the Joint War Plans Committee stated in a proposal that the problem of acquiring oil was so basic that "conventional warfare between the United States and the USSR would inevitably result in a material stalemate as both sides depleted their reserves."[34]

In contingency plans developed in 1949 for war between the United States and the USSR, American military planners believed it to be essential to the successful conduct of such a war that the oil fields of the Middle East be in the hands of the West. They maintained that, "if the Near and Middle East oil areas are not retained, or retaken in the early phases of the war, the oil position of the Allies would necessitate subsequent actions to retake it." The Allies estimated that in order to recapture the oil-producing areas at the head of the Persian Gulf from Soviet ground forces, they would need five divisions (three infantry, one armored, and one airborne), five fighter groups, and one light bomber group.[35]

Gradually, however, concerns about oil supplies receded, as the importance of nuclear weapons grew. Then, too, it was seen that oil shortages never materialized as a serious problem during the Korean War, in part because of major new production from North America and the Middle East. After Korea, the introduction of the fusion weapon (hydrogen bomb) and long-range jet bombers into the strategic arsenal meant that a global war with the Soviet Union would now be fought and "won" in days, so the preferred strategy became deterrence rather than protracted war. However, access to oil, coal, bauxite, and other raw materials continued to be given high priority by the U.S. military even as the world entered the era of the intercontinental ballistic missile (ICBM).

While the military establishment worried about "oiling another war," a different set of priorities emerged on the civilian front. Once the war was over, the end of gas-rationing in the United States and an economic boom led to a massive increase in gasoline use, primarily because of the explosion in the number of automobiles in service (20 million in 1945; 40 million in 1950). Increasing consumption served to raise prices and price increases stimulated greater oil exploration in North America, so that by 1950 U.S. reserves were 21 percent higher than they had been in 1945.[36]

Although the United States could continue to rely on its own oil to meet the growing demand, the European situation was very different. A postwar energy crisis developed that dramatically increased Europe's dependence on Middle East oil. The combination of wartime destruction and social and labor chaos coupled with the coldest weather of the century resulted in a near breakdown of coal supplies. (One effect of the cold was to literally freeze the rail junctions and switch points and prevent the transport of coal.)

It was the energy crisis of 1946–47 that drove the United States to reassume its leadership, which had been on the wane in the immediate aftermath of victory and the rapid demobilization of U.S. military forces. The clearest manifestations of new American involvement were the decision to replace Britain as the protector of Greece and Turkey (the Truman Doctrine) and the launching of the European Recovery Program (the Marshall Plan).

One of the effects of the Marshall Plan was to accelerate Europe's shift from coal to oil as the major fuel for industry and power generation. Plenty of oil was now available but maintaining the flow meant increasing dependence on the Middle East. Thus began the contemporary Middle East oil boom with a rapid expansion of fields and distribution systems, including new pipelines to the Mediterranean. However, the increasing importance of Middle East oil coincided with the rise of Iranian and Arab nationalism and the creation of the State of Israel and the intensification of the Arab-Israeli conflict.

During the 1950s two Middle East crises were precipitated over the control of oil and its distribution. In 1951 the Iranian government nationalized the Anglo-Iranian Oil Company, which had been jointly owned by Britain and Iran. Tension over control of the company had been building for several years because Britain was getting more revenue from taxes on company profits than Iran received in royalties. When prime ministers Ali Razmara and Hussein Ala failed to satisfy the Iranian public's demand for nationalization of the company, they were forced aside. Violent demonstrations were held in support of the Majlis's demand that the Shah appoint Mohammed Mossadegh, a leading proponent of the nationalization, as premier. Relations between Iran and the West became increasingly embittered. Mossadegh became prime minister on April 29, 1951, and implemented the Anglo-Iranian takeover and other policies unpopular in the West. A retaliatory oil embargo against Iran observed by most European shippers had a crippling effect on the Iranian economy. In August 1953 the United States and Britain, fearful that a Soviet coup and takeover of Iran was brewing, helped to overthrow Mossadegh and restore the Shah to full power.[37]

The second crisis occurred in 1956 when Egyptian president Gamal Abdel Nasser nationalized the Suez Canal as retaliation against a decision by the United States, supported by Britain, to cancel a World Bank loan to help Egypt build a dam at Aswan on the Nile in southern Egypt.

The repercussions of Nasser's action were to lead to the worst period in postwar relations between Britain and France, on the one hand, and the United States, on the other. The two European countries regarded the nationalization of the canal as a mortal threat to Western oil supplies, especially given the close evolving ties between Nasser and the Soviet Union. (France was also angered by Egyptian support of the insurgents in Algeria.) They were prepared to use force. The United States, while sympathetic to the British and French position, totally rejected any plans for military intervention to regain control of the canal. Britain and France engaged in war preparations during the summer and early fall of 1956 and were joined by Israel, which wished to expel Egyptian forces from the Sinai Desert in order to put an end to the guerrilla raids being launched from there. The subsequent military campaign produced mixed results. In a lightning offensive Israel succeeded in ejecting Egypt from the Sinai but under American pressure had to withdraw its forces in 1957. The Anglo-French operation was a disaster because the United States refused to support Britain and France in the financial markets. The subsequent run on the pound and franc forced both countries to back down. To compound the fiasco the Soviet Union used the crisis as a cover to march into Hungary and brutally suppress an anti-Communist uprising.

Western Basing in the Middle East and the Soviet Drive for Access Early in the Cold War

Very early in the cold war, the United States and its Western allies strove to develop and maintain a system of air and naval basing access all across North Africa and the Middle East and in and around the region's key bodies of water, the Mediterranean, the Red Sea, and the northwest Indian Ocean. This was obviously reflective of a rimland containment strategy directed against Soviet expansionism, which was predicated on such a basing system being constructed all around the Sino-Soviet bloc and its Eurasian heartland. The bases were a correlate of the alliance structure developed during the 1950s, featuring NATO in Europe (including Turkey) and the CENTO arrangements further east that had Turkey and Pakistan, respectively, as their western and eastern hinges. Later, beginning slowly in the late 1950s, but accelerating in the subsequent two decades, the USSR penetrated the region on the basis of ideological client states (Egypt, Syria, Iraq, India, North and South Yemen, Algeria, and Libya) undergirded by vast quantities of weapons supplies, and a competition for basing access developed between the two superpowers throughout the region.[38]

In point of fact neither the United States nor the USSR had significant basing points in the greater Middle East in the immediate aftermath of World War II, just as neither had had much access prior to that war (Britain and the USSR had to withdraw from their joint occupation of Iran in 1946).[39] But the United States and its allies were availed of an elaborate network of bases throughout the region because of Britain's continuing hold on its empire and related protectorates; to a lesser but significant degree France's remnant empire in North Africa served the same purpose. Well after World War II, but constituting a gradually dwindling asset, the United States and its British "host" were provided air and naval access to, among others, Gibraltar, Malta, Libya (Wheelus Air Base), Egypt (including a massive web of bases in the Suez Canal Zone), Crete, Cyprus, Jordan, Iraq (until the Baathist revolution), Oman, Aden, Bahrain, Pakistan, the Maldives Islands (Gan), and Ceylon; there were also bases on the periphery of the region in Kenya, the Seychelles, Mauritius, and more. Indeed Britain used many of its intraregional bases in an increasingly futile effort to suppress various anticolonial revolts in places such as the Suez Canal Zone, Cyprus, Aden, and Kenya. The United States, meanwhile, was also provided with air bases in Morocco while it was still a French possession, which proved to be essential for the forward basing of B-47 "Reflex" force bombers in the late 1950s.

Aside from the access provided by its NATO allies, the United States acquired a few bases on its own in the region in return for extensive security assistance. In the late 1940s it acquired rights to the Saudi air base at Dhahran. In the 1950s an independent Pakistan gave it access to a key intelligence air base at Peshawar, which it used to stage U-2 flights over the Soviet Union. Haile Selassie's Ethiopia came to provide the United States with access of various sorts, including the key intelligence and communications post at Kagnew. Most important of all, in the 1950s Turkey came to host a variety of U.S. facilities, including air and naval bases, stations for signals intelligence (SIGINT) radar systems, and seismological listening posts. Later on the Shah's Iran would provide the United States with additional intelligence facilities in the northern part of that country adjacent to Soviet Central Asia.

Several political and technological trends evolved during the first decades of the postwar period that characterized the patterns of basing access. On the political side there was the gradual unraveling of the British and French colonial empires, which by the 1970s saw Western access voided or curtailed in such countries as Morocco, Algeria, Libya, Egypt, Iraq, South Yemen, Malta, and Sudan, though such access was retained in Pakistan (up to 1965), Cyprus, Kenya, Ethiopia, and Bahrain, among others. The United States used security assistance to Iran and Turkey to gain extensive access along a wide swathe of the Soviet southern frontier. At the same time the Soviets by the 1970s had acquired important air, naval, and other facilities in Algeria, Libya, Egypt (also Yugoslavia), Syria, Iraq, North and South Yemen, Somalia, India, and Sudan, so there was a virtual "base race" in the region, which was paralleled by, among other things, competitive naval deployments in the Mediterranean and the Indian Ocean.

However, ongoing technological developments—in weapons, aircraft, satellites, and ships—gradually changed basing requirements for the major powers. These technological advances had a major impact in relation to the key geographical variables of distance, location, and the relationship of the various layered media, that is, the underseas, land, atmosphere, and outer space. Space limitations here preclude more than a cursory discussion.

First, the ranges of transport and combat aircraft were extended (all the more so once aerial refueling was introduced by the United States in the mid-1950s), which reduced the requirements for access to transport staging bases such as those at Wheelus and Dhahran. As a matter of fact by the time of the 1973 war, the United States was able to stage an arms resupply operation on behalf of Israel with just one staging base in the Portuguese Azores, abetted by the quiet use of a tanker facility in Spain. Ironically, though the Americans were able by the early 1950s to mount two-way bomber missions from the continental United States to the USSR with the giant B-36 bombers, by the late 1950s its shorter-ranged B-47 bombers required bases in Morocco and Spain, as well as in the United Kingdom, to provide a credible forward deterrent vis-à-vis Moscow. The advent of nuclear propulsion and the development of improved at-sea refueling capabilities similarly lessened but by no means eliminated the need for naval bases. But the 1950s, 1960s, and 1970s saw a plethora of new requirements for technical facilities related to communications, command and control, intelligence operations of various sorts, and space-related activities. These in turn led to needs for new types of facilities: space surveillance, satellite downlinks, telemetry monitoring of missile tests, nuclear detonation detection, satellite control stations, and early warning radars, among others. Then, too, there were new requirements for staging U-2 and SR-71 reconnaissance flights and for the basing of ocean surveillance aircraft to monitor the movement of rival submarines in the ocean and to destroy them in the event of full-scale war.

During this early period the United States experienced particularly difficult problems in regard to communications in the greater Middle East: The region did not have an adequate basic infrastructure and the United States did not have sufficient access for communications facilities all around the Indian Ocean. For many purposes, especially in connection with the high-frequency communications used by surface ships and land installations, the United States relied on a network surrounding the region comprising major communications facilities at Nea Makri on Crete, Kagnew in Ethiopia, Diego Garcia (originally used by the Americans solely as a communications base), and the Northwest Cape in Australia, as well as an earlier smaller facility co-located with the U.S. air facility at Dhahran. Only later were first telephone cables and more recently a fiber optic network put in place. In the 1960s the United States began to use satellites for communications, but the location of its satellites in geosynchronous orbits provided much less coverage for the Middle East than for the Far East or Europe. These problems are gradually being solved with the passage of time.

By the early 1970s, at the height of U.S.-USSR competition in the Middle East, both sides had extensive air, naval, and technical basing facilities throughout the region. The United States had a wide variety of facilities in Turkey, including its air base at Incirlik, naval access at Izmir, and command, control, communications, and intelligence (C^3I) bases at Sinop, Belbasi, Karamursel, and Pirinclik, involving stations for radar, seismological arrays, telemetry, and SIGINT. It had Tackman I and II intelligence facilities in Iran, similar installations in Cyprus and Crete, and a Sound Surveillance System network throughout the Mediterranean to track the movements of Soviet submarines. Along the region's northern tier, that is, the borders of Turkey and Iran with the USSR, the United States flew a variety of intelligence aircraft such as electronic intelligence ferrets and reconnaissance craft equipped with side-angle photography capability. As a matter of fact up to about 1980 intelligence was probably its single most important basing function in the Middle East. Access to the contiguous rim on the USSR's southern flank was a major advantage for the United

States, one that was only partly counterbalanced by Soviet access to Cuba after 1961.

The U.S. Navy Mideast Force was based on Bahrain; P-3 Orion ASW flights were mounted out of Oman's Masirah Island, Kenya's Mombasa, and Iran's Bandar Abbas. Key intelligence facilities were maintained in Morocco. The loss of its Iranian facilities in 1979 led, in the context of the United States' quasi-alliance with China vis-à-vis Moscow, to their replacement by similar facilities in China's Xingiang Province adjacent to Soviet Central Asia and its key nuclear and missile-launching installations. In the late 1950s U-2 flights were staged out of Incirlik and Peshawar; later there was a major satellite control station in the Seychelles related to the U.S. Defense Support Program for monitoring Soviet missile launches. In the late 1970s the twin events of the Iranian Revolution and the Soviet invasion of Afghanistan prompted the United States to enunciate the Carter Doctrine, create its Rapid Deployment Force (RDF), and seek additional air and naval access in key places such as Morocco, Egypt, Oman, Bahrain, Kenya, and Somalia, as well as to upgrade its capabilities in Diego Garcia. The last later came to accommodate prepositioned materiel for use in the Gulf and was also a base for, among other things, P-3 Orion ocean surveillance flights and the transiting of B-52 bombers (in a war with the USSR the B-52s could attack from the south, which would create omnidirectional defense problems). This new security architecture was designed primarily for a confrontation with the USSR and featured in-theater facilities to support an enhanced U.S. presence, as well as providing en route access for transports, ferried fighter bombers, and ships. But it became very useful in altered circumstances in 1990–91. By that time, of course, in the wake of the Camp David Accords, Egypt had become a key U.S. regional ally and provided crucial access during Desert Storm and the Somalia operation.

Beginning with the RDF in the late 1970s, the United States came to experience the serious problems that it has faced ever since in its defense planning for the region: the immense distance from the United States to the Middle East; the amount of materiel needed to support a massive, that is, an adequate military presence; and the politics of the countries needed to provide access. The first difficulty was, ironically, eased earlier by the proximity of American forces in Europe, but before 1990, these could not be moved elsewhere without leaving a void in Central Europe. In-theater access remained problematic up to the late 1980s (and to a lesser degree beyond that time), leaving the United States precariously reliant on distant Diego Garcia, and perhaps Israel in some circumstances, which led to endless discussions about the possible need for "forced entry" into the Persian Gulf area. Further, in the earlier period prior to the military technical revolution, there were fewer prospects for force multipliers that might ease the logistical problems in such areas as aircraft munitions, artillery, and tank ammunition—fewer of which are required when there is a higher rate of accuracy.

By the early 1970s the Soviets had acquired extensive access throughout the greater Middle East that entailed important bases for its Black Sea fleet (surface ships and submarines) in Egypt, Syria, South Yemen, and Algeria, and to a lesser degree in Libya, Yugoslavia, Iraq, and India. Soviet submarines patrolling the Mediterranean were afforded access to Tartus in Syria, Port Said and Alexandria in Egypt, and Annaba in Algeria. Soviet aircraft used bases in all of those countries. When Moscow conducted large-scale arms resupply airlifts to its friends in Angola in 1975 and Ethiopia in 1977, it had staging bases and overflight corridors in all of the countries listed, as well as in others such as Sudan, Chad, Benin, and the Congo. The old containment rim had been breached and the Soviets had viable north-south air logistics routes through and over the Middle East en route to sub-Saharan Africa.

The 1973 Arab-Israeli War and the Oil Crisis of the 1970s

When Egypt and Syria launched their surprise attack against Israel in October 1973, they changed the ground rules for post-World War II international relations. United States support for Israel triggered an Arab oil embargo and American car owners found themselves standing in line for gasoline. The fighting stopped on October 26, 1973. Two days later Egyptian and Israeli military representatives met for direct talks, but the embargo was not lifted. Under the terms of the embargo, Arab exporters imposed restraints on oil production and a total ban on the export of oil to certain countries. The ban initially targeted only the United States and the Netherlands, but was later directed against Portugal, South Africa, and Rhodesia as well.

The embargo triggered a sharp increase in oil prices, which allowed the Arab producers to cut exports without losing revenues. Resulting shortages in the market evoked fear and anxiety among consumers; oil companies started to purchase oil in a panic, which increased inventories and created heightened demand on the market. Price increases of 40 percent and "gas lines" soon became the most prominent symbols of America's loss of independence over its oil supplies.

All the major industrial countries suffered an economic downturn as a result of the increases in the price of oil. The U.S. gross national product decreased by 6 percent between 1973 and 1975, and unemployment doubled, reaching 9 percent. However, the effect of the embargo was felt most strongly by the poorest developing countries, which had no oil of their own and not enough hard currency to import the valuable commodity. The developed countries quickly devised new energy policies in order to lessen their dangerous dependence on the volatile Middle East producers. In February 1974 the United States convened an energy conference in Washington to develop a consensus among oil consumers to establish common policies in times of emergency. Among the conference's lasting achievements was the establishment of the Paris-based International Energy Agency (IEA). As a result of IEA initiatives targets for government-controlled strategic oil reserves were established. This led to the creation in the mid-1970s of the U.S. Strategic Petroleum Reserve. The United States built the Trans-Alaskan pipeline and gained access to the huge Alaskan oil deposits. In 1975 it adopted fuel efficiency standards for the automobile industry. Conservation efforts and searches for alternative sources of oil increased worldwide.[40]

The most significant new sources of oil were Alaska, Mexico, and the North Sea. Although the contribution of these deposits to the global market gradually increased in significance, the Organization of Petroleum Exporting Countries (OPEC) continued to be dominant through the 1970s, accounting for 65 percent of the total "free world" oil production in 1973 and 62 percent in 1978.[41] The political impact of OPEC's strength during this period was dramatic. For the next ten years, influence over oil prices by key Middle East states had a profound effect on the global economy and the security of the Middle East. Not since the late 1940s had the Middle East assumed such strategic importance. The real price of oil began to decline by 1975 because of high worldwide inflation. As demand fell OPEC's position became increasingly shaky and was only saved by the second oil shock of the Iranian Revolution.

Virtually overnight new or long-submerged hypotheses about international relations sprung into vogue with a focus on the impact of resource scarcity on political behavior and the role of force. Oil was the most visible scarce resource, but soon other energy-related minerals including coal and natural gas were added to the list as were metals such as platinum, minerals such as bauxite, and food products and water.

The concern about resource scarcity was paralleled with concern that the West's dependence on maritime trade and sea routes would make it vulnerable to the growing power of Soviet projection forces, especially the Soviet navy. Of particular importance was the West's dependence on Middle East oil and the proximity of that oil to the Soviet Union and Soviet-backed surrogates. It was feared that if the oil fell under Soviet control, Moscow would be in a position to dictate economic terms to Europe and Japan and significantly change the global balance of power.

The Suez Canal was closed as a result of the 1973 war and in any case the new supertankers coming into service were too big to use it, so increasing quantities of oil from the Persian Gulf were shipped to Western Europe and the United States via the Cape. As shown in maps 12 and 13, the daily shipment of oil through the Strait of Hormuz grew from about 2,400,000 barrels a day in 1957 to nearly 18,000,000 barrels a day in 1975, reflecting the staggering increase in overall demand for Persian Gulf oil as the economies of the OECD (Organization for Economic Cooperation and Development) countries grew.

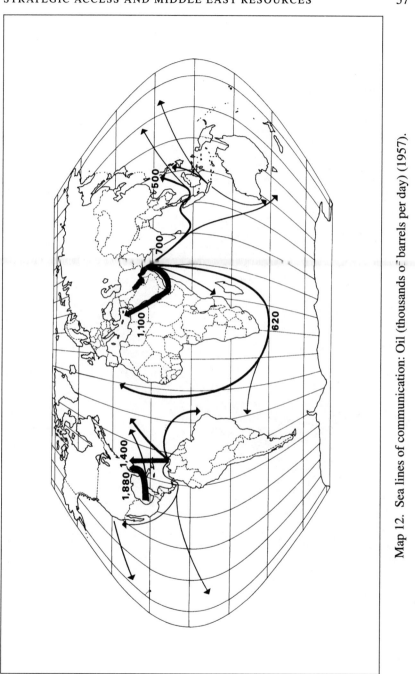

Map 12. Sea lines of communication: Oil (thousands of barrels per day) (1957).

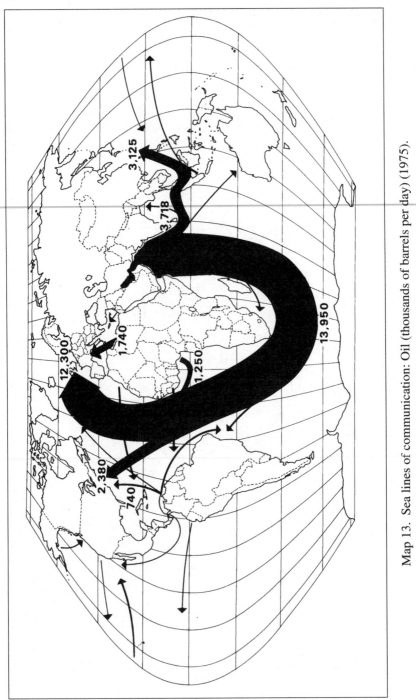

Map 13. Sea lines of communication: Oil (thousands of barrels per day) (1975).

To ensure that this flow of oil would not be disrupted, the United States and its allies developed an elaborate strategy to defend the Persian Gulf in the event of hostilities with the Soviet Union. The key pillar of this regional strategy was the pro-Western regime of Iran, ruled by Shah Mohammad Reza Pahlavi. By the 1970s strategy in connection with the Gulf was focused on the ability of the United States to project enough military power into the region and to deploy sufficiently forward to defeat any Soviet invasion. This ability was contingent upon a front-line defense of the Zagros Mountains in Central Iran, on the assumption that Soviet ground forces would run into major geographic obstacles if they attempted an invasion from Azerbaijan or from Turkmenistan. As long as the United States had access to major air bases in Turkey and Saudi Arabia, Soviet supply lines would be vulnerable to interdiction, and a Soviet advance could be sufficiently slowed to allow time for the Zagros deployment to take place.

This strategy assumed that Iran would be a U.S. ally in the event of a Soviet invasion, that Soviet airpower could be neutralized by a formidable U.S. air defense of the region; and that the Arab Gulf nations, especially Saudi Arabia, would provide base facilities. To some extent the U.S. strategy for defending the Gulf against a Soviet attack had similar features to the plan developed by Britain in 1942 to protect the region in the event that the German offensive in the Caucasus succeeded and was followed by an attack through Iran and Iraq.

The Iranian Revolution and the Soviet Invasion of Afghanistan

The geostrategic balance in the Gulf was rudely upset by two momentous events at the end of the decade. Antimonarchy political turmoil in Iran in 1978 was accompanied by strikes that halted oil production. By December 25, 1978, Iranian petroleum exports had come to a standstill. The unrest in Iran immediately drove oil prices up; in Europe they rose 20 percent above the official prices. In February 1979 the Shah of Iran was overthrown by an opposition alliance headed by the religious leader Ayatollah Khomeini. This caused another oil shock, the second of the decade.

Among oil consumers, the Iranian Revolution was met with fear and pessimism. Afraid that the panic of 1973 would be repeated, buyers made the situation worse by accumulating inventories. Fearful that prices would rise any day, companies inadvertently pushed prices higher by stocking up on cheap oil. British Petroleum, which had been getting 40 percent of its imports from Iran, invoked *force majeure* and stopped its deliveries to buyers; for example, it canceled its supply contracts with Exxon. Exporters also invoked *force majeure* to cancel previously concluded contracts or to raise prices. The shortage of oil on the world market, compounded by rising prices, underscored the vulnerability of Western consumers. Many observers declared the end of Western economic prosperity and the renewed dominance of OPEC. This impression was strengthened by the inability of the United States to resolve the Iranian hostage crisis of 1979.

At the same time Saudi Arabia started lobbying against the constant increases in oil prices. The Saudis were afraid that the increases would push world economies into recession and even depression, shaking future oil demand, as happened after the crisis in the early 1970s. In fact in 1980 the market did begin to sag, signs of recession appeared, and demand for oil started to drop. In response a number of OPEC countries decided to cut back on their production to stabilize prices.[42]

Following its 1979 revolution Iran became an adversary of the United States. While the new regime did not embrace the Soviet Union, its unpredictable behavior and strident rhetoric changed the balance of power in the region, and the United States and its allies now had to worry about confrontation with both Iran and the Soviet Union. The Soviet threat became more dangerous in December 1979 when USSR forces invaded Afghanistan. It was feared that a successful campaign in Afghanistan would radically improve Soviet military access to the Persian Gulf; for instance, it would bring Soviet combat aircraft, tactical missiles, and ground strike forces at least 600 miles closer.

As we noted earlier, in response to the invasion of Afghanistan U.S. president Jimmy Carter enunciated the "Carter Doctrine," which declared that the United States would regard any Soviet threat to the oil supplies of the Gulf as casus belli requiring an American military response. Carter ordered the creation of a Rapid Deployment Joint Task Force (RDJTF) to provide the forces to protect the Gulf. The Carter policy concerning the defense of the Gulf was adopted by the incoming Reagan administration in January 1981 and was supplemented with a more aggressive policy of supporting the Afghan resistance movement, which was waging guerrilla war against the Soviet army of occupation.

The Iran-Iraq War of 1980–88

In September 1980 oil markets were again seriously shaken. Saddam Hussein, eager to achieve Iraqi dominance over the Persian Gulf region and to preempt an expected ideological offensive by Khomeini, attacked Iran. He targeted the refinery at Abadan, which was the biggest in the world, and also severely damaged oil facilities as well as the port itself. In response Iran attacked and damaged Iraqi installations, shutting off Iraqi exports through the Gulf. Furthermore, by 1982 Iran had also convinced Syria to enter into a strategic alliance and to close its pipeline from Iraq. As a result Iraq could export oil only through a pipeline across Turkey.

In the initial stages of the war, almost 4 million barrels of oil a day, or 15 percent of OPEC's total output, were removed from world markets. Prices went up, with the price of Arab light oil reaching $42 a barrel. By July 1990 Kuwait had cut production from 2.5 million to its quota of 2 million barrels a day, while Iraq had increased its output to 3.5 million barrels a day. Their combined consumption was about 650,000 barrels a day. Total OPEC production was around 22.5 million barrels a day. Having learned from the two earlier oil crises, however, the Western governments seemed better prepared and unlikely to repeat their earlier mistakes. They persuaded the oil companies not to buy in panic—a practice that pushes prices up—but instead to use their reserves from 1979. The IEA also proved successful in persuading the oil companies to draw from their stocks, a policy that later turned out to be profitable. In addition the Saudis, whose strategy was to stabilize prices by creating an oil glut, declared at the outset that they would increase their production by 900,000 barrels a day. This increase compensated for almost one-quarter of the war-inflicted loss. Along with Saudi Arabia, Mexico, Britain, Norway, and other non-OPEC producers including Alaska continued to increase their production capacities. Non-OPEC countries, eager to gain markets, started cutting their prices. As a result an oil glut began and consumers resisted higher prices. This situation finally led to the 1981 meeting of OPEC, at which prices were somewhat standardized. That meeting marked the last time that OPEC prices went up; ever since market forces have driven prices steadily downward.

Thus the long war between the two major Gulf producers—Iran and Iraq—did not result in another oil crisis. Moreover, it coincided with a time when "the collapse in demand, the relentless buildup of non-OPEC supply, and the Great Inventory Dump" together produced a massive glut in the world oil market.[43]

Despite a sixfold increase in nominal prices during the 1970s, total world demand for oil increased 15 percent (from 2,700 to 3,039 million tons of oil equivalent) between 1973 and 1989. The increase in demand, however, came solely from non-OECD countries; OECD oil demand decreased during this period. In the developing world energy demand continued to grow even after the second oil shock, though at a slower pace in the 1980s than in the 1970s. Oil demand growth was concentrated mainly in the Middle East, followed by Africa, China, and Asian-Pacific states. At the same time, consumers in the communist countries of the Soviet Union and Eastern Europe—where demand was weakest—were sheltered from the oil crises by protective state policies. Nonetheless, in absolute terms the Soviet Union accounted for the largest share of total non-OECD demand—one-third of the total in 1989.[44]

The 1991 Gulf War

In 1991 the United States fought a major land war in Southwest Asia, primarily to prevent the energy resources of the region from falling under the control of Iraq's Saddam Hussein. Had Saddam Hussein been permitted to consolidate his annexation of Kuwait, most analysts believe it would only have been a matter of time before he exercised de facto political control over the supply—and therefore the price—of Persian Gulf oil, which then, as now, was largely determined by Saudi Arabia. The Saudi government would have been unable and unwilling to challenge him, and with the resulting increase in oil revenues, he would have been able to pay off his enormous debt to Western banks—accumulated during the Iran-Iraq war—and also complete his nuclear weapons program in a matter of a few years.[45] From his domination of the Gulf, he would have been able to pose a very serious threat to other countries in the region, including key American allies such as Israel. Under such circumstances the balance of power in the Middle East would have changed radically, to the disadvantage of the United States and its allies.

However, the Gulf War was a success, especially in regard to speed, efficiency, and the low number of casualties on the American side. That success, combined with the memory of past false alarms regarding oil shortages, has led to a certain amount of complacency. Neither during the Gulf crisis in 1990–91 nor during the eight-year Iran-Iraq war were oil supplies from the Gulf interrupted. Despite ominous warnings of potential catastrophes during the 1970s, when the oil crisis first loomed large on the strategic horizon, the 1980s proved calm in regard to oil. The escalation of the Iran-Iraq war and the disruption of oil supplies owing to attacks on tankers in the Gulf notwithstanding, the 1980s brought a net oil glut, and it is important to recall the reasons that oil prices fell rather than rose during that period of turmoil. First, the speed with which oil prices rose in the 1970s forced the industrial world to adopt radically new measures to conserve energy, which improved energy efficiency and reduced the demand for oil.[46] OECD countries, which accounted for 70 percent of world consumption of oil in 1973, accounted for only 56 percent of the demand in 1989. At the same time large, new supplies were beginning to come from the North Sea and Alaska. Second, although the economies of the West boomed in the early 1980s, they slid into a recession toward the end of the decade. The recession led to a further reduction in demand, and oil prices fell just as oil tankers were being sunk and interdicted in the Gulf as a result of the fighting.

In the immediate aftermath of the Gulf War, the Western industrial powers remained in a recession and the former Soviet Union and Eastern Europe were in economic turmoil, so there was no immediate urgency concerning oil supplies. Moreover the defeat of Saddam Hussein combined with the end of the cold war and the beginning of the Arab-Israeli peace process induced a sense of security in regard to available sources.

However, there is little room for complacency by the major oil-consuming nations. Although Saddam Hussein and the Iraqi military establishment remain crippled, there are still military threats to the Gulf and political instability is endemic to the region. More important, world oil demand is again growing rapidly and may need to be met increasingly from Persian Gulf sources in the next decade. The Gulf War showed that the West had learned from the crises of the 1970s and now knows how to handle emergency energy situations. Nonetheless, the war caused increasing numbers of observers to recognize the volatility of the Middle East and the danger to the West of becoming dependent on its oil reserves again.

Part Two

SOURCES OF CONFLICT AND ENERGY SECURITY

Chapter 3 considers major political and economic factors in the greater Middle East region that contribute to conflict, including long-term territorial disputes, population dynamics, and the scarcity of fresh water. Chapter 4 addresses the specific issue of regional energy resources, arguing that for the foreseeable future Persian Gulf resources will remain essential to the world economy but that the Caspian Basin can also provide large quantities of oil and natural gas for the market, provided that ways can be found to overcome the prevailing intense geopolitical rivalries.

THREE

The Contemporary Middle East: The End of the Cold War and Continuing Regional Conflict

FOR THE PAST several centuries, as we outlined in chapter 2, the key strategic developments in the Middle East were largely orchestrated by the great powers. What makes the contemporary Middle East different from the Middle East of the past is that there is presently only one major power—the United States —capable of exercising control and influence over the region. However, even the continued U.S. presence is dependent upon the goodwill and cooperation of regional players. For this reason the dynamics of regional and local conflicts now comands a more important role than it has in the recent past. We begin this chapter with an overview of how the breakup of the Soviet Union has affected the key countries in the Middle East. We then consider how the interaction of population dynamics, civil wars, and ethnic and religious conflicts creates a highly unstable region. The chapter concludes with a review of outstanding territorial disputes and the potential for conflict over access to fresh water. The strategic issues relating to oil and natural gas are examined separately in chapter 4.

The Impact of the Breakup of the Soviet Union

The breakup of the Soviet Union was the seminal geopolitical event of the second half of the twentieth century. Between 1989 and 1991 the largest and possibly the most authoritarian and tightly controlled empire the world has ever seen collapsed. Comparable in scope to the dissolution of the Austro-Hungarian and Ottoman empires at the end of World War I and the defeat of the

Axis powers in 1945, the collapse of the Soviet Union will have a profound impact on the international environment for decades to come. However, unlike the situation in 1919 and 1945 when the victorious powers were able to impose some sort of order on the reemerging political boundaries, the events following the demise of the Soviet Union have been chaotic.

The primary difference is that the Soviet Union imploded. It was not defeated in a classic military encounter after which the victors could dictate terms to the vanquished. Some would argue that there was a logic to the breakup of the Soviet Union dictated, in part, by geography. Many of the "new" republics that are now shown on the map were in fact old countries, for example, Latvia, Lithuania, Estonia, and Ukraine. What happened is that the *internal* borders that demarcated the republics within the USSR almost overnight became *international* borders.

There was, nevertheless, hope that a new world order would emerge that would reflect the basic Western ideals that had been upheld in the fight against communism. Peace, democracy, and economic development were to be the triumvirate of the future. The cornerstone of the new world order would be a more cooperative security environment in which major alliances armed and prepared to fight each other would be replaced by a mutually reinforcing defense concept—an expanded NATO without the Warsaw Pact countries as enemies. The goal would be for each participant to contribute to overall security and for the body itself to agree on rules of engagement to resist actions of aggressor countries deviating from the norm. A new emphasis on the United Nations and other international organizations would be paralleled by a new focus on economic competition and cooperation as the primary ingredients for global growth and political stability. But this was not to be. The legacy of the Soviet Union's demise has so far been confusion and regional conflict. Over time a new more stable regional order may emerge. This is certainly the hope of Russia's leaders, who are extremely concerned about the strategic consequences of losing control of what they call "the Near Abroad." Their efforts to revitalize the Commonwealth of Independent States (CIS), which was established in December 1991 following the breakup of the Soviet Union, were based on the logic that the ties that bind the former republics (economic cooperation) are as significant as the diversities that separate them (ethnicity and nationalism). So far, however, the forces of instability are dominant.

The reason is clear. Peace, democracy, and free-market economic development presuppose a high degree of order within and among societies. Order, in turn, presupposes clear, predictable rules concerning the exercise of power. Peaceful, orderly societies are, by definition, stable societies in which it is usually possible to predict with some degree of certainty what the outcomes will be if any group attempts to create disorder and instability. Order will then be restored according to accepted, predictable methods.

For the last thirty years the Warsaw Pact and NATO countries enjoyed a stable but putatively dangerous relationship. This stability was the result of several conditions: the awesome destructive capacity of thermonuclear weapons; the total integration of nuclear weapons into the force structures and military doctrines of the opposing alliance; clearly demarcated and accepted borders separating the forces; explicit policies regarding the use of force should either side invade the other; virtual consensus that there could be no victor in the event of a full-scale war; an understanding that escalation to nuclear war would be highly likely if war broke out; clear lines of communication between the adversaries and clear acceptance of Soviet and American dominance among the Pact and NATO members; the ability of the adversaries to contain their competition in regional conflicts outside the central theater and to resist military intervention in the conflicts within the alliances; and stable regimes in both of the opposing camps.

More specifically, the Soviet Union exercised such total control within its own empire and around its borders that of the nine points of direct contact between the empire and the outside world (Finland, the NATO countries, Austria, Yugoslavia, Iran, Afghanistan, China, Mongolia, and North Korea) only the border dispute with China was a serious source of potential conflict. This is in stark contrast to the situation today. The disintegration of the Soviet Empire has seen the outbreak of dozens of border disputes between newly established or reestablished states. Until stability along the borders of this multitude of countries is ensured, it is certain that fighting and conflict will continue. In sum, the breakup of the Soviet Union has undermined two key elements of the cold war balance of power—stable borders and stable regimes.

Map 14 shows the nine points of friction between the Soviet Union and the rest of the world and the situation today. Russia now has eleven points of contact with its new neighbors and several of the borders are unstable. On its southern front Russia now has borders with Ukraine, Georgia, Azerbaijan, and Kazakstan. Further south but not contiguous to the new Russian border are Moldova, Armenia, Turkmenistan, Uzbekistan, Tajikistan, and Kyrgyzstan. In other words, Russia suddenly found itself confronted with ten independent southern republics that until 1991 had been controlled and dominated by Moscow. Given the rapidity with which the breakup of the Soviet Union occurred, it is not surprising that conflict and chaos have been endemic in this region. Furthermore, if we add the conflicts taking place to the west in the former Warsaw Pact countries and *within* Russia, especially Chechnya, it is not difficult to see why from a historical Russian perspective the stability of its borders, especially its southern border, has once more assumed such importance and why, in turn, the new republics are primarily concerned with their own security problems and the activities of the giant to their east and north. These changes have already had a serious impact on the major countries even further south in what might be called the traditional Middle East. The foreign

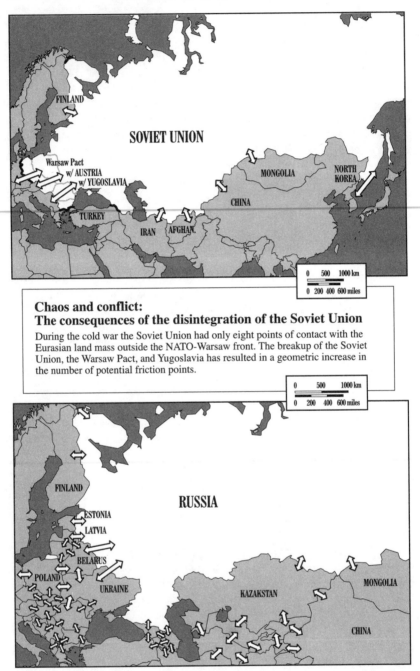

Chaos and conflict:
The consequences of the disintegration of the Soviet Union

During the cold war the Soviet Union had only eight points of contact with the Eurasian land mass outside the NATO-Warsaw front. The breakup of the Soviet Union, the Warsaw Pact, and Yugoslavia has resulted in a geometric increase in the number of potential friction points.

Map 14

and security policies of Turkey, Egypt, Israel, Syria, Iraq, Iran, the Arab Gulf, Saudi Arabia, India, and Pakistan have all been influenced by these events.

Turkey

Throughout the cold war, Turkey played a key role in Western strategy because of its location on the southeastern flanks of NATO. Its primary importance was that its geography would permit the West to contain Soviet maritime access from the Black Sea to the Mediterranean in time of war, and require that the Soviet Union deploy large forces to protect its southern flanks. The United States had a large array of C^3I (command, control, communications, and intelligence) bases in Turkey that were vital because of their location, and all the more so after the United States lost access to Iranian bases in 1979. During the cold war Turkey emerged as a respected member of the NATO community, and is using this experience to make its case for eventual membership in the European Union. Its ongoing conflict with Greece has always been a flashpoint within NATO and a source of considerable embarrassment to the United States and Europe, particularly over the Cyprus crisis. However, the United States remained committed to Turkey's integrity and provided considerable military assistance in exchange for access rights, including base facilities, which became very important during the Gulf War. Without access to Turkey's bases military operations in northern Iraq, including the rescue of the Kurds (Operation Provide Comfort) after the fighting was over in 1991, would have been impossible.

However, the changes that brought about the end of the cold war and the breakup of the Soviet Union have radically shifted the parameters of European and Middle East strategic front lines. Turkey now finds itself at the center, rather than at the periphery, of a changing environment. It remains a key eastern Mediterranean power with a very important role in the Balkans. Its continuing conflict with Greece over a number of issues ranging from Cyprus to disputes over air traffic control and offshore resources is a major source of concern and could lead to a worsening of its relations with the European Union. It is seen by the Arab countries and Israel as a key player, both for good and ill. Its conflict with Syria over both water issues and Kurdish terrorists is one of the major stumbling blocks to regional conflict resolution. On the other hand, Turkey's relationship with Israel is of great importance to both Israel and the United States. If peaceful conflict resolution prevails, Turkey could once more become a major crossroad for the physical link between Europe and the Middle East. Whether Turkey is finally admitted to the European Union or not will in large measure determine whether it can counter the appeal of radical Islam and remain secular and pro-Western. Alternatively, if its leadership increasingly embraces Islam, it will never be fully accepted as a member of the European Union.

Turkey's relationship with the Caucasus is also extremely important. Indeed, Turkish involvement in the southern Caucasus, notably Georgia and Azerbaijan, together with its growing economic ties with Central Asia, is reflected in the new road and sea links from Turkey across the Caspian Sea. Possible pipeline routes from the Caspian through Turkey to the Mediterranean would further cement these close economic ties.

These reasons, together with the continuing strategic importance of the Black Sea, bear directly on relations between Turkey and Russia. Turkish-Russian competition over these and other issues is likely to grow rather than decrease in the years ahead and has been compared to the geopolitical competition between Russia and the Ottoman Empire during the nineteenth century. Whether this relationship becomes adversarial with military dimensions remains to be seen. There is no doubt that Russia views Turkey's activity—both actual and potential—very seriously, particularly in light of its own chaotic policies in the north Caucasus, including the disastrous war in Chechnya. In Central Asia secular, democratic Turkey has been seen as a countering factor to Islamic fundamentalist appeal. While the importance of this aspect can be exaggerated, there is no doubt that Turkey, with its European connections and investors, offers important access for the hard-pressed economies of Central Asia.

Thus in all directions, there has to be a new emphasis on Turkey's role. This is best illustrated by pointing out how disastrous it would be for the United States if there were to be a collapse of secular rule in Turkey with the concomitant shift to a more revolutionary, radical state. This possibility became more real in 1996 when the virulently anti-Western, pro-Islamic politician Necmettin Erbakan was named to head a coalition government. If Erbakan and his Refah party eventually assume power in their own right it would torpedo any chance of Turkey joining the European Union. It would fundamentally change Turkey's relations with Israel and its attitude toward the Arab-Israeli peace process, and could lead it to play a dangerous, spoiler role in Central Asia. In that event confrontation and conflict with both the United States and Russia might be on the agenda. Alternatively, if Turkey continues to follow a cautious, pragmatic policy consistent with its geographic and economic importance and resolves its conflicts with Greece, its influence in the decade ahead could well be a strong positive factor in encouraging peace and stability in the region. Turkey could become a source of fresh water and a conduit for oil as well as a road link among Central Asia, Europe, and the Middle East.[1]

Egypt

The breakup of the Soviet Union has resulted in benefits for Egypt. The removal of the most powerful opponent of the Camp David peace process initially enhanced Egyptian prestige throughout the Arab world. As Gorbachev, and then Yeltsin, assumed power in Moscow, Egypt's centrality in Arab politics regained much of its former glory. It has since become much easier for Egypt to work with its former Arab enemies, including Syria. At the time of the Gulf War in 1990–91, Egypt was fully integrated into mainstream Arab politics. Furthermore, the breakup of the Soviet Union and the collapse of socialism throughout the Soviet Empire made it much easier for Egyptian leaders to introduce radical market reforms and abandon the heavyhanded socialism of Nasser and other former Egyptian leaders.

Yet there has been a downside for Egypt. Egyptians have found themselves more and more tied to Israel and the United States and marginalized on key issues concerning Arab-Israeli diplomacy. This may have changed since the last Israeli election in May 1996. The coming to power of a right-wing Israeli government headed by Benjamin Netanyahu has ironically given Egypt a greater role as a broker in the peace process. Though close ties with the United States continue to serve its interests, Egypt enjoys greater flexibility for more independent policy activity in the region. Generous American military aid to Egypt is based almost entirely on the continuation of the Arab-Israeli peace process. In the past Egypt was looked upon as a prize that the United States had pried away from the Soviet Union. Now it is seen primarily in the context of regional politics. Furthermore, the breakdown of communism has led to a surge of Western interest in democratic institution-building in the Middle East, and some of Egypt's more authoritarian methods have increasingly come under criticism from human rights and democracy advocates. However, the United States has overlooked abuses in these spheres in view of the danger to its strategic interests if the Mubarak regime were to fall.

Israel

One immediate benefit to Israel of the breakup of the Soviet Union was the outflow of new immigrants from Russia and other CIS states, which has important implications for the long-term growth of the Israeli economy. In addition, relaxed relations between Israel and Moscow have been paralleled by a dramatic increase in diplomatic recognition of Israel and acceptance throughout the world, including the return of Israeli diplomats, envoys, and businessmen to many African countries that had expelled them after the 1967 war. Israel's new acceptance has been symbolized by the establishment of diplomatic relations with the CIS, including the Muslim countries, as well as the Vatican. This could be set back if the Arab-Israeli peace process is stalled, but absent a complete reversal of the process there is unlikely to be a return to the old days when Israel was considered a pariah state by a majority of the members of the United Nations.

On the negative side some Israelis consider that Israel is no longer seen as a strategic asset by the United States and that it will be harder to persuade

Congress to grant the aid for military activities that Israel regards as still necessary in view of the ongoing conflict with the Arabs and other Muslim countries such as Iran and Pakistan. Furthermore, the highlighting of human rights issues now that the cold war is over has put into focus some of Israel's policies toward the Palestinians, which have included significant infringements of human rights. If the peace negotiations with the Palestinians continue to make progress under the new Israeli government, the issue will diminish in importance.

Syria

The most obvious impact on Syria of the breakup of the Soviet Union was the weakening of Hafez al-Assad's regime vis-à-vis Israel and Egypt, primarily because the rug was pulled out from under the Syrian military establishment and it was suddenly forced to pay in hard cash for arms supplies from Russia. Thus Syria has fewer cards to play in its negotiations with the West and with Israel. On the positive side Syria has become part of the peace process and, although a reluctant partner, over time may benefit economically through the normalization of relations with the West and the increased flow of investments into the country. There is no doubt that the end of Soviet military support was a dramatic strategic setback for Syria and it is unlikely to regain the influence it had in Arab politics in the 1970s and 1980s unless the peace process aborts and Israel is once more regarded as a prime enemy of the Arab world. This possibility cannot be ruled out in view of the setback to the peace process in 1996.

Iraq

Although Iraq and the Soviet Union had good relations at the end of the 1980s, Iraqi leader Saddam Hussein regarded the breakup of the Soviet Union as an opportunity for his armies to dominate the Middle East. With the cold war over he did not believe the United States would intervene to challenge him militarily. With the Soviet Union gone, Iraq would assume a regional hegemony. These ambitions ended with the Gulf War, and today Saddam Hussein finds himself isolated, with little access to any of his former friends in Russia. Though this might change if the sanctions are lifted, Iraq is certainly much weaker today than it has been at any time in the recent past.

Iran

Iran, like Turkey, finds itself with new neighbors at a time when its geo-graphical location is becoming more important. As we have stressed throughout this study, one of the most promising major repositories of oil and gas after the Persian Gulf is the Caspian Basin. This fact has not gone unnoticed by the mullahs in Teheran. Iran literally straddles these two huge energy regions and has tried for the last few years to capitalize on its geography and negotiate agreements with its Caspian neighbors to move gas and oil across its territory by pipelines to ports on the Mediterranean and Persian Gulf. However, it has run up against the opposition of the United States, whose relations with the regime in Teheran remain adversarial. While it might be in America's long-term interests to see Iran become the conduit for oil and gas from the Caspian Sea—to ensure that alternative egress routes for energy are available—this is unlikely in the short term. Meanwhile Iran has developed rail links with Turkmenistan, and direct train service between Central Asia and the Gulf is now a reality. With more cordial relations between Iran and its neighbors this could become an important alternative to the road routes into Central Asia that Turkey is currently developing.

Iran plays an important role vis-à-vis its neighbors in all directions. Its relations with Russia have political, military, and economic overtones and, indeed, Russian-Iranian decisions on the future of the Caspian Sea may be determinative for many of the oil agreements that will be worked out with Azerbaijan, Turkmenistan, and Kazakstan in the future. Iran's political involvement with the competing factions in the continuing Afghanistan civil war is significant because it will remain a key player regardless of the outcome of the fighting. If Iran's efforts to improve its relations with Pakistan and India are successful it will have geostrategic importance for the region and for the United States. Further, its growing involvement with China in regard to military concerns and energy is an indicator that the latter's interests in the Middle East transcend weapons sales and are related to its long-term energy requirements.

Iran's highly complex relations with the Arab Gulf states are of direct concern to the United States and potentially the most volatile source of conflict. Owing to its long coastline, Iran's maritime potential features prominently in the thinking of U.S. military planners. This naval capability together with its nuclear and terrorist activity represent a more serious threat to American interests than its conventional land forces. Iran's opposition to the Arab-Israeli peace process is one of the most serious complaints that the United States and its allies have against the regime. If Syria and Israel ever resume negotiations and consummate a peace treaty residual Iranian opposition might or might not be an important factor, but there is no doubt that Iran is capable of playing a spoiler role, including the use of terrorism, throughout the Gulf as well as in the Arab-Israeli conflict.

The debate about America's relationship with Iran goes on, but Iran's strategic importance is undeniable. It is also a foregone conclusion that virtually every concerned party, particularly the United States, would benefit if relations were to improve, but this is only likely to happen if there is a change of regime or radical change in leadership in Teheran.

The Arab Gulf

The Arab Gulf countries have been, as least in the short term, beneficiaries of the end of the cold war. They have found it much easier to work with the United States. They now have established strong ties with the new republics of Central Asia. Over time these links could develop into fruitful economic relations (though this could lead to competition over energy exports). The Arab Gulf states are far less coy about dealing with or talking to Israel, and while there is still no formal peace between the Arab Gulf and Israel, the zero sum game no longer applies—namely American military assistance to the Gulf Cooperation Council (GCC) is no longer seen as a threat to Israel and vice versa. On the other hand, the GCC states are highly vulnerable to conflicts that have broken out around their periphery in the past few years, including the instability in Iraq and Iran and the ongoing wars in the Caucasus and Central Asia.

Another downside for all the GCC countries has been the new international focus on democracy and civil rights. Some of the GCC countries have glaringly repressive systems, and each of them has responded in a different way to the pressures regarding, for example, their treatment of women and religious minorities. It remains to be seen whether they can adjust to the scrutiny of an intrusive world and still keep their basic traditions intact.

Saudi Arabia

Saudi Arabia is considered the most important of the GCC countries because of its geography and its resources. The Saudi Arabian littoral straddles the Gulf of Aqaba, the Red Sea, and the Persian Gulf. An improvement of relations between Saudi Arabia and Israel would be a powerful factor in increasing the chances for a comprehensive Middle East peace. Alternatively, a collapse of the Saudi monarchy, or the emergence of a more radical one, would have most deleterious effects for both the peace process and for America's strategic posture in the region. Without the access and basing provided by Saudi Arabia, the United States would be hard pressed to mount another major military operation against a resurgent Iraq or to deal with any other military contingency in the region that required the use of extensive ground forces and land-based airpower.

Saudi Arabia's other crucial role for the foreseeable future is a function of its capacity to increase or decrease its oil production on relatively short notice. Saudi Arabia is the one country in the world today whose oil production can significantly affect market conditions and therefore the price of oil. As long as Saudi Arabia remains unique in this respect—and there is not likely to be any competitor for that role in the next five to ten years, though some would argue that Iraq could ultimately be in the same position—concern about its future will remain high on national agendas, in the East as well as the West. Asian countries are increasingly interested in the region's energy resources, so they too will have to pay attention to the nuances of the Saudi royal family and the sociological pressures that affect the entire Arab Gulf. Thus the continued stability of the Saudi kingdom is of great importance to the geostrategic balance in the region.

India

Over time the breakup of the Soviet Union will likely be seen as an important factor helping to propel India toward modernization. However, in the short run India was adversely affected by the collapse of the Soviet Union. Indian-Soviet relations were very close, especially in the area of military cooperation. Many of the weapons systems in the Indian army and air force were produced jointly with the Soviet Union, at very low cost. Now it is becoming more expensive for India to equip its army with Russian goods, and it has little foreign currency to buy weapons on the open market. On the other hand, the end of communism in the Soviet Union has had a salutary effect upon Indian economic development and there are signs that the Indian economy is on the verge of major growth. While India has always nurtured great power ambitions, its decision to bid for such status was not taken seriously until the end of the cold war and the 1990–91 Persian Gulf crisis forced its leaders to modernize the Indian economy. While the Western allies gloried over the defeat of Iraq, India, even though it played almost no part in the war, had to deal with the problems and traumas that grew out of it. The temporary rise in oil prices had severely drained its limited reserves of foreign currency; during his occupation of Kuwait, Saddam Hussein had taken about 140,000 Indian expatriate workers as hostages; and India's closest ally, the Soviet Union, was siding with the United States. Finally, the stunning success of the American military operation drove home to India just how obsolete its own military equipment was, how costly it would be to be to modernize its armed forces, and how irrelevant its leaders were to world events, even those close to its own borders.

Though precipitated by the Gulf War, these crises had been a long time in the making—some would say as early as the 1950s, when India's Congress Party chose socialism and a centrally controlled economy that became one of the most protected and autarkic in the world. Although India became self-sufficient in food production and was relatively untouched by the fluctuations in world trade, the price of this rigorous centralized control was a huge bureaucracy, infuriating red tape, very few incentives for private industry to compete in the marketplace, and an extremely hostile attitude to foreign investment.

By June 1991 India was on the verge of bankruptcy, with barely enough foreign currency to pay for a month's worth of necessary imports, such as oil and high-technology items. To meet this crisis the Indian government had to turn to the International Monetary Fund for emergency assistance, and that help was conditional: India would at last have to be serious about the reform of its antiquated economy. On July 24, 1991, Finance Minister Manmohan Singh unveiled the New Industrial Policy. The statement it made was blunt:

> The government has decided to take a series of measures to unshackle the Indian industrial economy from the cobwebs of unnecessary bureaucratic control. These measures complement the other series of measures being taken by the government in the areas of trade policy, exchange rate management, fiscal policy, financial sector reform and overall macroeconomic management.[2]

Since then there has been speculation on the implications of modernization for India. Optimists believe that reform is unstoppable, that India is on the way to major power status, and could become a worthy competitor to China early in the next century. Others agree that India has all the attributes needed for success but that much depends upon what happens in the next few years. Investors must still be convinced that reform will continue and that real profits can be made and taken home. Confirmed pessimists still believe that India is incapable of the necessary modernization and that it is too insular and caste-ridden to adapt to long-term change. A few critical observers point to the sharp contrast between India and China, lauding the latter, although as China enters a period of political uncertainty and economic dislocation this comparison may be short-sighted.[3] As in many other countries, human rights issues in India have come under more scrutiny in recent years, particularly in connection with the treatment of Muslims in Kashmir.

Pakistan

The Soviet Union's demise has had a most profound effect on Pakistan. With the Soviet defeat in Afghanistan, Pakistan lost its most persuasive leverage with the United States, which had provided Pakistan with political and military support during the war. No longer seen as a strategic asset challenging communism along its common border with Afghanistan, Pakistan has been subject to punitive measures by the U.S. Congress as a result of the implementation of the Pressler amendment, which cuts off major military assistance to Pakistan so long as it continues to produce nuclear weapons grade fissionable material. Pakistanis feel betrayed by the United States and no longer regard it as a natural ally against communism. Furthermore, Pakistanis are aware that with the end of the Soviet-Indian relationship there is a growing body of opinion in the United States that regards India, not Pakistan, as the most important country to woo in South Asia.

The Geographic Component of Regional Conflicts

Many factors contribute to the plethora of ongoing conflicts in the Middle East. Here we consider four related categories that have distinct geographical components: demographic trends, including migration; ethnicity, religion, and civil war; unresolved territorial disputes; and scarcity of fresh water.[4] We note that the human geography of the greater Middle East is in a state of flux as a consequence of political factors within and outside the region. Ethnic conflicts—intensifying with the end of the cold war and new international support for the principle of self-determination for ethnic groups—spill over national frontiers and challenge the authority of local regimes in the Middle East and on its periphery. As states on the periphery fragment and refugees flee, they often carry ideas, attachment to social movements, and fidelity to Islamic militancy that strike a responsive chord in—and effectively transform the human geography of—neighboring states in the Gulf and elsewhere in the region. This is particularly problematic for unstable states in the region that are near the locus of conflict, especially those with substantial minority populations with cultural or religious ties to groups involved in violent political transformation elsewhere.[5]

When these changes are juxtaposed against the backdrop of historical territorial disputes and the political and economic consequences of fresh water scarcity, it becomes clear why the greater Middle East region remains a powderkeg and why military force remains such an important element in regional geopolitics.

Demographic Trends and Asymmetries

Demographers predict that the population of the traditional Middle East will increase from 274 million in mid-1995 to almost 500 million in 2025. In Iraq, Jordan, Kuwait, Oman, Syria, the West Bank and Gaza Strip, and Yemen it is expected to more than double by 2025.[6] In the wider Middle East region, the population of the states in Central Asia and the Caucasus is projected to increase from 70.7 million in mid-1995 to 111 million in 2025, while that of India and Pakistan will jump from 1.16 billion in mid-1995 to over 1.6 billion by 2025. In particular, the populations of Tajikistan, Turkmenistan, Uzbekistan, Kyrgyzstan, and Pakistan are projected to almost double by 2025.

Such predictions are tentative, change frequently as revised data become available, and rely on a host of demographic and economic assumptions.

However, they do provide some sense of the rapid growth in population throughout the region. Given the continued failings of regional governments to provide for the daily needs of hundreds of millions of people, a population increase of several hundred million over the next thirty years is a daunting prospect. Demographic trends are yet another reminder that the Persian Gulf, Levant, and South Asia are difficult to isolate on a growing number of issues. Migration, including labor migration, and the spillover of population problems from one country to the next and beyond defy solely local approaches to demographic issues.

A number of variables could affect long-term forecasts of regional population growth, though they are unlikely to be significant in the short and medium term. They include factors ranging from government intervention in family planning to changes in behavior and birth rates due to economic growth or recession. The accurate prediction of the rate and timing of declines in fertility is an important factor in projections of population growth. The UN prediction of a total population of 490–560 million in 2025, for example, assumes that the fertility rate will decline to nearly half of current levels.[7] Obviously, a slower decline could lead to more serious population dilemmas as the problems described below would be exacerbated.

The composition in terms of age of Middle Eastern and Central and South Asian populations is another factor influencing future growth, and population growth is accelerated by the age ratios in many parts of the region. While only about 20 percent of the population in the developed world is under the age of fifteen, the figure soars to 40 percent in many parts of the greater Middle East. In 1995 approximately 45 percent of the population of Iran, Iraq, Jordan, Syria, Saudi Arabia, Pakistan, and Tajikistan were children under fifteen. In the West Bank, Gaza, and Yemen, 50 percent of the population is below the age of fifteen. As these children enter their childbearing years, the population growth rate will continue to increase because even as the fertility rate drops, a boom in the number of childbearers could offset any downward trend. Also, the relative "youth" of society is a significant factor in political instability. Young people in large numbers without jobs or meaningful careers are susceptible to becoming political recruits for social movements that promise better opportunities through a change in the political regime. Some observers have suggested a strong causal link between the relative youth of a society and its propensity to engage in violence and revolutionary behavior.[8]

Social and educational issues related to women will also modify population figures. Among women the literacy rate, level of employment, and average marriage age are important factors in population trends. Higher education levels, for example, generally lead to greater use of birth control. The place of women in Middle East society is controversial and, in many countries, women are denied access to the types of jobs and educational opportunities that have traditionally resulted in lowered fertility rates. There is a wide discrepancy in the region between literacy rates for men and for women. In Egypt 34 percent of the women as against 63 percent of the men are literate; in Iran, 43 percent of women as opposed to 64 percent of men; in Saudi Arabia, 48 percent of women as opposed to 73 percent of men.

Many countries in the region are experiencing enormous strains on social services as a result of rapid population growth. In areas such as education, housing, health services, labor, and basic food supplies, many governments are unable to adequately meet the needs and—as a result of modern communications, especially television—the expectations of their people. At best, some states simply maintain a limited number of services for an ever-expanding population, and their people live in greater poverty, with more frustration, and, overall, a lower quality of life. The countries that have been able to avoid this trap are better able to meet the rising expectations of their inhabitants.

The young age of the population also exacerbates the shortage of services. While the industrialized world has a 3:1 ratio of working adults to children, the ratio in the Middle East today is only 1:1. In a sense each working adult in the region is responsible for supporting, either directly or with the help of government programs, three times as many children as in the developed world. With such a ratio many parts of the region can hardly get ahead.

Currently the states of the Middle East can be grouped into two demographic categories. Many of the countries noted for rapid population growth have managed to slow the rate. Egypt and Iran both have overwhelming population-related problems, but forecasts call for a decrease in the growth rate over the next thirty years. In Egypt, as well as in all the Arab Mediterranean countries, the rate of population growth has started to decline.[9] In many other areas the growth rate remains high and the population is expected to increase by 94 percent (in Pakistan) to 175 percent (in the West Bank and Gaza). Afghanistan, Tajikistan, Iraq, Jordan, Kuwait, Oman, Syria, and Yemen all fall into this latter category.

Despite this catalogue of population ills, the Middle East has several notable exceptions: some of the energy-rich countries of the Persian Gulf that have relatively small populations and Israel. Oil and natural gas production has allowed most of the GCC countries to escape the cycle of expanding populations and inadequate services and resources, although Saudi Arabia, Kuwait, and Bahrain have recently experienced financial difficulties. Oil is not an automatic guarantee of salvation. A host of other factors can cause major shortfalls even in energy-rich states, and Iran, with its growing population, is a prime example of a country in such a situation.

Israel has avoided overpopulation problems.[10] An advanced economy, high education and literacy levels, and a more Westernized approach to most women's issues have kept population growth relatively low, and Israel actually would welcome a higher fertility rate. Ultraorthodox (and some orthodox)

Jewish families do continue to have much larger families, but secular Israeli Jews have followed a similar pattern to Western Europe and the United States.

Migration issues are often closely related to those concerning population. In the Middle East there are numerous examples of migration issues that are linked to military or political events. The wealthy Gulf states rely heavily on Asian labor, especially in the aftermath of the Gulf War when many Palestinian workers were expelled and took up residence in Jordan. Palestinian refugees from earlier Arab-Israeli wars, many living in underdeveloped refugee camps, remain a central stumbling block in Israeli-Palestinian negotiations. In general the presence of foreign workers raises citizenship questions and often skews the age structure in both labor-exporting and labor-importing countries. It also fosters a superior-inferior relationship between the upper-class native population and the lower-class foreign workers that can contribute to instability within the host country as well as cause tension between the migrant labor's nation of origin and its host country. The dramatic statistics of population composition in the GCC countries are shown in tables 3-1 and 3-2.

The characteristics of a given population can also serve as indicators on the military front. Asymmetries in population size, literacy, and ethnic composition influence military power calculations and therefore determine the type of military doctrines that are best suited for the security posture of a particular country. In areas such as the Gulf, the number of expatriates and members of the indigenous population that can serve in the armed forces constitute a key factor. Countries with small populations such as Israel will tend to rely more heavily on deterrent strategies or on preemptive strategies that would lead to a speedy end to the conflict. On the other hand, countries such as Iran with larger populations may, depending on the political climate, be prepared to engage in longer-term conflicts with greater emphasis on ground combat because of their ability to absorb higher numbers of casualties. It is difficult to imagine any war between Israel and the Arab states lasting for eight years as did the Iran-Iraq war.

Table 3.1. GCC Population Data

Country	Population	Percent nationals	Percent foreign	
Bahrain	575,925	68	32	(13 percent Asian)
Kuwait	1,800,000	39	61	(9 percent South Asian)
Oman	2,100,000	73	27	. . .
Qatar	533,916	25	75	(36 percent South Asian)
Saudi Arabia	18,700,000	69	31	(10 percent Afro-Asian)
UAE	2,900,000	24	76	(50 percent South Asian)

Source: *The World Fact Book 1995* (Washington, D.C.: Central Intelligence Agency, 1995).

Table 3.2. GCC National and Nonnational Labor Force
Thousands

Country of residence	1985				1990			
	National	Non-national	Total	Percent of non-national of total	National	Non-national	Total	Percent of non-national of total
UAE	71.8	612.0	683.8	89.5	96.0	805.0	901.0	89.3
Bahrain	72.8	100.5	173.3	58.0	127.0	132.0	259.0	51.0
Saudi Arabia	1440.1	2661.8	4101.9	64.9	1934.0	2878.0	4812.0	59.8
Oman	149.8	335.7	485.5	69.1	189.0	442.0	631.0	70.0
Qatar	17.7	155.6	173.3	89.8	21.0	230.0	251.0	91.6
Kuwait	127.2	551.7	678.9	81.3	118.0	731.0	849.0	86.1
Total GCC	1879.4	4417.3	6296.7	70.2	2485.0	5218.0	7703.0	67.7

Source: Arab Labour Migration: Arab Population Conference, Economic and Social Commission for Western Asians, Amman April 4–8, 1993, p. 4.

Some of the more obvious links between population and security issues have been mentioned above. Population issues are relevant in discussions of the strain on natural and human resources, the size of the armed forces, and the strength of opposition groups. But the wider issue of population and conflict also deserves greater scrutiny. A number of states believe that they face an existential demographic threat. To what extent do population issues foster military conflict? How deeply are population issues affected by conflict?

Demographic threats manifest themselves in several different ways in different countries throughout the region. Relevant factors include the impact of asymmetric birth rates and the effect of legal and illegal immigration and emigration. With respect to Israelis and Palestinians, for example, asymmetric rates of population growth in favor of the Palestinians at one point put Israel in an increasingly vulnerable position. This demographic dynamic changed, however, when the Soviet Union, and then Russia, Ukraine, and other successor states opened the gates for Jews to emigrate to Israel. With numbers often translating into political power, fear of demographic changes exists on all sides of the Arab-Israeli conflict.

Ethnicity, Religion, and Civil War

Fragmentation and civil war generally have spillover effects that threaten the stability of states outside the epicenter of conflict, as is most vividly demonstrated in the Balkans with the breakup of Yugoslavia. The flow of refugees into neighboring states not only strains local government resources, but can create instability because of tensions with the resident population.

Refugees can also introduce political ideas that threaten local norms and, more important, can carry with them a commitment to their cause that can threaten their country of origin and spark interstate conflict. Civil war is particularly threatening to neighboring states that are home to substantial populations that identify ethnically or religiously with one or more of the groups involved in the nearby violent confrontation. There is a real fear that civil unrest will spill over into neighboring countries, and that segments of those societies that identify with protagonists in the conflict will make it impossible to avoid international involvement.

The geographical perspective is critical here because proximity to the conflict epicenter is an important factor in determining the destabilizing impact of regional and peripheral conflict. Also important, however, is the degree of instability already present in countries close to the zone of conflict. An explosion next door cannot but worsen an already volatile situation. The emergence of a revolutionary or hostile regime nearby can threaten national unity and stability, particularly if it activates existing cleavages within neighboring societies.

The stability of the Middle East is threatened by the potential fragmentation of Afghanistan, Iraq, and Azerbaijan and the spillover effect of regional conflicts in North Africa and the Indian subcontinent. The most important cases are reviewed briefly (see table 3-3).

Table 3.3. Typology of Middle East Conflicts in the 1990s

Interstate	Causes/issues
Arab-Israel	Territory/ideology
Armenia-Azerbaijan	Territory (Nagorno-Karabakh)
Egypt-Sudan	Water
India-Pakistan	Territory/religion
Iran-Iraq	Territory/ideology
Turkey-Syria	Territory/water

Internal/cross-border	Causes/issues
Afghanistan	Civil war/tribal
Algeria	Civil war/unrest (government-Islamists-Berbers)
Bahrain	Unrest (Sunni-Shia)
Egypt	Insurgency (government-Islamists-Copts)
India	Insurgency (government-Sikhs-Muslims)
Iraq	Insurgency (government-Kurds-Shias)
Iran	Insurgency (government-Exile groups)
Israel	Terrorism (government-Palestinian-Lebanese Islamists)
Lebanon	Insurgency (Christians-Hezbollah)
Somalia	Civil war (tribal)
Sudan	Civil war (ethnic-religion)
Turkey	Insurgency (government-Kurds)

Since the Najibullah regime fell in April 1992, Afghanistan has been ravaged by civil war and violent conflict among factions of the Mujahideen. Iran is not the only regional state with a stake in the war's outcome. Pakistan, Uzbekistan, Tajikistan, and Saudi Arabia also support different factions in the ongoing struggle. The periodic cease-fire agreements have fallen victim to renewed eruptions of hostilities, though in the closing months of 1996 it appeared that the Pakistani-backed Taliban forces were on the brink of victory.

The Kurdish issue has been of great importance to Turkey since the birth of the modern Turkish nation. The 1920 Treaty of Sèvres called for the creation of a weak state from the ruins of the Ottoman Empire, and included the provision for an autonomous region for the Kurds. However, Turkish nationalists founded a modern and strong state and the treaty became irrelevant. Many Turkish leaders became convinced that the creation of a Kurdish state would inevitably weaken Turkey. The first major challenge to the new Turkish state came from a Kurdish religious leader, whose forces occupied one-third of Kurdish Anatolia. This first attempt and two subsequent Kurdish revolts were ruthlessly suppressed by the government. The fact that the three major armed rebellions against the state were led by Kurds, and originated in the Kurdish region, has firmly established the Kurds in Turkish minds as the originators of the primary challenge to their independent existence.[11]

In a unique conjunction of circumstances, the Kurds in Iraq were given autonomy in northern Iraq at the end of the Gulf War by coalition leaders. Although the current Kurdish leadership is divided, all the factions are sensitive to the vulnerability of Kurdistan, and none is now pressing for independence. On the contrary, they constantly reiterate that a solution for the Kurds need not threaten any existing national borders. Turkey would prefer a unified, democratic Iraq, in large part because the future of the Kurds can best be addressed within the existing borders of an Iraq that is democratic and respects human rights. However, neither Turkey nor any other major power is confident that a democratic regime will emerge as the successor to Saddam Hussein, nor do they have a political road map or the leverage to advance this agenda.

The Kurds spill across the frontiers of Turkey, Iran, Iraq, and Syria, complicating relationships among the four. Turkish, Iranian, and Arab commitment to the Kurds is marginal and that of the West is uncertain. The coalition action on behalf of the Kurds at the end of the Gulf War has internationalized the issue, and the future of the Kurds, hostage to the future of Iraq and the successor to Saddam Hussein, has serious implications for Turkey, Iran, and Iraq and their relationships. For instance, if the Kurds of northern Iraq eventually establish an independent state in the wake of a collapse of the Baghdad regime, Turkey would face a serious dilemma concerning its own Kurdish population and the insurgency in the east. Similarly, Iran has reason to fear an independent Kurdish state. Iranian leaders talk openly of the Kurdish enclave in northern

Iraq as "another Israel" and fear that its success will encourage its own Kurds to seek independence or the right to join their counterparts in Iraq. Clearly Turkey, Iran, Iraq, and Syria would all be opposed to an independent Kurdistan. Yet if they tried to crush a fledgling state with military force, they could face strong opposition from the West, especially if that state were seen to be democratic. In these circumstances a war over the independence of Kurdistan could escalate, creating grave problems for both regional and external powers, especially the United States.

In the Caucasus Turkey and Iran are engaged in a diplomatic competition in the conflict between Armenia and Azerbaijan, while the fighting has edged closer to Iran and hundreds of thousands of refugees have fled the expanding zone of conflict. The proximity of the fighting and the flow of refugees create serious problems for Iran. In August 1993 an Iranian foreign ministry statement seriously warned "the Armenian forces for their repeated aggressions" close to Iran's borders.[12] Some 12 million Azeris live in northern Iran near the border with Azerbaijan.[13] The proximity to Azerbaijan could theoretically lead to a situation in which the external Azeri community supports Azeris within Iran against the central Iranian government. In reality, however, Azerbaijan is enmeshed in a war with Armenia and has no resources to extend. Iran, already home to one of the largest refugee populations in the world, has admitted thousands of refugees from the fighting.[14]

There are at least 600,000 Baluchis in southeast Iran. Other than the existence in the 1970s of the Baghdad-based Baluch Liberation Front, Baluchi nationalism has been quiescent. The Baluchi are often involved in smuggling across the Afghani and Pakistani borders despite the presence of Iranian Revolutionary Guards. They are able to do so because of the extraordinary weakness of the Iranian government in the Baluchi-inhabited provinces.[15] In early 1994 scattered attacks and violence rocked Iranian Baluchistan.[16] There are an estimated 4.6 million Baluchi in Pakistan, mostly in the southwest near the Iranian border.[17]

The violent confrontation between Russia and the insurgent movement in the Chechnya district between 1994 and 1996 was another indicator of the turmoil and instability that persist in the Caucasus and all along Russia's southern borders. The possibility of protracted conflict in this strategic area is bound to affect Russia's policies toward both Turkey and Iran in the coming decade. Furthermore, as we discuss in chapter 4, oil pipelines from the Caspian Basin run through the Caucasus to the Black Sea, and these disturbances make it sensible to develop alternative supply routes either to the north through Russia proper or possibly through Turkey, Iran, or Pakistan. However, all these alternatives have various problems that make them risky ventures and capital for their development is unlikely to be found in the near future.

On the Indian subcontinent the continuing crisis between India and Pakistan over Kashmir has led both countries to seek ways to broaden their regional alliances. In recent years Iran and Pakistan have often found common ground arising from geographical proximity, mutual opposition to Soviet influence, their Islamic constitutions, and their troubled relations with the United States. Leaders in both countries have called for stronger ties and increased cooperation on regional issues, which could be further enhanced if India's nationalist Hindu parties achieve greater political power in Delhi. Both countries are founding members of the Economic Cooperation Organization that now includes Central Asian countries. Both have been deeply affected by the protracted conflict in Afghanistan with refugees and militants creating havoc in their border areas. Though Iran and Pakistan see themselves as mediators in the conflict, they have different approaches and allies in Afghanistan, and this has occasionally caused friction. There have been rumors of a military alliance—including cooperation on nuclear weapons—between them, a possibility that is clearly making India apprehensive.[18] However, Iran's policies toward India and Pakistan are a careful attempt at balance to preserve its options. Iran regards India as an important potential partner. Indian and Pakistani interest in Iran nevertheless attests to its strategic importance to both and the interconnections between the conflict on the subcontinent and in the Gulf.

The Muslim population in India—120 million strong and scattered throughout the subcontinent—is the country's largest minority and the world's second largest national Muslim community, second only to that of Indonesia. Hindu-Muslim tensions, generated through the years of aspiration toward independence and during the clashes and resettlement after partition,[19] have been most prevalent in India's northern states, particularly Gujarat and Uttar Pradesh, although Muslims can be found in significant numbers throughout the country.

The social and economic lives of most of India's Muslims differ little from those of Hindus in their community and most of India's Muslims consider the language of their community to be their mother tongue. Many Muslims join in local Hindu celebrations and worship some local Hindu gods.[20] The political ascendancy of Hindu nationalists in recent years, however, has been a source of anxiety for many of India's Muslims in part because the Hindu-nationalist Bharatiya Janata Party (BJP), which emphasizes a national identity based fundamentally upon the Hindu religion, gets much of its support from militant Hindu extremist groups such as the Rashtriya Swayamsevak Sangh (RSS) and is immutably associated with the destruction of the oldest Muslim shrine in India—the Ayodhya mosque in Uttar Pradesh—and the subsequent riots that killed 3,000 people.[21]

Muslims and other minority groups fear that the BJP political mantra, "Hindutva," which to many signifies the restoration of Hindu primacy in India, will lead to the criminalization of non-Hindu sacred practices and result in an upsurge in violence should the Hindu nationalists retain power. The BJP manifesto calls for the removal of special constitutional provisions for the

Muslim majority state of Jammu and Kashmir, the abolition of the special code of civil laws for Muslims that have been part of Indian law since British colonial times, and the construction of a Hindu shrine on the site of the Ayodhya mosque.[22]

As previously noted, Muslims in the state of Jammu and Kashmir—the only Muslim majority state in India—have been a major source of instability for India as well as an important factor in India-Pakistan tensions. The Indian government's approach to the Kashmir crisis has continued to involve a tough military response combined with generally unsuccessful attempts at dialogue.[23] Any significant and lasting diffusion of the Hindu-Muslim tensions, which are rooted in the fundamental conception of Indian national identity, must await resolution of the Kashmir crisis, a context in which Muslims constitute a continuing threat to the integrity and stability of India.

Sikhs, a distinct religious group accounting for about 2 percent of the Indian population, live primarily in the states of Haryana and Punjab and adjacent areas. Years of agitation won the Sikhs a Punjabi-speaking majority state in 1966, but in the 1970s they began to struggle for autonomy, an aspiration based in part on regionalism (greater autonomy and more economic benefits for the state of Punjab) and in part on the desire to preserve the cultural integrity of the Sikhs as a religious community.[24] Violence and terrorism in Punjab eventually led, in 1983, to the removal of the elected government in that state and the institution of President's Rule, which ended in 1985 only to be reinstated between 1987 and 1992. Today Hindu-Sikh tensions are lower than they were during their peak in 1984, when the assassination of Prime Minister Indira Ghandi by her Sikh bodyguards led to a tragic massacre. Nonetheless, this bloody conflict, based on religious and cultural differences, regionalism, and years of violence, is far from resolved.

Iraq, Saudi Arabia, Lebanon, and Bahrain all face problems with their Shia populations. In Iraq Shias, numbering some 10 or 11 million people, constitute a majority. As several other groups in that country, the Shias are oppressed; their leaders have been arrested and executed by the hundreds. In March 1991 their uprising against Saddam Hussein was brutally crushed. The Iraqi government continues to target the Shias through its draining and shelling of the country's southern marshlands. The Shias in Iraq have received little international support. Kuwait and Saudi Arabia are overtly hostile and even Iran has not done very much. Most of the countries in the region fear the fragmentation of Iraq and the possibility of an Iranian-dominated Shia entity in southern Iraq.

In Saudi Arabia the Shias, who account for about 6 percent of the population, live mostly in the eastern Province along the Persian Gulf. They suffer from general discrimination and unfair employment practices that are entrenched features of Saudi society and are systematically harassed as well as denied full religious freedom. Incidents of unexplained sabotage of oil facilities in the early 1980s and periodic disturbances have been attributed to disgruntled Shias.[25] Some changes in Saudi policy in 1993 did lead to marginal improvements for the Shias. Bahrain has faced a very different problem, namely unrest among its Shia majority. The minority Sunnis believe Iran is behind the unrest but disparity of income and opportunity is also a factor.

The Shias in Lebanon are probably the largest single ethnic or religious group in the country, numbering at least 1 million.[26] Hezbollah, a militant Shia party backed by Iran, continues to resist the Israeli occupation of southern Lebanon. Future Hezbollah military actions against Israel are partly hostage to Iranian-Syrian relations and to ongoing Syrian negotiations with Israel on the terms of a peace agreement.

On the African continent many civil conflicts threaten regional stability. In the Sudan a particularly violent civil war between the predominantly northern Arabs, who are Muslims, and the southern black African Christians has raged for over twenty years. Until 1995 the southern Sudan People's Liberation Army was plagued with internecine tribal conflict that limited its offensive capabilities against the north. However, these quarrels were set aside in 1995 and it was once more able to challenge the extremely unpopular and isolated Islamic government in Khartoum.[27]

In Somalia twenty-one years of one-man rule ended in January 1991 with the flight of General Muhammed Siyad Barrah from the capital. Fighting between rival factions led to 40,000 casualties, and by mid-1992 the civil war, drought, and banditry combined to produce a famine that threatened some 1.5 million people with starvation. Sporadic clashes among members of rival Somali clans continue and the hard political decisions have not yet been faced.[28]

The most serious crisis in North Africa is in Algeria. In January 1992 the Algerian government, anticipating a near certain and overwhelming victory of Islamist political parties, canceled the second round of national elections, and the country was plunged into civil war. Led by the military the government continues to battle Islamic militants, and there have been over 30,000 deaths since the fighting began. Opposition forces have attacked not only the military and security forces but politicians and foreigners as well.[29] No mediated resolution to the conflict appears likely: either the military government will intensify its repression, creating hundreds of thousands of refugees, or Islamist forces, further angered and alienated by the failure of traditional Arab governments to support their cause, will prevail. Algeria cannot but be a source of instability for both Europe and the Middle East.

Complicating the Algerian problem is the fact that approximately 20 percent of the country's population are Berbers, who have remained relatively unassimilated in the majority Arab culture and retain their own language and traditions. In the late 1960s when the Algerian government declared Arabization as a national goal and began to promote Arabic in the bureaucracy and in schools, the Berbers, particularly the Kabyles in the mountains of Kabylia, east of Algiers, began to mobilize against the Arab-speaking majority's "cultural imperialism" by campaigning, through strikes and demonstrations in support of the Berber Cultural Movement, for official recognition of their language and culture.[30] The unrest continued until 1995, when, after continued demonstrations that included an eight-month strike by Berber students, the government finally initiated a pilot project for the teaching of Tamazight, one Berber dialect, and set up an official body responsible for promoting Berber language in school curricula and in the media.

This represents a dramatic policy shift for the government, which has tried to steer Algeria toward an Arabic-based culture and discourage the use of regional non-Arab dialects, but the Berbers, many of whom oppose the Islamic fundamentalists, have become important government allies in the center-stage struggle against Muslim militants.[31] The Berbers began arming themselves in 1993 in response to Islamist attacks,[32] and by 1995 vigilante Berber groups were being armed by the government as part of its strategy to annihilate the Islamists.[33] There is no doubt that government appeasement of the Berbers has been accelerated by the violent clash between army-backed government forces and militant Muslims.

The Egyptian government is also involved in a serious battle with Islamic militants, who have targeted both tourists and the minority Copts. At times the conflict has been responsible for an alarming drop in Egypt's earnings from its tourist industry, the country's primary source of hard currency, which in turn has been the cause of serious pressure on the government of President Hosni Mubarak. Moderate intellectuals and policy analysts in Egypt are concerned about the political paralysis in the government and its single-minded focus on Islamic militants, though during 1995 tough and sometimes harsh measures against the fundamentalists seemed to curb their more extreme activity and tourism rebounded.

The regional balance of power would shift dramatically were Islamists to take power in Egypt. Egypt's pivotal role in promoting a peaceful resolution of the Arab-Israeli conflict would most probably end, Israel would be seriously alarmed, the PLO might be fatally weakened, and the Gulf states would be threatened. Iran would welcome another fundamentalist regime and the opportunity to end its isolation through a new alliance. The political map of the Middle East and the Gulf would be dramatically altered as Islamist forces were strengthened throughout the region.

Unresolved Territorial Disputes

Disputes over the control and ownership of territory are among the most usual causes of conflict and war. The genesis of the Arab-Israeli conflict, which has dominated Middle East geopolitics for the past forty years, is about territory, and the main issues are well known. Until such time as the remaining territorial disputes between Israel and its immediate neighbors, especially the Palestinians, Syria, and Lebanon are resolved, an Arab-Israeli peace will remain elusive (see map 15). It is noteworthy that peace between Israel and Egypt and Jordan did not happen until most of the territorial issues were negotiated, and even today this peace may be precarious if present regimes and governments change.

The remaining territorial disputes in the Arab-Israeli conflict all involve multiple factors and will be difficult to resolve. In the last twenty years Israel has settled territorial and border disputes with Egypt and Jordan; the Israeli-Jordanian treaty resorted to novel methods of leasing to overcome final differences on sovereignty and land use for agriculture. Despite this progress three bitter disputes remain. The Israeli-Palestinian, Israeli-Syrian, and Israeli-Lebanese territorial issues all involve both economic and strategic considerations. All three contested areas serve as military buffers for Israel. With the West Bank resolution is further complicated by the strong religious claims made by some Israelis.

ISRAEL-PALESTINIANS. In 1967 Israel captured both the West Bank (from Jordan) and the Gaza Strip (from Egypt) as well as the Sinai Desert and the Golan Heights. Concerning the West Bank, no Israeli government will return to the pre-1967 lines. Israel has long argued that the West Bank is a key strategic asset and the settler movement adds religious overtones with allusions to biblical Israel. At the same time Palestinians regard the return of the West Bank as the necessary condition for full peace with Israel. They believe that Israeli settlements are an instrument for the creeping annexation of the West Bank; as Israeli settlements grow Israel will claim that more and more land cannot be relinquished. More Jewish settlers make the return of Palestinian land more difficult.

In addition to military security, water security plays a central role in the Israeli desire for control of the West Bank. The West Bank mountain aquifer provides approximately 25 billion cubic feet of water a year, of which Israel currently takes 80 percent.[34] This "provides a third of Israel's water consumption, 40 percent of its drinking water and 50 percent of its agricultural water."[35] Consequently Israel has prohibited the drilling of Palestinian wells since 1967. As with all water distribution problems, the dispute over the West Bank involves both control of the land and the legal discussion of who has the right to its water. Israel, for instance, might be willing to cede the land over most of the aquifer if water agreements still guaranteed that it has the right to pump much of the aquifer's water.

Gaza is less important to Israel. In the past Israel made a strategic argument that justified the need for Gaza to prevent Arab armies from driving up the coastline to Tel Aviv. Though there are a few thousand Israeli settlers in the Gaza Strip, some of whom live in settlements that are positioned to enhance the

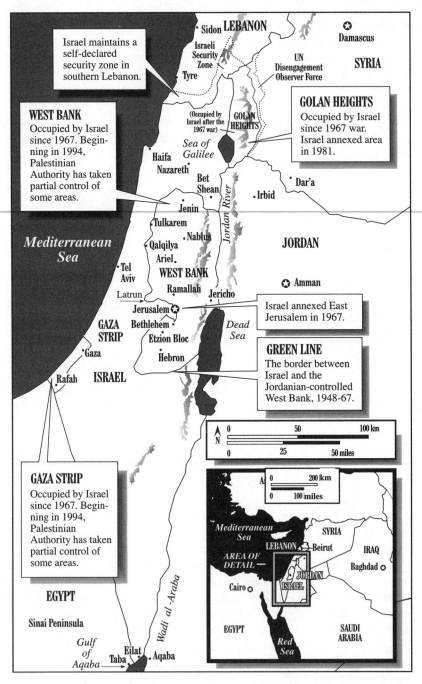

Israel maintains a self-declared security zone in southern Lebanon.

WEST BANK
Occupied by Israel since 1967. Beginning in 1994, Palestinian Authority has taken partial control of some areas.

GOLAN HEIGHTS
Occupied by Israel since 1967 war. Israel annexed area in 1981.

Israel annexed East Jerusalem in 1967.

GREEN LINE
The border between Israel and the Jordanian-controlled West Bank, 1948-67.

GAZA STRIP
Occupied by Israel since 1967. Beginning in 1994, Palestinian Authority has taken partial control of some areas.

Map 15

security of all of Israel, even many right-wing Israelis are not interested in retaining the overcrowded, poverty-stricken area.

ISRAEL-SYRIA. In the 1967 war Israel captured the Golan Heights from Syria and has held the territory until today. In 1981 it extended Israeli law to the Golan, a move some considered tantamount to annexation. However, since Syrians believe the Golan has been Syrian territory from time immemorial, it would be political suicide for any Syrian leader to yield to anything less than a full Israeli withdrawal from the territory.[36] Continued Israeli occupation of the Golan poses a direct and immediate military threat to Damascus, which is only thirty-seven miles from the cease-fire line.[37] For Israel, however, control of the Golan Heights not only bolsters its strategic position but enhances its control of area water sources, as the Golan currently provides Israel with 30 percent of its drinking water.[38]

ISRAEL-LEBANON. Israel has maintained a self-declared security zone in southern Lebanon since the withdrawal of Israeli forces from most of Lebanon after the 1982 Israeli invasion, and this has de facto moved the border a few miles north of the international line. The northern border of the security zone includes barbed wire fences, limited access, and observation posts. In many ways the security zone is run by Israel and the Southern Lebanese Army as an entity separate from the rest of Lebanon.

This security zone provides a buffer for Israel as it attempts to ward off Hezbollah and other anti-Israel guerrillas. Lebanon rejects the Israeli occupation of the southern strip and charges that Israel also has an economic motive for the seizure. Once again the key economic factor is access to water, in this case allegedly the Hasbani and Litani rivers. Loss of the security zone in Lebanon might make an equally wide strip of northern Israel vulnerable. From the Lebanese perspective an Israeli attack on Lebanon would start from a more forward position if Israel maintained control of the territory. However, for large-scale conventional military engagements, the zone, unlike the Golan and West Bank, does not add a great deal to Israeli military capabilities against Lebanon.

However, Syrian military planners have expressed concern that Israel might use its presence in southern Lebanon to outflank the Syrian army and avoid a war on the Golan. Syria could do the same in reverse, which would be an especially appealing option as long as Israel retains control of the Golan Heights. Talks would likely lead to an Israeli withdrawal from the security zone and a restoration of the international border as the dividing line between Israel and Lebanon as neither side rejects the existing international boundary.

PERSIAN GULF DISPUTES. In the Persian Gulf, where no peace process is under way, the disputes are equally serious (see map 10). The Shatt-al-Arab has a long history as the meeting point of the Arab and Persian worlds. In more recent years it has been a vital waterway for the export of oil from both Iraqi and Iranian ports. After many years of conflict, Iraq agreed, in the 1975 Algiers

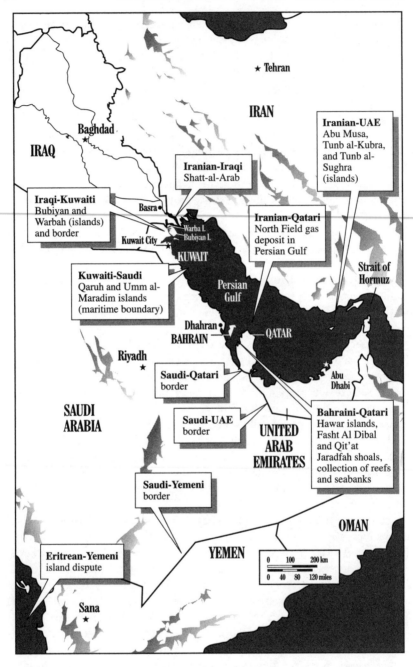

Map 16

Communique, to a delineation of their mutual border that was more favorable to Iran in exchange for an end to Iranian support for Iraqi Kurdish rebels. In 1980 Iraq abrogated the agreement and the eight-year Iran-Iraq war began, with much of the fighting taking place in and around the Shatt-al-Arab. In 1990 two weeks after the Iraqi invasion of Kuwait, Saddam Hussein reversed his position and agreed to abide by the 1975 agreement.

Iraq-Kuwait. Iraq has ongoing territorial disputes with Kuwait. The border between them has caused friction as far back as 1962 because of Iraq's consistent claim of sovereignty over the whole of Kuwait, previously a part of the Basra governorate under the Ottoman Empire before the British separated it in 1932. Territorial claims over the islands of Bubiyan and Warbah, which occupy a strategic position at the entrance to the Iraqi port of Umm Qasr, and a dispute over three miles of the border near the Rumaila oil field brought the two countries to the brink of armed conflict in March 1973. These two disputes figured prominently in Baghdad's justification for the invasion of Kuwait in August 1990.[39] Iraq continues to seek sovereignty over the islands because of their strategic importance as well as for the economic benefit they would provide through an increased share of seabed oil.[40] At various times Iraq has offered Kuwait an exchange of Iraqi territory for either or both of the islands, an exchange of fresh water for Kuwait from the Shatt-al-Arab, or straightforward leasing. Kuwait has refused all these offers.

The findings of the UN commission that was set up after the 1991 Gulf War to demarcate the Iraq-Kuwait border were approved unanimously by the UN Security Council on May 27, 1993. Basing its rulings on a 1963 agreement in which Iraq and Kuwait agreed to respect a common border described in a 1932 exchange of letters, the commission moved the prewar de facto boundary in both directions, favoring Kuwait in several critical areas.[41] The commission put the official boundary 2,000 feet north of the previous frontier, effectively handing Kuwait an extra strip of territory and a greater share of the Rumaila oil field, eleven oil wells that Iraq had drilled in the Ratga oil field prior to the invasion,[42] and a portion of the Umm Qasr port, including some naval facilities.[43] Iraq refused to accept the decisions, being especially opposed to the ruling on Umm Qasr, which remains its only outlet to the sea, its only port, and its only naval base following the closure of the Shatt-al-Arab waterway after the Gulf War. Paradoxically Kuwait has never claimed Umm Qasr, and official Kuwaiti government maps show the boundary line as south of the port even in mid-1992.

The Iraqi government sent a letter to the UN secretary-general on May 21, 1993, calling into question the finality of the commission's decision, denouncing the commission's mandate, composition, and working methods, and suggesting that Iraq may dispute the very existence of Kuwait.[44] The demarcation undoubtedly has reinforced Iraq's sense of grievance and increased the probability of irredentist action in the future. Although Iraq formally recognized the independence of Kuwait in 1994, the recognition was grudgingly given under intense international pressure in the hope that it would lead to the lifting of international sanctions. It seems clear that even if Saddam Hussein's rule is replaced by a more conciliatory leadership, Iraq's antipathy toward Kuwait, grounded in the issues of borders and access rights, will continue. Iraq will remain a revanchist state until a formal Iraqi-Kuwaiti border agreement is openly and freely negotiated.

Iran–United Arab Emirates. Iran has a major territorial dispute with the United Arab Emirates (UAE) over the three strategically located islands of the Greater and Lesser Tunbs and Abu Musa on a tanker route at the entrance to the Strait of Hormuz. The location of the islands and Iran's military presence highlight the strategic importance of the dispute.

Claiming that the islands were Persian before Britain transferred their sovereignty to the sheikhs of the Trucial States almost a century ago, Iran landed troops on Abu Musa and the neighboring Tunbs in 1971, on the eve of the British withdrawal. The Tunbs, only fifteen miles from the Iranian island of Qeshm and located on the Iranian side of the median line,[45] were historically part of the Ras al-Khaimah emirate, while the Sharjah emirate claims sovereignty over Abu Musa, located on the Arab side of the median line thirty-five miles from the UAE coast.[46]

Iran and Sharjah signed a memorandum of understanding in 1971 that gave Iran full jurisdiction over the range of hills on the northern side of Abu Musa, where it could maintain a military presence, while allowing Sharjah political control over the rest of the island, including its small population.[47] In 1992 that arrangement proved unacceptable to Iran in the wake of the Gulf War and the new American military presence in the region. In an effort to flex its muscles, Iran ordered the expulsion of all non-Sharjah nationals, including teachers, engineers in charge of water facilities, and health workers providing essential services for the UAE government.[48] Iran never denounced the 1971 agreement governing administration or made any additional claim to the island. Nonetheless, the UAE contends that the Iranian government is expansionist and seeks to annex the island. The crisis reached a peak in August 1992, when Iranian officials prevented 100 residents, mostly Egyptian Arabic teachers and their families, from landing on the island. Subsequent talks held in Abu Dhabi in late September 1992 fell apart when the UAE delegation insisted on placing the Tunbs on the agenda for discussion.[49] The GCC, in the Abu Dhabi Declaration issued on December 23, 1992, rejected the acquisition of territory by force and emphasized UAE sovereignty over Abu Musa and the Tunbs.[50] There have also been some oil claims related to the islands.[51]

Despite commercial ties between Iran and the UAE, the conflict over Abu Musa makes the bilateral relations somewhat unstable. In February 1993 the UAE president, Zayid bin Sultan al Nuhayyan, vowed to regain control of the three disputed islands. Moreover, he stated that the "development of friendly ties between the two countries hinges on boosting confidence and measures that show Iran's adherence to international law and respect of the UAE sovereignty."[52] In 1994 both the GCC and the Arab League supported the UAE position on the islands and toward the end of the year the UAE proposed that the matter be referred to the International Court of Justice (ICJ). Iran has so far refused to take this step and the increasingly bitter dispute continues. Teheran is cognizant of the significant strategic benefit that accrues from control of the islands as well as of the fact that overall Iran is far stronger than the UAE. Especially in view of the possibility of rights to oil fields, both sides are likely to continue to press their claims with vehemence.

Arabian Peninsular Border Disputes. Many borders on the Arabian peninsula, particularly those in open desert areas, were never clearly defined by the European colonial powers that created them. In recent decades the competition for control of oil and natural gas reserves has fueled numerous conflicts. Owing to the economic significance of many areas in this oil-rich region, national boundaries are widely disputed, with periodic escalations into violence. During the December 1994 GCC summit the member states determined to try to resolve all their regional border disputes by the time of the next meeting, scheduled for the following year in Musqat,[53] but they did not succeed, and most of the potentially explosive territorial disputes still remain unresolved.

Saudi Arabia and Qatar have a long-standing border dispute over a ten-square-mile strip of territory said to be rich in oil and gas deposits.[54] The quarrel erupted into violence in September 1992 at a border outpost that has been claimed by both sides for over two decades.[55] An Egyptian-brokered border demarcation accord, signed in December 1992 to ease tensions and to end Qatar's boycott of the GCC,[56] set up a framework that was to have led to a final agreement within one year, but little has been done to implement it. Meanwhile there was a series of shooting incidents between March and October 1994 that resulted in several deaths, including that of a Qatari civilian.[57] Both countries have expressed their desire to resolve the dispute, but the Saudi kingdom remains at odds with its small neighbor over a maverick foreign policy that has permitted a significant U.S. presence on Qatari territory, and aims to improve ties with Iran and Iraq while normalizing relations with Israel.[58]

A dispute over the potentially oil-rich Hawar islands and shoals of Fasht Al Dibal and Jarada, as well as a collection of reefs and seabanks, has been the cause of tension between Qatar and Bahrain since they gained their independence in the early 1970s. Britain officially accepted Bahrain's claim to the cluster of islands and reefs as early as the 1930s,[59] but Qatar has always disputed Bahrain's sovereignty over the territory. Saudi mediation since the late 1980s and GCC efforts to keep the dispute from becoming explosive have not resolved the tensions, which continue unabated. Qatar brought the dispute before the ICJ for the second time in November 1994 without the consent of Bahrain, which continues to control the territory in question.[60] The ICJ ruled that the case is within its jurisdiction and that it would hear it on the basis of unilateral application by Qatar. The emir of Qatar, Hamad bin Khalifah al Thani, has explained, however, that "if the two parties reach acceptable results through the Saudi mediator, then the case will be withdrawn from the ICJ. We hope to reach a brotherly solution through the mediator that will satisfy the two parties. Otherwise, we will have to wait until the ICJ decides on this issue."[61]

There is also no demarcated border between Saudi Arabia and Yemen, and there are several disputed areas, coveted by both countries, partially because of posited oil deposits. Once again economics is the key irritant that blocks a smooth resolution of the dispute. In 1992 Saudi Arabia warned several international oil companies to stop searching for oil in disputed territories. The oil companies argued that they were operating on Yemeni, not Saudi, land.[62]

The dispute began in the early 1930s when Saudi Arabia defeated the Yemeni Imamate in a border war over the provinces of Asir, Najran, and Jizan. In addition to periodic reports of border skirmishes, such as in early 1995, various Yemeni leaders have also alleged over the years that Saudi Arabia has provided arms to factions within Yemen.[63] Several rounds of talks between the two governments have failed to produce an agreement. Saudi Arabia and Yemen signed a memorandum of understanding in February 1995 to demarcate their borders on the basis of the 1934 Taif agreement, and technical committees have met to discuss the issue.[64]

North and South Yemen have had a long-standing border dispute involving the Shabwah oil field straddling their common border, which was never fully demarcated. After two border wars during the 1970s and subsequent border clashes which were denied by both states, an agreement was concluded in 1988 to replace existing border posts and allow for joint exploitation of the cross-border Marib-Shabura oil fields.[65] Even after unification of the conservative north and the formerly Marxist south in 1990, the ownership and distribution of oil wealth continued to be a source of friction between them, particularly after new oil was discovered in the south.

South Yemen, home to one-fifth of the total Yemeni population, began providing united Yemen with one-third of its oil resources. Bickering over the distribution of the country's oil income, disagreement over the political and social structure of the new country and power-sharing between the north and south, and southern dissatisfaction with the corrupt tribalism of the northern administration plunged the nation into civil war in May 1994.[66] Northern victory over the southern secessionists has not resolved the underlying disputes and current political unity is likened by some analysts to a form of colonization of the south by the north.[67]

Kuwait and Saudi Arabia have been progressing toward a resolution of their long-standing territorial dispute over the islands of Qaruh and Umm al Maradim, situated, respectively, twenty-three miles and sixteen miles off the coast of the northern part of the neutral zone that was created by the British in 1922. During the 1960s, the zone was partitioned through a series of agreements that applied to the land boundaries and extended six marine miles into the seabed and subsoil.[68] The offshore territory has not yet been partitioned and is the focus of current concerns. Ownership of the islands is intricately linked to critical maritime boundary delimitation, a process that involves the division of oil-rich Gulf waters between the two nations.

Kuwait annexed the northern portion of the neutral zone after the December 1968 demarcation agreement, and claims the islands on the basis of an exchange of letters with Iraq in 1923 and 1932. In 1961 it offered Saudi Arabia a share of any proceeds that might accrue from oil taken out of the islands in return for Saudi acknowledgment of Kuwaiti sovereignty over them. Saudi Arabia refused, and Kuwaiti ownership of the islands is still being contested by Riyadh.[69]

The arguments began as early as September 1949, when the Kuwaiti government granted the American Oil Company of California a concession to conduct oil exploration on the two disputed islands. Following protests from Saudi Arabia, the firm stopped its operations. Kuwaiti foreign minister Sabah al-Ahmad al-Jabir Al Sabah was in Saudi Arabia on July 18, 1995, discussing border demarcation.[70] He announced that the talks had made progress toward demarcating the maritime boundary between the two nations, and had led to "agreement on specific points," adding that "experts from the two countries would follow up on those points in upcoming meetings."[71] Ownership of the islands is sure to be at the heart of negotiations to follow.

CAUCASUS AND CENTRAL ASIA. There is also conflict over territory throughout Central Asia and the Caucasus. Azerbaijan continues to war with neighboring Armenia over Nagorno-Karabakh, whose population is 75 percent ethnically Armenian. In May 1992 two Azerbaijan towns, Shusha and Lachin, were occupied by Armenia, and this occupation—which provides a corridor between Armenia proper and Nagorno-Karabakh—permits Armenia to maintain a logistics route into Nagorno-Karabakh, whose legal status remains undetermined.

Various groups within Georgia are also struggling to gain autonomous control over territory. The Ossetians, a separate ethnic group speaking a language based on Persian, have long sought to unify South Ossetia (a region largely populated by Ossetians) with the North Ossetian Autonomous Republic. The autonomous region of South Ossetia, which was created in 1922, was abolished by the Georgian government in 1990 and reinstated in 1992. Also in Georgia the Abkhazians have been struggling to wrest control of their territory, formerly an autonomous Soviet republic, out of the hands of the government in

Tbilisi. Tensions in Abkhazia have led to open warfare on a much larger scale than in South Ossetia.

Of the many territorial problems among the Central Asian states of the former Soviet Union, those between Uzbekistan and Tajikistan are among the most complex. The disputes stem from Soviet border-drawing in the 1920s that included in Uzbekistan several historic Tajik-populated cities of the Zeravshan Valley, including Samarkand and Bokhara, with their many important cultural and religious sites. This was considered by many Tajiks to have been a great injustice that has yet to rectified. Some Tajiks also claim Uzbekistan's Surkhandarya region, west of the existing border. The Uzbeks, in turn, aspire to the "recovery" of Fergana Valley towns such as Khodjend, Isfara, and Kanibadam.[72]

SOUTH ASIA. South Asia also has its share of territorial disputes. Indian-Pakistani tensions preceded statehood and are most apparent in the contested area of Jammu and Kashmir, where territorial dispute is anchored in ideological and religious claims. Today India, China, and Pakistan all control some part of the disputed territory. In late 1947 Muslim guerrillas began invading Kashmir and Indian troops soon arrived to repel the invaders. The current line of control divides Jammu and Kashmir (India) from Azad Kashmir (Pakistan). To date the two governments have been unable to agree on how to move toward a final resolution and settlement in Kashmir, and no change in this stalemate is expected in the near future.

The dispute over Kashmir is also about the rival conceptions of statehood in South Asia since the time of independence. India sees itself as a multinational state with room for all communities, and considers that there is little need for countries such as Pakistan that are based on one religion or ethnicity. In contrast, Pakistan was created as a Muslim state because its founders saw the need for statehood for both peoples of the subcontinent, Hindus in India and Muslims in Pakistan. While most territories were divided between India and Pakistan through civilian and military means in a manner that ultimately proved acceptable to both sides, Kashmir was left split. Each side sees control of Kashmir as part of the fifty-year-old debate about independence. With such a strong ideological base, the dispute is very difficult to resolve and has continued to this day.

EGYPT, SUDAN, AND NORTH AFRICA. Since 1902 both Egypt and Sudan have administered land across their shared border in order to facilitate tribal grazing. These areas, located both north and south of the international boundary, are now considered by each country to be part of its sovereign territory. Sudan lays claim to an area known as the Halaib Triangle, control of which could enlarge its claims to the Red Sea seabed and its resources, as well as to a smaller area, the Wadi Halfa salient, that is now mostly under the waters of Lake Nasser.[73] Fighting erupted in the disputed areas in June 1995 after an assassination attempt against Egyptian president Hosni Mubarak led Egyptian officials to point the finger at Sudan. Although the territorial dispute may be fueled by economic factors, it is primarily a legal issue, which is highlighted whenever these two regimes clash on other bilateral fronts, such as over the Nile waters or support for Islamic fundamentalism. As long as these other differences remain unresolved, the territorial dispute will continue to contribute to the deterioration of relations between the two countries.

A dispute between Chad and Libya over the Aozou strip, which was annexed by Libya in 1973, was finally resolved after a February 1994 ruling by the ICJ in favor of Chad.[74] The boundary between Libya and Chad was first defined in 1898 under a convention signed by France and England, in an attempt to delimit their respective spheres of influence east of the Niger River. In 1955, four years after Libya's declaration of independence, a French-Libyan treaty recognized the boundary between Libya and French Equatorial Africa as fixed.[75] Libya, however, argued that this treaty had not established a definitive border and in 1973 annexed the Aozou strip, thought to contain uranium deposits that could help Libya achieve nuclear independence. Violence along the Libya-Chad border erupted repeatedly during the 1970s and 1980s because of rival claims to this uranium-rich territory, with Libya arguing on the basis of cultural and ethnic ties.

Conflicts over Water

Records of conflict over access to fresh water are found throughout the history of the Middle East. In the seventh century B.C. Ashurbanipal of Assyria seized water wells in his war against Arabia. In the same century Sennacherib of Assyria destroyed the city of Babylon and made certain to destroy the canals that supplied water to the city in the process. In recent years fighting broke out along the demilitarized zone between Israel and Syria in the 1950s when Syria tried to stop Israel from building its national water carrier, an aqueduct to provide water to southern Israel that draws water from upstream sources in Syria.[76] In the 1960s Israel used force, including air strikes against water-diversion facilities, to prevent Syria from diverting the headwaters of the Jordan River (the Hasbani Springs and the Yarmuk River). These military actions contributed to the regional tensions that eventually led to the 1967 war.[77] Between 1967 and 1970 Israeli forces repeatedly shelled the east Ghor (now known as the King Abdallah) Canal in the Jordan Valley in response to PLO raids, and water installations were regarded as strategic targets. In 1974 Iraq massed troops along the border and threatened to bomb Syria's Al Thawra Dam on the Euphrates, alleging that the dam had reduced its flow of water.[78]

Dams, desalination plants, and water-conveyance systems were targeted by both sides during the Gulf War. Retreating Iraqis destroyed most of Kuwait's extensive desalination capacity. In mid-1992 the Iraqis continued to suffer

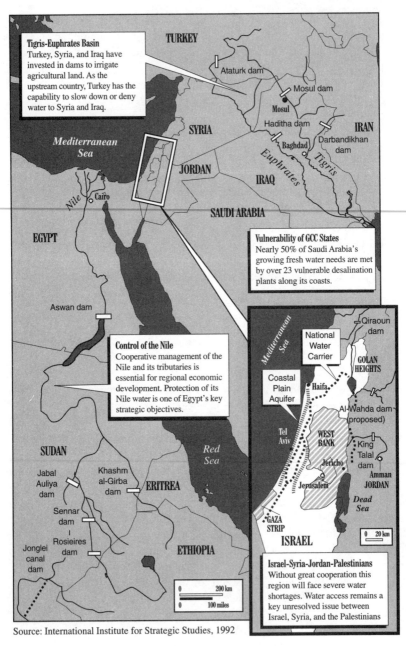

Tigris-Euphrates Basin
Turkey, Syria, and Iraq have invested in dams to irrigate agricultural land. As the upstream country, Turkey has the capability to slow down or deny water to Syria and Iraq.

Vulnerability of GCC States
Nearly 50% of Saudi Arabia's growing fresh water needs are met by over 23 vulnerable desalination plants along its coasts.

Control of the Nile
Cooperative management of the Nile and its tributaries is essential for regional economic development. Protection of its Nile water is one of Egypt's key strategic objectives.

Israel-Syria-Jordan-Palestinians
Without great cooperation this region will face severe water shortages. Water access remains a key unresolved issue between Israel, Syria, and the Palestinians

Source: International Institute for Strategic Studies, 1992

Map 17

severe water shortages because Baghdad's modern water-supply and sanitation system had been destroyed by the U.S.-led coalition forces. The threat of the "water weapon" was made clear during the early days of the Gulf War when there were discussions about using Turkey's Euphrates dams to cut the flow of fresh water to Iraq in retaliation for its invasion of Kuwait.[79]

The growing demand for fresh water in the Arab-Israeli arena and the Arabian peninsula poses a particularly sensitive set of issues that can contribute to conflict (see map 17). The looming "water crisis" in the region—an element of its physical geography—coupled with troubling population dynamics in many areas—changing human geography—has increased the strategic importance of water and its impact on power relations among the countries of the Middle East. The water problems of Egypt, Iraq, and Syria, for example, directly affect their relations with neighboring countries that control the sources of the Nile, Tigris, and Euphrates rivers.

According to the World Bank *Annual Report 1994* the Middle East and North Africa are moving toward a "water crisis." Owing to rapid population growth and urbanization, the demand for water has reached a point where withdrawal now exceeds replenishment; the *Report* predicts that "by the year 2000, withdrawals in most countries will exceed total fresh water potential."[80] Competition for water resources gives rise to several problems. First, the distribution of water throughout the region creates new categories of rich and poor countries—water-rich upstream states in the greater Middle East region will gain increasing leverage over their downstream coriparian neighbors— dramatically affecting the regional balance of power in a potentially destabilizing way. Second, countries may engage in direct conflict over control or allocation of water or both.

A third problem that stems from the shortage of fresh water, particularly on the Arabian peninsula, relates to the fact that access to new technology is essential for efforts to exploit natural resources in the region and to develop alternative water sources. The region's acute water shortages could be overcome through the development of joint projects such as canals, water pipelines, and nuclear desalination plants in key areas. However, increasing reliance on high technology, involving as it does capital-intensive investment, increases the vulnerability of the region to devastating damage in the event of war. Just as infrastructural development transforms the target landscape—an aspect of the military geography of the region—parallel improvements in the range and accuracy of conventional weapons have increased their effectiveness in neutralizing infrastructural targets with minimal collateral damage where desirable.[81] With respect to the energy-rich countries of the Arabian peninsula, the physical geography of the region, in terms of the scarcity of water, is transforming the military geography of individual nations, since efforts to mitigate water shortages through the construction of large-scale, dual-purpose water-desalination and power-production plants are leading to the creation of strategic targets that are highly vulnerable to modern conventional weaponry.

The Jordan Basin

The Johnson plan of 1953 would have divided the waters of the Jordan River, among Jordan (46.7 percent), Israel (38.5 percent), Syria (11.7 percent), and Lebanon (3.1 percent).[82] The plan was never implemented because the Arabs wanted more water for Jordan and Syria, while Israel felt that it could obtain larger quantities of water through its own diversion scheme. Israel's unilateral diversion of the Jordan's headwaters (including the Hasbani River and Dan Spring in Lebanon, the Baniyas River on the Golan Heights, and the Wazzani Springs in southern Lebanon) to its national water carrier, which distributes fresh water right down to the Negev, has long been a point of bitterness for the Arabs in the Jordan Basin and throughout the Middle East. Syria's taking of water from the Yarmuk River, a tributary of the Jordan that rises in Syria, has also been problematic, particularly for Jordan, which has been the weakest state-actor in the region with regard to Jordan Basin water-sharing. The 1994 treaty between Jordan and Israel included a water annex, which, based upon the provisions of the Johnson plan for allocation of the waters of the Yarmuk, secured Jordan's water share from Israel, increasing its intake from the Yarmuk by 1.6 billion cubic feet.[83] Syria, which has been taking 7.8 billion cubic feet annually from the Yarmuk since the 1950s, has yet to engage in negotiations over its use of these waters.[84] The Johnson plan only allocated 4.4 billion cubic feet to Syria and 883 million cubic feet to Israel, the level allocated to the latter in the recent peace treaty.

For decades Israel has been drawing 80 percent of the 24 billion cubic feet of water provided annually by the underground water basin (a mountain aquifer running north to south) located mainly under the West Bank.[85] As we have pointed out, the aquifer provides Israel with a third of its water supply—40 percent of its drinking water and 50 percent of its agricultural water. Israeli military occupation orders, in place since 1967, include a prohibition on drilling of new wells, which prevents Palestinians from getting better access to the aquifer. Forty percent of Israel's fresh water comes from aquifers beneath the West Bank and Gaza Strip, which represent 95 percent of the known underground reserves in those areas.[86]

Syria and Jordan have requested funding from the World Bank for their projected Unity Dam on the Upper Yarmuk, which will regulate the water supply and hence ensure sorely needed water for the Jordan Valley and vital municipal and industrial water for the Amman-Zarqa urban complex. The World Bank has not financed the project since it needs the approval of all the riparians, including Israel, which depends on the Yarmuk for 3 percent of its national water supply and has withheld its approval.[87]

If there is a land-for-peace agreement between Israel and Syria it will have to resolve critical water issues. Water from the Golan, with its many rich water springs, provides 30 percent of Israel's drinking water.[88] In addition, the Wazzani Springs, located within Lebanon two miles away from the Lebanon-Israel border, feeds almost 4 billion cubic feet of fresh water annually into the Hasbani River, which flows, along with the Baniyas River, into the Sea of Tiberias in northern Israel. Israel secured the Baniyas Spring—the source of the Baniyas River—when it captured the Golan Heights in the 1967 war, and during the 1978 invasion of Lebanon it took the Wazzani source.[89] The waters of the Baniyas and the Wazzani have played a central role in the reclamation of arid land for Jewish settlers, and Israel will surely be loathe to return them to Lebanon and Syria unless some water-sharing agreement can be negotiated. Israel's military presence in southern Lebanon since 1978 serves as a potential bargaining chip to ensure the uninterrupted flow of the Wazzani Springs and the Hasbani River into Israel.[90]

Syria continues to maintain a strong military presence in Lebanon in part because it seeks to ensure that the headwaters of the Orontes River, which rise in the Bekaa Valley, will not be seized by hostile forces.[91] The Orontes flows from Lebanon, into Syria, and then into the Turkish province of Hatay, an area that is claimed by Syria. As the de facto upstream state, Syria has been taking 90 percent of the average annual discharge of the river to support its national irrigation schemes.[92]

The Euphrates-Tigris Basin

The Southeast Anatolian Development Project (known as GAP, its initials in Turkish) could, in theory, divert up to 90 percent of Iraq's intake from the Euphrates River and 40 percent of Syria's through the construction of a system of dams that will bring irrigation and hydroelectric power to one of Turkey's poorest regions. Since April 11, 1995 Euphrates water has been diverted into the fields of Turkey's sun-scorched Harran-Urfa Plain. Currently only 2 percent of the river's flow has left its natural course into Iraq and Syria. However, in the years to come Turkey plans to keep at least a third of the river water within its national boundaries for its own agriculture and energy needs.[93] Over half of the estimated cost of the GAP has already been spent, and half of the twenty-two planned dams have been completed or are presently under construction. Minister of State Kamran Inan has insisted that Syria and Iraq will always have enough water to meet their needs. However, in spite of Iraqi and Syrian protest, the Turkish government reduced the flow of the Euphrates for 30 days in January 1990 in order to fill the reservoir behind the dam. At the inauguration of the Ataturk Dam on the Euphrates in July 1992, Prime Minister Suleyman Demirel further sparked the anxieties of Turkey's downstream neighbors when he asserted that Turkey had the right to do whatever it wants with its waters.[94]

The average annual discharge of the Tigris at the Turkish border is 600 million cubic feet; sources in Turkey contribute 45 percent of the flow of the river, while tributaries in Iraq whose headwaters lie in Iran (the Adhaim, the Great and Little Zab, and the Diyala rivers) add a further 940 million cubic feet to the flow. Turkey's GAP will eventually reduce the flow of the Tigris in the interest of Turkish agricultural development. In part to offset Arab criticism of its projects, Turkey proposed between 1987 and 1991 a water "peace pipeline" to the arid areas of the Middle East, which we discuss in chapter 10.[95]

Adding to the complexity of the situation is the fact that the three protagonists in this water dispute have unresolved historical animosities. Mutual suspicion rooted in half a century of Turkish domination of the Arab world, Turkey's membership in NATO, its improved relations with Israel, and irredentist claims to disputed territory all complicate Ankara's relations with its Arab neighbors.[96] In 1984 as the massive Ataturk Dam was being built, Damascus began to support the Kurdish Workers party (PKK) to punish the Turkish government for embarking on a path that was potentially threatening to Syrian national security. The subsequent signing in 1987 of the protocol between Turkey and Syria led to the displacement of Kurdish rebel camps away from the Turkish border[97] and the transfer of PKK headquarters from Syria to the Bekaa Valley in Lebanon.[98] In 1992, however, Turkish officials acknowledged that the Iraqi and Syrian regimes were still supporting the PKK as a form of retribution for Turkey's current policy with regard to the Euphrates.[99]

The Nile Basin

Egypt depends on the Nile for 97 percent of its water, but 95 percent of the Nile's runoff originates in Sudan, Ethiopia, Kenya, Uganda, Rwanda, Burundi, Tanzania, and Zaire.[100] Egypt's per capita water use is far higher than the country's potential per capita internal renewable resources. In addition, while regional water supplies are diminishing as a consequence of climatic changes, the Egyptian population is projected to reach 75 million by the year 2000.[101]

Although Egyptians depend mainly on Nile water flowing from other countries,[102] Egypt's agreement with Sudan (Egypt and Sudan are the largest consumers of Nile waters) regarding allocation of the Nile waters was signed without reference to the demands of the other upstream users. This agreement does not take into account heightened tensions resulting from additional water development in upstream countries that would reduce the supply of water available to Egypt. Ethiopia controls 85 percent of the headwaters of the Nile,[103] and current plans to develop its hydroelectric infrastructure could reduce the Blue Nile flow by almost 9 percent.[104]

At least 50 percent of the discharge of the White Nile is lost in the Sudd swamplands of southern Sudan.[105] Attempts to build the Jonglei Canal through the Sudd to accelerate the northern flow of river waters out of the swamp heightened the internal tensions within Sudan, which led to the 1983 renewal of the civil war. The partially constructed canal, seen as a symbol of northern Sudanese exploitation, was destroyed, and there are no plans to rebuild it, as the war continues to rage throughout southern Sudan.

Former Egyptian under-secretary of foreign affairs Boutros Boutros-Ghali has said that the only thing which could make Egypt ever go to war again would be an attempt by Ethiopia or another power to divert the Nile River.[106]

The Arabian Peninsula

The Arabian peninsula receives infrequent and small amounts of rainfall, implying little, if any, surface runoff. Northern and central Saudi Arabia, most of Kuwait, Bahrain, Qatar, and the northern UAE have almost no surface runoff whatsoever. Southwestern Saudi Arabia, most of Yemen, southern UAE, and southern Oman receive enough rainfall to generate runoffs that contribute to alluvial (shallow) aquifer recharge and flood irrigation.[107]

The relative wealth of most states on the Arabian peninsula has made it possible for them to build and operate extensive desalination facilities to supplement their peacefully shared (for the time being) groundwater and surface runoff resources. In the future, however, as domestic water use continues to increase, groundwater resources become gradually depleted, and oil revenues cease to be sufficient to foot the bill for improved desalination technology and the maintenance of existing plants, the situation may change, increasing the possibility of regional tensions around water resources.

Agricultural development on the peninsula is either extremely water-intensive (and hence costly) or else requires expensive drip-irrigation technology, which is affordable in the short run but may not be feasible in the future given the projected depletion of oil reserves. Domestic food security, one possible motivation for costly water and agricultural policies, will become an increasingly difficult goal for this arid region.

The reluctance of the nations of the Arabian peninsula to engage in serious discussions regarding the importation of water from other countries in the region (e.g., Turkey through the proposed peace pipeline) is based partly on the need for internal water security. The fear that a simple turn of a spigot could cut them off has forced Saudi Arabia, Kuwait, and the UAE to invest heavily in desalination technology, which, unlike water importation, leaves control of the vital resource within the nation. Desalination, however, is not necessarily the more strategically secure option, given the vulnerability to attack of the desalting facilities.

Saudi Arabia, followed by Kuwait and the UAE, is the world's leader in installed desalination capacity. Prices per cubic meter of desalinized water range from $0.81 to $1.74, not including the cost of pumping the water from sea level to the place where it is used.[108] Almost all of the Saudi desalination plants are dual purpose power and desalination facilities, and most of the water used for Gulf petrochemical production comes from desalination facilities.[109] This makes them even more likely to be strategic targets in a conflict.

Conclusions: The Role of Economic Uncertainty

Against this backdrop of multiple conflicts with multiple causes, the relative economic deprivation of different population groups can transform nagging disputes into full-scale violent encounters. Thus the state of the regional economies is a vital factor in the overall assessment of conflict potential.

Much of the concern about the future is uncertainty about the economic well-being of the key countries. Some observers consider that the best hope for conflict resolution is the growth of regional economies, a belief that is especially prevalent in regard to the Caspian region and among individuals most wedded to the Arab-Israeli peace process. In contrast, economic troubles in the Gulf countries, particularly in Iran and Iraq, will limit the capacity of these two countries to dominate the region and expand their military capabilities, at least for the next few years. Among the oil-producing countries, especially the GCC group, the major concern is that the economic boom of the 1970s and 1980s is over and that budgetary crises will have a destabilizing impact on their traditional societies. Although demand for oil and gas will grow, the price of energy may not keep pace with the needs of their societies. The combination of population growth, high defense expenditure, and generous entitlement programs has severely constrained the budgets of even the richest countries, such as Saudi Arabia. Furthermore, the budget squeeze comes at a time when expectations, especially among the youth, have been rising owing to better health, better education, and more awareness of the wider world. For other major countries in the region, including Turkey, Egypt, Pakistan, and India, poor economic performance may be directly tied to the appeal of Muslim and Hindu radicalism.

In sum, while this study is not specifically focused on economics, the subject is of immense importance and must be regarded as one of the major backdrops to any analysis of conflict and cooperation in the greater Middle East region. While we have focused here on more tangible geographic issues such as territory, population dynamics, ethnic conflict, and water scarcity, the underlying importance of economic factors must be constantly emphasized and is discussed again in chapter 10. Although an active Middle East peace process is under way, with the potential for bringing economic benefits to the entire region, it has not yet stood the test of time, nor has it had much impact on the Persian Gulf and South Asia. For these reasons most of the countries in the region will continue to regard military force as the principal ingredient of national power and will continue to prepare for possible military conflict with the neighboring states and against dissident groups within their borders.

FOUR

The Strategic Energy Ellipse: The Persian Gulf and Caspian Basin

OVER THE NEXT fifteen years world demand for energy will reach unprecedented levels: It is expected to increase by between 34 and 46 percent between 1993 and 2010.[1] The increased demand will be met primarily by oil, natural gas, and coal. The energy needs of the newly industrializing countries (not including those in the Organization of Economic Cooperation and Development [OECD], Eastern Europe, and the former Soviet Union) could more than double by the year 2010, at which time OECD consumption could represent less than half of world energy demand. The principal factors in this projected dramatic growth are the booming economies of East Asia and South Asia.

World demand for oil will grow in parallel with these expectations, and the estimate is that it will increase from 70 million barrels a day (mb/d) in 1995 to between 92 and 97 mb/d by 2010.[2] Estimating the future supply and demand for oil is subject to many uncertainties—as indeed are all long-term economic projections—so it is misleading to use precise figures for the period up to the year 2010. Yet based on current trends—and assuming no catastrophic upheaval in the world economy—it is possible to suggest the general trend lines for the next fifteen years.[3] Furthermore, there is enough oil in the ground and under the seabed in the Middle East, Russia, and Asia to meet global demands into the indefinite future. The problem is getting the oil out of the ground and guaranteeing its distribution to the marketplace at an acceptable price. The primary obstacles to increased access to this oil are political, economic, and logistical rather than geological, and are tied into questions of sovereignty and regional stability. The world is awash in oil (and coal and natural gas), but the most promising sources for further production are located in two of the most politically unstable regions on earth—the Persian Gulf and the Caspian Basin. Of these two regions the Persian Gulf remains the most important.

Using numbers generated by the International Energy Agency (IAE) and other reliable sources, one can see clearly that Persian Gulf oil will probably become more rather than less important. It is estimated that by the year 2010 Gulf oil suppliers will be called upon to provide between forty and forty-five mb/d, a little under half of the world's requirements. The exact quantity of oil that will have to come from the Persian Gulf will depend on the rate of world economic growth, the availability of alternative sources, capacity constraints on further production, and price. Low-price oil and high-growth but inefficient economies will lead to increased demands for Persian Gulf oil because, at the margin, low prices make Gulf supplies more cost effective than the more expensive oil from other regions. Whatever the final numbers, two dramatic points stand out in all the authoritative projections:

—There is no realistic alternative to Persian Gulf oil to meet increased world supply in this period.

—The Asian countries will require increasing quantities of oil from the Persian Gulf.

One of the primary uncertainties during this period will be the amount of oil (and to a lesser extent, natural gas) that can be exported from the area adjacent to the Persian Gulf—the energy-rich Caspian Basin. This region has the potential to supplement Persian Gulf oil, yet the political and economic problems of transporting Caspian oil and gas to market remain unresolved. The three major oil and gas producers of the region—Azerbaijan, Kazakstan, and Turkmenistan—cannot get their products to market without going through another country's territory. This raises extremely sensitive geopolitical issues, which will be discussed at length in this chapter. While the Persian Gulf remains the repository of the greatest reserves in the world, it, too, is a region bedeviled with conflict. Until there is stability in the Gulf and more particularly a resolution of the crisis involving Iran and Iraq, there can be no assurance of uninterrupted oil supplies.

There is the possibility of oil becoming available from the other major reserve regions, including Siberia and western China. However, political obstacles and economic costs involved in developing these regions are also likely to limit their contribution to the market within the time frame of this study. Finally, technologies are being developed for more cost-effective fuel cells and electric vehicles, and there is the possibility that alternatives to the internal combustion engine for transportation will become generally available. However, few believe that these technologies will seriously reduce the demand for gasoline in the next fifteen years. The issue of alternative energy sources is examined in appendix 1.

The Enduring Importance of the Persian Gulf

Approximately two-thirds of the world's proven oil and one-third of its natural gas reserves are controlled by the Persian Gulf states (see figures 4-1 and 4-2).[4] If the estimated reserves of the Caspian Basin are added to these figures, the respective percentage for reserves goes higher, perhaps 70 percent for oil and over 40 percent for natural gas.[5] For this reason the Gulf-Caspian Energy Ellipse outlined in map 18 is one of the most significant geostrategic realities of our time.

The continued demand for Persian Gulf oil is related to its abundance and to the growth of the world economy. If sustained economic growth continues in North America, Europe, and Asia over the coming decade, the world will be faced with a sharp increase in demand for energy. The increase in Asian demand for Middle East oil will probably be the single most important development, and it will have major economic and geopolitical implications. If the countries of the former Soviet Union and Eastern Europe manage to get their economies in order, energy demand will grow at an even faster pace. It is

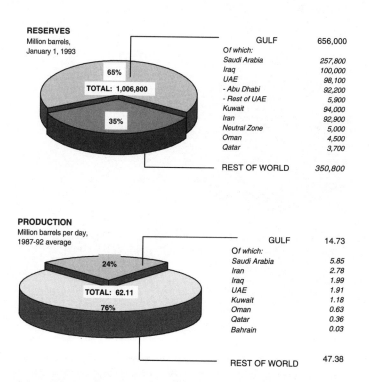

Figure 4-1. World oil reserves and production.
Source: Energy Map of the Gulf, *Petroleum Economist,* 1993.

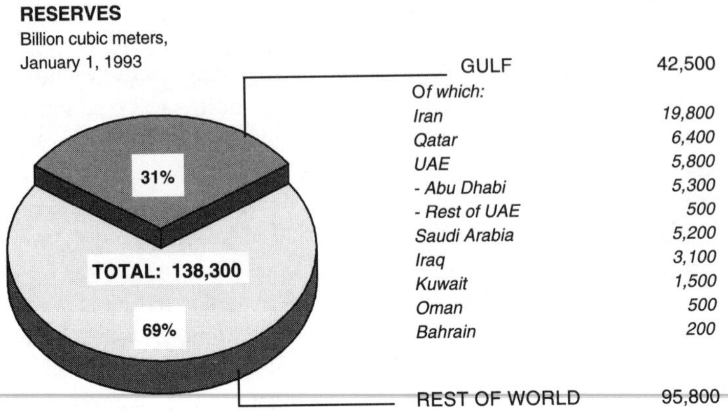

RESERVES
Billion cubic meters,
January 1, 1993

	GULF	42,500
	Of which:	
	Iran	19,800
	Qatar	6,400
	UAE	5,800
	- Abu Dhabi	5,300
	- Rest of UAE	500
	Saudi Arabia	5,200
31%	Iraq	3,100
TOTAL: 138,300	Kuwait	1,500
	Oman	500
69%	Bahrain	200
	REST OF WORLD	95,800

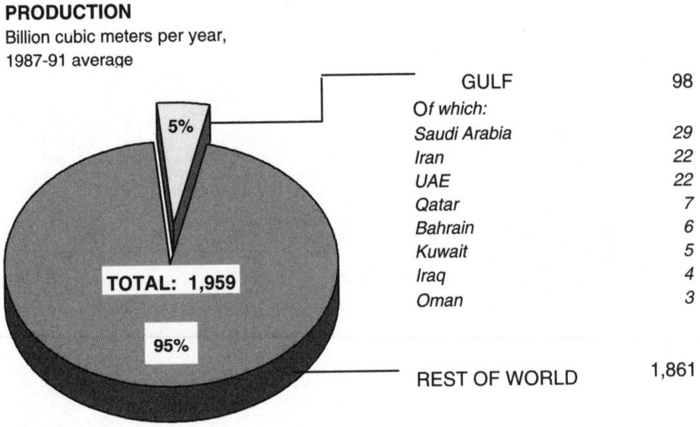

PRODUCTION
Billion cubic meters per year,
1987-91 average

	GULF	98
	Of which:	
5%	Saudi Arabia	29
	Iran	22
	UAE	22
	Qatar	7
	Bahrain	6
	Kuwait	5
TOTAL: 1,959	Iraq	4
	Oman	3
95%		
	REST OF WORLD	1,861

Figure 4-2. World gas reserves and production.
Source: Energy Map of the Gulf, *Petroleum Economist,* 1993.

possible that wild-card events such as global economic catastrophe or major war will hold down demand, but barring such scenarios energy needs are bound to rise.

In practical terms this means that for the next fifteen years the industrial world will remain critically dependent on energy from a region that remains vulnerable and unstable. Furthermore, given the inability of the small countries

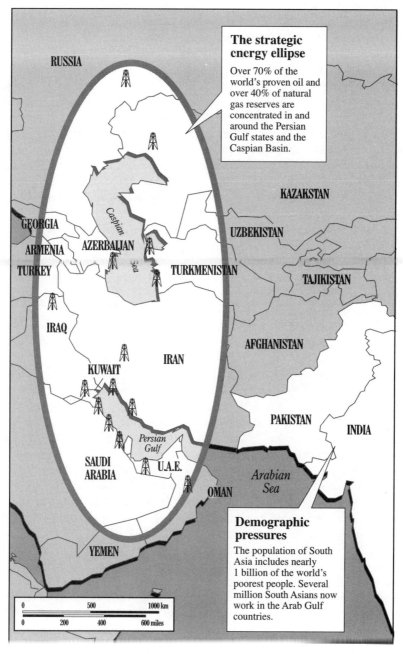

The strategic energy ellipse

Over 70% of the world's proven oil and over 40% of natural gas reserves are concentrated in and around the Persian Gulf states and the Caspian Basin.

RUSSIA

KAZAKSTAN

GEORGIA

UZBEKISTAN

ARMENIA

AZERBAIJAN

TURKEY

Caspian Sea

TURKMENISTAN

TAJIKISTAN

IRAQ

AFGHANISTAN

KUWAIT

IRAN

PAKISTAN

INDIA

Persian Gulf

SAUDI ARABIA

U.A.E.

Arabian Sea

OMAN

YEMEN

| 0 | 500 | 1000 km |
| 0 | 200 | 400 | 600 miles |

Demographic pressures

The population of South Asia includes nearly 1 billion of the world's poorest people. Several million South Asians now work in the Arab Gulf countries.

Map 18

in the Gulf, notably the Gulf Cooperative Council (GCC) states, to defend themselves in the near future against a rearmed Iran or Iraq, there is no alternative to a continued American military presence in the region. The Persian Gulf will thus remain a key U.S. military concern, perhaps long after the American land presence in Europe and Asia—where other options for regional deterrence exist—has diminished. However, this is not inevitable. Political questions as to why the United States has to continue to assume this commitment will probably be raised with increasing frequency in coming years. While the United States imports more oil from the Gulf than it did at the time of Operation Desert Storm, the percentage of total U.S. oil imports from the Gulf has fallen in the past two years. This has important geopolitical consequences as it has been happening at a time when European and Asian imports from the Gulf have been increasing.

Despite the obvious dangers of conflict in the Persian Gulf, demand for its energy continues to grow and the reasons are simple: Gulf oil and natural gas are plentiful; they can be extracted from the ground at a relatively low cost; and they are found along effective and well-developed transport routes. This reality has a powerful impact on the political fortunes of the major democracies, including the United States.

Consequently, even in the aftermath of the Gulf War, when the popularity of the Bush administration was at unprecedented heights, the United States made no attempt to intervene in the oil market and use the experience of the war to reduce American dependence on Gulf oil. The most persuasive arguments against such an initiative at that time revolved around the business cycle. In 1991 the United States was entering into a recessionary period. Imposing a tax on Gulf oil to reduce demand would have worsened the economic situation less than two years before the 1992 election. The political reality is that American presidents are hostage to the business cycle and a unique American sensibility to energy costs, especially gasoline prices. Had the end of the Gulf War coincided with an upturn in the economy or been in the middle of a boom, such a step might have been possible. Alternatively, had the Gulf War been more costly to the United States in terms of casualties, a radical approach to foreign oil dependence might have been politically feasible.

This scenario highlights a problem that confronts democratic governments, all of which are very sensitive to short-term economic trends if they wish to retain power. Since these trends are closely linked to day-to-day prices, which, in turn, are umbilically linked to energy costs, there is an inevitable incentive to hold these costs as low as possible. Historically, cheap energy has been linked to American prosperity. To adapt the economy to rely less on cheap oil will be difficult and painful and may be possible only in a national emergency.

In light of these truisms the contrast between the United States and Europe is worth noting. All the Western European countries have imposed a very high tax on energy, particularly petroleum. This overall strategy was adopted years ago, at a time when government-imposed price controls were the norm after the stringencies of the 1930s and World War II and as economic recovery permitted governments to capture this share of economic rent relatively painlessly. Much less of this tradition exists in the United States, and it would take a major new crisis to change American behavior in this regard. Thus the key to the U.S. energy policies is the high sensitivity of the American political system to radical changes in energy prices. There are also other fundamental structural differences between the United States and Western Europe. The latter, for reasons of geography, history, and a much longer tradition of central government control, has instituted far-reaching and relatively efficient public transportation systems. However, because of greater distances in the United States, many of the European systems would be uncompetitive. In many regions of the United States gas-consuming cars, trucks, and aircraft are the most efficient means of transportation.

There is also an important cultural issue. The United States, more than any other country in the world, is wedded to the private automobile. Until a form of transportation that can compete in terms of comfort, convenience, and cost with the private automobile in a free market with free choice, people will drive their own cars. Today Americans are back to driving larger cars, especially sports utility vehicles, and gas consumption is rising despite more efficient engines. Modern automobiles are more comfortable, reliable, safer, and pleasant to drive than their predecessors. These preferences will not change radically over the next ten to fifteen years. Quite to the contrary, it is certain that other countries wish to follow the American example. There will be more cars throughout the world as the middle classes of Asia discover the pleasures of air-conditioned automobiles with stereophonic sound.

Increasing Energy Demand (1995–2010)

Table 4-1 shows the 1996 estimates of world oil demand and supply between 1995 and 2010.[6] Beyond the year 2010, however, innovative policies to develop alternatives and reduce dependence on the Persian Gulf could begin to have a significant impact on demand. For example, by the year 2010, competitively priced alternatives to fossil-fueled automobiles could be available and increasingly in use, particularly if legislation is introduced to make these alternatives more competitive (see appendix 1).

As we noted above total world energy demand is expected to continue to increase over the next fifteen years at an average annual rate of 2.2 percent, with growth concentrated in non-OECD countries. The rate of increased demand in OECD countries is projected to be just 1.4 percent, and only 0.8 percent for the former Soviet Union (FSU) and Central and Eastern Europe. Growth in energy demand will therefore come mainly from the Rest of the World (ROW) states. The latter's share in world energy demand is expected to

Table 4-1. World Oil Demand and Supply (Capacity Constraints Case)
Million barrels a day

Area	1995	2000	2010
North America	21.7	23.8	26.1
Western Europe	13.9	14.6	15.9
Central and Eastern Europe	1.4	1.7	2.2
Former Soviet Union	4.7	5.0	6.4
Africa	2.2	2.9	4.2
Asia	18.1	21.1	29.8
South and Central America	4.1	4.8	6.2
Middle East	4.1	4.6	6.0
Stock changes	0.2	0.3	0.3
Total demand	70.3	78.6	97.1
OECD	21.0	22.5	21.2
FSU/CEE	7.4	8.1	10.6
Non-OPEC ROW	12.1	14.2	14.8
OPEC	28.3	32.1	48.6
Processing gains	1.5	1.7	1.9
Total supply	70.3	78.6	97.1
OPEC's share (per cent)	40	41	50

Source: *World Energy Outlook 1996* (International Energy Agency, IEA/OECD, 1996, Paris)
See endnote 6 for a discussion on IEA methodology for estimating demand and supply.

rise from 28 percent to about 40 percent, while the share consumed by OECD and FSU countries will decrease. The energy demand for OECD countries is expected to fall from 55 percent of total world demand in 1991 to less than 50 percent in 2010; in Eastern Europe and the FSU the projected drop is from 20 percent to about 13 percent of total world demand.[7]

The share of oil in primary energy demand is expected to decrease from 42 percent in 1991 to 38 percent in 2010 for the OECD countries, and from 41.6 percent to less than 40 percent for the ROW states, a downward trend that began in 1973. In Eastern Europe and the FSU, the drop is projected to be from 16 to 12 percent. In contrast, the share of ROW oil demand follows an opposing trend, rising from 28 to about 41 percent.[8] Globally the decrease in the share of oil will be paralleled by an increase in the shares of gas and solid fuels. The projected relative increase in natural gas use is especially significant for the ROW; it is expected to rise from 13 percent in 1991 to 17 percent in 2010. But in spite of the decreased share of oil in the total mix, IEA projects that until 2010 *oil will remain the most used fuel in the energy mix.*[9]

The IEA forecast ties the increase in energy demand, and particularly in oil demand, to a projected increase in economic growth over the next fifteen years.

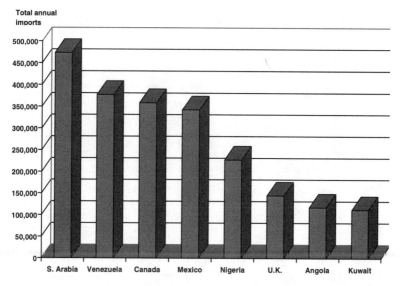

Figure 4-3. Country of origin of U.S. imports of crude oil, 1994 (thousands of barrels).
Source: Derived from Energy Information Administration, *Monthly Energy Review,* August 1995.

The developed economies (OECD) are expected to grow at an average rate of 2.6 percent in North America, 2.5 percent in Europe, and 2.5 percent in the Pacific industrialized countries, mainly Japan. The Chinese economy is projected to grow at an annual rate of 7.8 percent, while East Asia's economies will grow at a yearly rate of 4.1 percent. The average growth rate for the ROW group is projected to be 5.4 percent.[10]

U.S. reliance on imported petroleum reached about 45 percent in 1994 (see figure 4-3).[11] In 1991 Americans consumed about 17 mb/d and accounted for over 25 percent of total world oil consumption. Since 1983 U.S. oil consumption has increased by about 14 percent. At the same time U.S. domestic production has fallen sharply, resulting in a rise in imports. The Energy Information Agency (EIA) predicts that U.S. oil consumption will increase from 17 mb/d in 1992 to more than 21 mb/d in 2010. Thus it can be expected that the United States will rely on foreign suppliers for up to 60 percent of its oil needs by 2010, assuming continued low prices.[12]

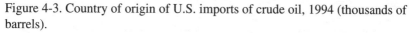

Between 1985 and 1993, some 58 percent of the increase in gross U.S. crude and petroleum product imports was supplied by the Persian Gulf countries. In 1990 more than 30 percent of U.S. oil imports came from the Gulf states, mainly from Saudi Arabia and Iraq.[13] However, the recent trend is toward greater reliance on "short-haul" oil from within the Western Hemisphere be-

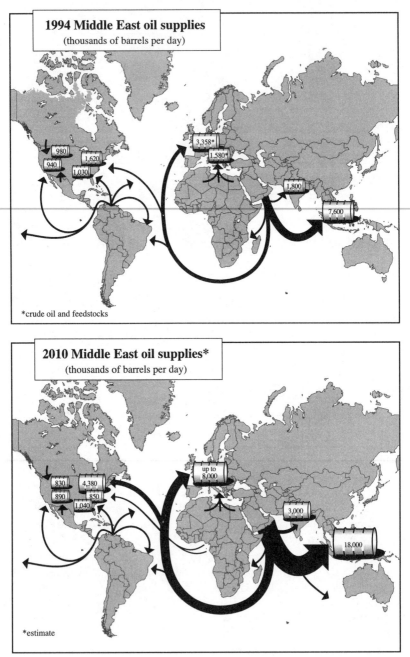

Map 19

cause of a massive change in oil company inventory policy, and to a lesser extent from the North Sea and west Africa. Most of the European Union's oil nccds havc also historically bccn supplied by the Middle East. Though their oil consumption and share of oil imports declined through most of 1980s, the oil needs of Europeans remained significant. The European Union, which consumes 18 percent of total world oil, is responsible for less than 4 percent of world production. In 1992 almost 4 mb/d—more than 25 percent of total Middle East oil production—were exported to the European continent.[14]

The incremental demand for oil will be highest in the ROW.[15] A World Bank study of the eight countries (including India and China) that consume half of the energy and one-third of the oil used by the entire developing world examines why economic expansion in the less developed world is usually accompanied by a rise in energy demand despite unfavorable prices on the world market. The study concludes that energy intensities increase when countries are in the early stages of industrialization and urbanization and peak when economies mature. High energy consumption in developing countries is due to structural economic changes that occur during sustained growth from a low industrial base to a higher one.[16]

In this connection there is a close correlation between demand for oil and transportation activity. Most of the incremental demand for oil in the industrialized world in the next fifteen years is expected to come from the transport sector. In developing countries one of the prime factors driving higher oil demand is the "unstoppable need for mobility" and the resulting increase in the use of cars and trucks, which, at least initially, have lower fuel efficiencies than those in OECD countries. By 1990 transportation accounted for 40 percent and industry for 17 percent of the total oil use in the eight countries covered by the study.[17] Hence even in the developing countries, oil will increasingly be used for transportation and residential needs rather than for heavy industry. The Shell Oil Company estimates that the number of cars worldwide could double to 1 billion by the turn of the century. Another survey projects that there could be 70 million motorcycles, 30 million trucks, and 100 million cars in China by the year 2015, with room for more.[18]

As suggested the Asia-Pacific area will become more dependent upon Middle East crude oil over the next fifteen years. In 1993 the Asia-Pacific countries imported more petroleum products (3.77 mb/d) than they exported (2.1 mb/d of products).[19] By the year 2000 the oil trade imbalance is expected to become even more serious, especially since increased demand for diesel fuel will put strong pressure on the market. As a result the Asia-Pacific dependence on imports of crude oil is projected to increase from 56 percent in 1993 to 72 percent in 2005. Imports of crude are expected to come primarily from the Middle East, which will account for 92 percent of total regional oil imports.[20]

If the predictions about future demand are even approximately accurate, the impact on oil flows from the Gulf are likely to be dramatic. Map 19 shows oil flows from the Gulf in 1994 and 2010. In 1994 the amount of Gulf oil going to both North America and Europe was only about one-third of that in 1975, at the height of the oil crisis (see map 13). Part of the falloff was due to Iraq's removal as a key exporter but other reasons were increased supplies from the North Sea and the Americas, conservation, and economic recession. Yet in 1994 Asian daily imports were more than double those in 1975. If we project ahead to 2010 the situation changes again. According to current estimates European and American demand for Gulf oil will be up to around 12 mb/d. Yet what is truly important is the estimated figure for Asia of 18 mb/d, or more than Europe and the United States combined. This is a most significant statistic and one that will have a profound impact on the geopolitics of the region.

If the projections for oil demand show a steady increase based upon expectations of economic growth and a greater use in transport, the supply side of the ledger is showing a number of equally clear trends. The most important of these are the continuing dominance of the Persian Gulf oil fields and the decline of older oil fields in North America and Europe. Big questions remain as to the availability of oil from the Caspian Basin, the South China Sea, China, Central Asia, and Russia. A look at prospective oil suppliers indicates that the two primary regions to supply the industrialized world with oil in the future will be South America and the Middle East, which will need to double production by the year 2010 in order to satisfy the growing demand for oil. Of these by far the most important is the Persian Gulf.

The Growing Importance of Natural Gas; Comparisons with Oil

While the geopolitical importance of the Middle East is associated with petroleum supplies, the significance of its huge natural gas reserves is likely to increase in the coming decades. As figure 4-2 shows, the Persian Gulf contains over 30 percent of proven world natural gas reserves. If the reserves of the Caspian Basin are included, the figure rises to over 40 percent. Worldwide demand for natural gas is growing. One estimate is that by 2015 world gas use could equal coal consumption, owing to the improved efficiency of natural-gas-fired electricity generating plants and the relative cleanness of gas compared to coal and oil.[21] Over the longer run, new technologies being developed to reduce the high costs associated with gas transportation would further increase the desirability of natural gas.

Although natural gas and petroleum are in some ways in competition with each other, the economics, infrastructure logistics, marketing, and geopolitics of these two fossil fuels are very different. Oil is an eminently fungible commodity but one found in relatively few locations. The high risks of investment in the oil industry occur during the exploration phase. Assuming its quality is acceptable, once crude oil has been extracted from the ground or from beneath the sea, it can be moved by a variety of means to dozens of markets. It can, for instance, be transported by pipeline, truck, train, ships of all sizes, or even aircraft to refineries and then sold to consumers all over the world. This results in a benchmark price for oil that determines global supply and demand in the marketplace.

The opposite is the case with natural gas. Although natural gas deposits are more geographically dispersed than oil deposits, the gas market is regionally priced and there are large differences between different markets. One reason is the difficulty of transporting gas to distant markets. There are only two basic ways to transport gas: by pipeline to dedicated markets or by liquified natural gas (LNG) tankers. The capital costs of installing LNG facilities are enormous. The start-up costs of an LNG terminal can reach $7 billion. For this reason it is very difficult to attract capital for such ventures unless there is a guaranteed, long-term, reliable market for the product and a belief that the producer will be a stable and reliable supplier. Thus the investment risks are spread over a larger time frame than is the case with oil. Furthermore, there is more uncertainty in estimating such risks.[22]

Another distinction in the respective marketabilities of oil and natural gas is the profitability of exports. Oil is virtually guaranteed an export market, and therefore it has high priority for most countries as a foreign exchange earner. Gas, on the other hand, being much more market specific than oil, will not necessarily be a profitable item for export, if the alternative is a boom in domestic demand.

This reality has significant implications for energy geopolitics in the Caspian-Gulf region. Consider the dilemma that faces Iran. Iran is blessed with the world's second largest reserves of natural gas. Its oil industry, on the other hand, is maturing and its oil fields are old and need major investment for modernization and huge amounts of gas injection to sustain pressure in these oil wells. At first sight it might seem logical for Iran to move from oil to gas export, since there is a worldwide demand for gas and gas fields are virtually untapped, but this is where the problems arise.

First, since the upfront costs of developing gas fields are so extraordinarily high, long-term markets must be assured before such investments are made. The most natural clients for Iranian gas are nearby neighbors, particularly Turkey, Pakistan, and India. Yet in order to deliver the gas, even to its closest neighbors, Iran would need to invest very heavily in modernization of its existing gas pipelines, or, in the case of Pakistan and India, build extremely expensive new pipelines. In addition, Iranian demand for energy at home is growing and there are very strong motives for supplementing and then re-placing petroleum consumption with gas consumption, leaving the oil available for the profitable export market and the earning of foreign exchange.[23]

Second, the location of Iran's gas presents a problem for domestic consumption. Most of Iran's gas fields are located in the south of the country, whereas the population that needs the gas is in the north. One alternative would be for

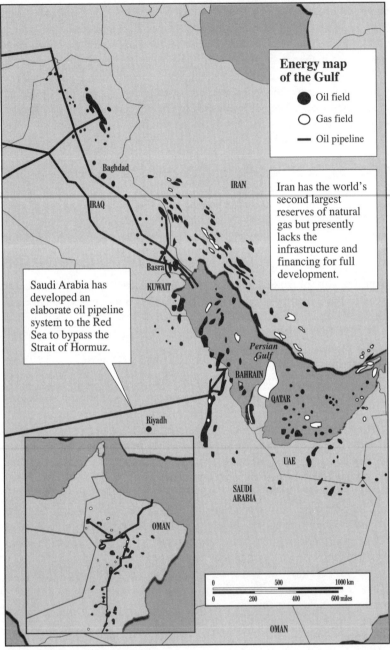

Energy map of the Gulf

- ● Oil field
- ○ Gas field
- — Oil pipeline

Iran has the world's second largest reserves of natural gas but presently lacks the infrastructure and financing for full development.

Saudi Arabia has developed an elaborate oil pipeline system to the Red Sea to bypass the Strait of Hormuz.

Baghdad

IRAN

IRAQ

Basra

KUWAIT

Persian Gulf

BAHRAIN

QATAR

Riyadh

UAE

SAUDI ARABIA

OMAN

OMAN

Source: *Petroleum Economist,* 1993

Map 20

Iran to import gas from Turkmenistan for its northern consumers and to export gas from its southern fields to market. However, given Iran's precarious political position and the United States' threat to impose sanctions on foreign companies that deal with Iran's energy industry, Iran may find it impossible to raise the necessary capital for its gas project absent a change in its politics.

In other words, the prospect of Iran's gas reserves finding their way to the world market in the near future is far from inevitable. While it is true that East and Southeast Asia have a need for gas, transshipping it from Iran—leaving aside the political problems with the United States—would be an extraordinarily complicated and expensive job. Obtaining gas from other regional sources, such as Australia, or even less politically controversial countries in the Gulf, such as Qatar and Oman, may be the preferred strategy. These political issues have led the Japanese, among others, to think that a land gas pipeline leading from Central Asia to China and then passing underwater onto Japan, despite all the obvious logistical problems, may be the most feasible solution in terms of national security interests.

Over the long term, one development that could significantly change the nature of the gas market and introduce more flexibility would be the creation of a major inter-Asian network of connecting gas pipelines, analogous to the system in the United States. This approach would enable gas to be redirected to target markets with much more speed and efficiency than is currently the case. It would also provide more fungibility in the overall gas business and help the industry become more competitive and less tied to specific market opportunities.

THE GEOGRAPHY OF PERSIAN GULF ENERGY. From a geographical perspective two issues concerning Persian Gulf oil and natural gas are of paramount importance: the location of the oil and gas fields and the associated distribution system, including the collecting system, oil and gas separation plants, storage tanks, refineries, loading facilities at ports, and pipelines. As map 20 shows, the geography of the energy has some very important features:

—Significant oil and gas fields lie offshore in the Persian Gulf itself. Most dramatic is the huge gas field (North Field and South Pars) which lies mostly in Qatar's exclusive economic zone, with a smaller but significant portion belonging to Iran.

—Most of Iran's oil and gas fields are located in southwest Iran and along the north littoral of the Gulf. Its gas supplies are a long way from its domestic market in the north of the country. Although Iran has huge natural gas reserves, its distribution system for gas is underdeveloped. Virtually all of Iran's oil egresses to market go via the Strait of Hormuz.

—All of Oman's oil and gas are located onshore, where many of the United Arab Emirates' (UAE's) oil assets are offshore.

—Saudi Arabia's giant field at Ghawar is approximately 250 miles long and between 20 and 25 miles across, by far the biggest field in the world. Saudi Arabia has also developed a sophisticated network of egress routes including oil and gas pipelines to terminals on the Red Sea, thus bypassing Hormuz. It also uses the Trans-Arabian Pipelines, which goes to Jordan and on to Syria.

—Iraq, on the other hand, has virtually all its oil and gas onshore located in the south and northeast of the country. Before the Gulf War Iraq exported oil by sea and by pipeline south through Saudi Arabia and oil and gas north and west through Turkey to the Mediterranean and Turkish cities.

—There are several proposals to build more pipelines, especially for natural gas, including one from southern Iran to Pakistan and India.

—Several big oil fields straddle international boundaries and maritime economic zones, including Rumaila-Ratgar (Iraq and Kuwait). Disputes over the extraction of oil from this field was one of the factors leading to the Iraqi invasion of Kuwait in August 1990.

CAN THE GULF MEET GLOBAL OIL NEEDS? Having the necessary proven oil and natural gas reserves in the ground is an important prerequisite, but not the only condition, for maintaining a steady energy supply in the future. It remains to be seen whether the Persian Gulf countries can expand their capacity to meet growing demand.

A great deal will depend upon the financial health of the key oil-producing countries and whether they can find the necessary capital to expand their production capacity. Iraq and Saudi Arabia have been producing substantially below capacity, Iraq because of the Gulf War and UN sanctions and Saudi Arabia because of OPEC quotas. (Saudi Arabia has excess capacity of 2.1 million barrels per day.) Plans to expand production in both countries have been

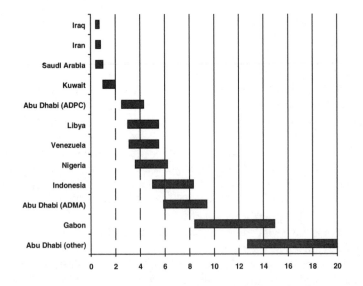

Figure 4-4. Oil exploration: costs of incremental OPEC production.

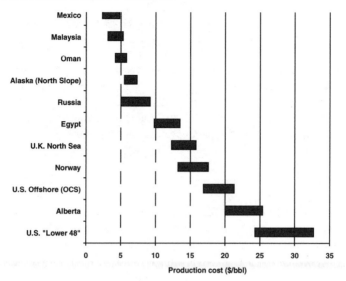

Figure 4-5. Oil exploration: costs of incremental non-OPEC production.

delayed because of financial constraints. All Gulf countries have been under pressure to fund other important and expensive programs, including defense, health, education, and housing initiatives, and this situation has resulted in lower investment rates for expanded oil production.[24] Thus, while in theory the Gulf is capable of producing considerably more oil, the estimated additional amounts will remain uncertain as long as Iraq remains a pariah state and the other producers face financial problems.

Oil Production Costs

Along with capacity expansion production costs are an important indicator of whether the world will continue to enjoy a steady supply of oil from the Gulf. If one compares the costs of oil production worldwide, "the 'Big Four' in the Gulf—Iran, Iraq, Kuwait, and Saudi Arabia—stand out as doubly blessed and uniquely low-cost producers: not only are their reserves the most abundant in the world, but their production is also by far the world's cheapest" (see figures 4-4 and 4-5).[25]

The costs of oil production in Iraq, Iran, Saudi Arabia, and Kuwait have increased over the last twenty years but are still under $2 a barrel. Low production costs in the leading Gulf exporters are linked to the rate of flow in Gulf wells. (The rate of flow is a function of natural or induced pressure.) The average flow rate for the "Big Four" is 4,000 barrels a day (b/d), which is 100 times greater than that of the average U.S. well. In Venezuela and Indonesia the flow rate is less than 200 b/d, much higher than that of wells in the United States or Canada but still far below Gulf rates.[26]

Production costs also depend on the location of wells, as well as on their depth. Deep wells in remote or hazardous locations are more costly than shallow wells in easily accessible locations. One important factor is whether the wells are on- or offshore. Offshore production costs are two to four times higher. Some older offshore wells in Abu Dhabi have production costs of up to $10 a barrel, and in fields like the Upper Zakum project and Al Bunduq costs are higher. The costs for the Middle East members of OPEC are not so high as to present a barrier for the future development of new wells, considering that the payback period of new wells is six months or less. Iran, however, may incur serious capital shortages—representing up to one year's gross oil export receipts—after financing its gas injection schemes. Such shortages would make new investment more difficult. In OPEC countries such as Venezuela, payout periods have been much longer.[27] Private investment in hydrocarbon development in Venezuela was prohibited by law until 1995, when the law was changed. Despite very demanding terms foreign investment interest is now booming—and so is Venezuelan production.

Estimating Oil Prices

The future supply of and demand for oil is directly related to price. Low prices encourage greater demand and reduce incentive to find alternative energy sources. However, they also deprive the producer of income and by extension the capital needed to expand production and to make it more efficient. High prices, on the other hand, increase incentive for consumers to reduce consumption and seek alternative energy sources. Sustained high prices provide the producers with capital to further exploit their resources and market niche. Clearly, then, estimates about the future price of oil have an important bearing on the geopolitics of the Persian Gulf.

Attempts to predict trends in oil prices began after the oil shock in 1973 when, for the first time in many years, prices were no longer stable. After the two oil shocks, at a time when most oil experts predicted rising prices and increasing dependence on Middle East oil, Israeli economist Eliyahu Kanovsky argued that prices would go down and OPEC's importance diminish. He was right: in the 1980s prices began to fall, and the major Middle East oil exporters experienced a sharp decline in oil revenues. The change in OPEC's policy originated with the Saudis, who decided that high prices were not in their best interests and undertook measures—moderate prices and increased exports—to gain back lost markets.[28]

However, an energy survey in the *Economist* in June 1994 suggested that if the Gulf states invest in their industries as little as possible, oil markets will be tight and the price of oil will certainly rise.[29] James Schlesinger, secretary for energy under former President Carter, stated that, "By the end of this decade we are likely to see substantial price increases, for which the consuming countries, and notably the American people, are unprepared."[30]

The question of future oil prices causes heated debate. Those, like Kanovsky, who believe price will remain low rest on the notion that supply will be plentiful in the 1990s, especially in the aftermath of the Gulf War, which stimulated exploration worldwide. Non-Gulf producers such as Venezuela, Libya, and Nigeria are planning increases in their production capacities and non-OPEC output is likely to rise significantly since Western companies, wary from the Gulf War, are, in theory, attracted to areas far from the Middle East. Yet as the next section outlines, the region getting the most attention is literally next door—the Caspian Basin. Non-OPEC investment will presumably grow, aided by new Western technology, which will decrease exploration costs. However, many of the potential sites for future expansion of oil production currently face political and economic difficulties that could hinder increased oil exports. Studies of existing oil exploration sites in the FSU and the South China Sea highlight some of the more acute political and logistical constraints on new oil production. The analysis shows that for at least the next decade or so, oil supplies from regions such as Russia, China, and North Africa are unlikely to become viable alternatives to Persian Gulf oil (see appendix 1).

The reality is that even if prices do stay depressed through the 1990s, the importance of Persian Gulf oil will increase. According to the IEA analysis, "under assumptions of low oil and gas prices, the world's dependence on the Middle East and Venezuela is even greater than under the high-growth assumptions."[31] Viewed in the domestic context, the oil price dilemma confounds: if prices on the world oil markets stay depressed or even drop, dependence on Gulf oil is likely to increase; if prices go up, the Gulf countries will get richer and the U.S. deficit will rise.

Can U.S. Dependence on Gulf Oil Be Reduced?

There have been many proposals designed to limit oil imports from, and thus dependence on, the Gulf. Some advocate encouraging further imports from South America and by 1995–96 market forces were orienting U.S. imports toward more Western Hemisphere supplies that are now becoming available. Others suggest increasing American domestic production of oil by juggling import fees, tariffs, and tax credits. Virtually all of these options entail major political costs and will result in either higher gas prices, higher taxes, or both, with concomitant damage to the economy and growth. Furthermore, within the continental United States and Alaska, there are now far-reaching environmental laws and political pressures that prevent the exploitation of remaining abundant oil sources in places such as the California coast or the northern slope of Alaska, though this could change with new political alliances in Congress. As a way to diversify the oil supply, the Office of Technology Assessment (OTA) proposed in 1991 "helping the Soviets to expand oil production in return for a share of it," and assisting "sister nations in the Western Hemisphere," by contributing to the development of oil reserves in Venezuela and other U.S. allies.[32] In theory the United States could reach an agreement with countries in Central and South America—particularly Venezuela, Colombia, and Mexico—to meet U.S. needs. In the short run, however, this would mean higher costs at the pump because billions of investment dollars would be necessary to transform Mexico, Colombia, and Venezuela into alternatives for the Persian Gulf.

The 1991 OTA report concluded that the greatest opportunities for oil replacement lie in the transportation sector, since it is the nation's largest oil user, accounting for 63 percent of total U.S oil use in 1991.[33] Major oil-replacement opportunities for light-duty vehicles in the 1990s include improved fuel efficiency, conversion to natural gas and other alternative fuels, better traffic management, and replacement of old cars with new, fuel-efficient ones. As the report explains, however, "the transportation system is virtually locked into petroleum use for all but the long-term, and efforts to shift to alternate energy sources face significant hurdles."[34]

There are, of course, other proposals designed to ensure that American dependence on Persian Gulf oil does not increase costs at the gas pump and that the United States is never again beholden to the powers of the OPEC cartel. One such proposal is competitive bidding, which prevents OPEC from imposing its will on the rest of the world by inflating oil prices. In theory such policies will work. The problem is that as long as they do and as long as Gulf oil is reasonably priced, there will be little incentive to develop alternative sources of oil, let alone alternative fuels to substitute for oil. The dependence will thus persist, if not grow. Under current conditions—with oil cheap and in ready supply—the U.S. oil-replacement capability, which relies on conservation, efficiency, and fuel-switching, has fallen significantly over the last seven years.[35]

There is no way out of this dilemma in the immediate future. Persian Gulf oil will remain a critical prerequisite for Western industrial growth and increasingly for the industrial growth of Asia and much of the developing world. This dependence on Gulf oil entails geopolitical risks. It does not take great wisdom to see that a rearmed Iraq is likely to emerge after the lifting of sanctions; another Gulf War can therefore not be ruled out. It may not be an Iraq armed with nuclear or chemical weapons, or with long-range missiles, but in the absence of an American security guarantee it could be an Iraq equipped with conventional capabilities quite sufficient to intimidate and thus control its neighbors.

On the other hand, the dependence between the United States and the GCC countries is increasingly a two-way street; oil is only part of the equation. Not only are the GCC countries more beholden than ever before to the United States for security, but since Desert Storm there has also been significant growth in U.S.-GCC economic cooperation. As the GCC countries become richer, and as they diversify and enlarge their economies, they will not only become more important to the world economy, but also more interdependent with it. GCC capital will increasingly fund development schemes in the West as well as in the Middle East itself. The GCC will have a major incentive to invest in the stability of the world markets and economy, since there will always be surplus capital that cannot be used directly for development within the countries themselves. This is not to say that the GCC countries cannot or will not further expand their own nonoil industries, but as they develop they will need to increase their imports in areas other than arms, which they are already purchasing in vast quantities. Economics aside the GCC may also ultimately embrace political values closer to those of the United States and the West. At that point a relationship more akin to relationships between the United States and the countries of Western Europe, Canada, and Israel may emerge, in which case the relationship between the West and the GCC will become even closer than it is today.

The Persian Gulf and Asia's Energy Needs: Geopolitical Implications

While forecasts of increased Western dependence on Persian Gulf oil merely reinforce current policies that put a premium on regional stability, the growing needs of Asia pose a new problem, and the economic dimensions of the energy equation will have a very important security component. If the estimates of Asian energy requirements are of the right order of magnitude, it is clear that this "Asian energy gap" will have profound consequences for the global economy and the geopolitics of the Middle East. Increasingly close ties among the Gulf, China, and the Indian subcontinent could transform the political and economic geography of the region. With respect to the Indian subcontinent, proximity and disparity in population size and per capita income make it certain that closer ties to the Persian Gulf will have major political implications.

By way of reference we consider the situation at the time of the last Gulf crisis in 1990–91. There was no doubt who was in charge. President George Bush was able to orchestrate a unique coalition against Saddam Hussein owing to the exceptional circumstances of that moment. Iraq's invasion of Kuwait was such a flagrant violation of the most basic principles of international relations that the votes at the UN condemning Iraq and applying sanctions were nearly unanimous. For the first time since World War II, the United States and the Soviet Union found themselves on the same side concerning the use of force in the Middle East. The oil market in 1990 was sufficiently buoyant that the crisis and war had little long-term impact on oil prices and there were no gas lines in the United States and other industrial countries. The war itself was conducted with speed and efficiency and very few allied casualties.

The chances of encountering such favorable conditions the next time there is a crisis in the Gulf are not high. A new crisis could be more ambiguous in regard to naming the culprit. This could erode hope for a consensus at the UN, or elsewhere, on what to do. While the United States presently has the military capability to deter high-level threats from countries such as Iran and Iraq, it has little control over the internal security problems that the Gulf countries will face in the coming decade. These threats have to do with social and religious factors, including autocracy, corruption, asymmetric income distribution, migration, human rights abuses, and Islamic extremism.[36] Given increased demand the oil market might be tighter when the next crisis develops and, unlike 1990, the key Asian countries might be less willing to accept American leadership in crisis management. It is also likely that Russia would be more likely to challenge American Middle East policies, especially given its increasing stakes in Central Asia and the Persian Gulf. Similarly, if China and India become major importers of Middle East energy, they will likely have their own, more independent and possibly more confrontational, policies toward the region. There have been suggestions that China's embryonic blue water navy may eventually be deployed in the Indian Ocean in part to assure the security of oil supplies from the Gulf.[37] If this happens China will be following the traditional behavior of maritime powers, who have always put a high premium on sea control missions. In this context much has been written about the relationship between China's energy needs and its assertive political-military activity in the potentially oil-rich South China Sea. More speculative, but potentially very important, is the possible link between China's growing needs for Persian Gulf oil and its plans to develop a blue water navy with the capability of protecting its sea-lanes in the Indian Ocean and South China Sea and thereby posing a putative challenge to other maritime powers in the region, especially India and the United States.[38] The key to the geopolitical linkage between China and the Middle East is oil. If China's economic growth continues and if its efforts to rapidly develop its own potential oil resources in the Tarim Basin and the South China Sea fail or are continually delayed, it will join the United States and Japan as one of the top three oil importers in the world. As one writer put it:

> China will be as robust as the United States in defending its access to oil supplies. . . . China regards the Middle East as a major supplier of oil for its domestic needs. China trades arms for oil and has recently been supplying Iran with missiles, tanks, and other military equipment. It has also sold missiles to Saudi Arabia and military equipment to Kuwait in exchange for oil supplies.[39]

It is impossible to imagine Chinese naval deployment in the Indian Ocean not soliciting a response from India. In sum, new permutations of power relationships will emerge in the next century that may bear little resemblance to those that existed in 1990–91.[40] What role the United States will play in these new relationships is a key question for the future.

Finally, big question marks remain as to who will be ruling the two most powerful countries in the Gulf—Iraq and Iran. In theory there could be new pro-Western regimes in both countries, with profound consequences for regional stability. If these changes in regime coincide with a successful Israeli-Syrian peace settlement, there will be increasing international pressures on the GCC countries to adopt more democratic institutions, and this could itself lead to instability and conflict. On the other hand, if Iran and Iraq remain adversarial states, the possibility of further interstate armed conflict cannot be ruled out. Sooner or later Iraq will begin exporting oil and will then have the money to rebuild its conventional armed forces.

The conclusion must be that the geostrategic importance of the Persian Gulf will grow. At the same time the adjacent Caspian Basin will also probably become a primary source for energy if political problems can be solved. Whatever happens, the many sources of conflict in the Persian Gulf region will mean that security concerns and military planning will continue to go hand in hand with economic development and efforts at conflict resolution.

The Caspian Basin: Energy and Geopolitics

A hundred years ago the Great Game between Russia and Britain involved geopolitical competition along a swathe of territory running from the Bosporus, through the Caucasus, to the Caspian Sea and Central Asia, and into the Hindu Kush Mountains. At that time Britain's preoccupation was Russia's colonization of Central Asia and its parallel construction of the Trans-Caspian railway to Bokhara and Samarkand. Today a new Great Game is evolving between the countries of the Caspian Basin and the key external powers, including the United States, and the reason is energy.

It is ironic that the most likely competitor for Persian Gulf oil producers is the region next door. The estimated oil and gas reserves of the Caspian Basin are very great and *could* be as high as 200 billion barrels of oil and 279 trillion cubic feet of natural gas, though most industry analysts use the figure of 90 billion for oil.[41] They represent a potential treasure trove for the struggling economies of the five littoral states and their immediate neighbors. Russia, Iran, Azerbaijan, Kazakstan, and Turkmenistan all have significant oil or natural gas deposits, while countries and companies, both regional and international, are hoping to control the access routes that will allow for the export of Caspian oil and natural gas.

There are three elements of Caspian energy geopolitics that are of paramount concern: the ownership of the resource, the amount of hydrocarbons available for extraction, and the production and distribution including direct and environmental costs. Whether or not there will be cooperation or conflict over access to Caspian energy will depend upon whether these issues are resolved in an adversarial or cooperative manner.

Indeed, what makes the developing Caspian energy rush unique is the question of ownership and geography. Until 1991 the Caspian Sea was controlled by the Soviet Union and Iran. Its legal status following the breakup of the Soviet Union has yet to be defined. All five of the littoral states face major economic and political obstacles limiting their capacity to export energy. Three of the major oil and gas producers—Azerbaijan, Kazakstan, and Turkmenistan—are surrounded by other countries, and as we noted earlier cannot get their energy to market without crossing territory in another country. It is interesting that in the long and turbulent history of the oil business this crossing of borders has been very rare. In the past all the great oil-and gas-exporting countries (Algeria, Angola, Bahrain, Brunei, Canada, Colombia, Egypt, Iran, Indonesia, Iraq, Kuwait, Libya, Mexico, Nigeria, Norway, Qatar, Oman, Romania, Russia, Saudi Arabia, United Arab Emirates, United States, and Venezuela) have had direct access to the world's shipping lanes.

For many years the big multinational oil companies were able to negotiate bilaterally with individual countries and market their products with little need to cooperate with neighboring states. Thus the cozy deals that American oil companies made with Saudi Arabia before World War II and a variety of companies with Libya in the 1960s reflected the way the oil business worked. But in the Caspian region, the major foreign energy companies, as well as the producer countries, must also negotiate with adjacent countries, especially Russia, and reach mutually acceptable agreements if the energy is to be exported. If all the parties do not cooperate the oil and gas will remain in the ground.

Owing to this geographic reality, the multiple routes being considered to transport the energy have taken on political and strategic, as well as economic and environmental, overtones. However, since the economic stakes for the winners are so high the principal participants are only reluctantly accepting the need to cooperate and compromise. Russia, for instance, has tried to use strong-arm tactics to ensure a monopoly on the key routes from Azerbaijan and Kazakstan. However, its ability to dictate terms has been limited because of its own financial problems and the chaos and conflict along its southern borders, especially the war in Chechnya. Iran's problems relate to its poor relations with the United States and its difficulties in attracting the investment needed to develop its huge, but underutilized, natural gas reserves and to exploit its geography as a transit route for the oil and gas of its neighbors.

Since the choices for oil and gas routes involve so many complicated decisions, many different options have been proposed involving countries as far apart as Bulgaria, Greece, and Turkey in the west, and India, China, and Pakistan in the east. Each of the many options has costs and benefits. For instance, Bulgaria and Greece, in cooperation with Russia, are proposing an oil transport route that would avoid Georgia, Turkey, and the Bosporus, but the plan would probably further aggravate Greek-Turkish tensions. One of the more intriguing proposals has been to build natural gas pipelines from Turkmenistan and Iran that run into Pakistan and then on into India. Apart from the economics of such a project, it raises some important questions about the relationship between the subcontinents' two nuclear-armed antagonists. Could the pressing need for natural gas and the benefits of a shared project help them work more closely together to resolve their outstanding differences, including Kashmir? Could it reduce the pressures on both countries, but particularly India, to expand their civilian nuclear plants to meet growing electricity needs? Or will the ongoing crisis over Kashmir and the mutual suspicion between the two countries put any prospects for energy cooperation on indefinite hold?

What happens concerning ownership and access will have major strategic consequences for the region, its neighbors, and the external powers. Competition over the oil and gas and related pipelines—an issue rooted in the geographic location of these natural resources—is helping to shape the relations among interested states. The choices pose dilemmas for all concerned, including the United States. If there is no agreement among the littoral states concerning ownership and access issues, the disputes could degenerate into serious political conflict which, in turn, will discourage international investors from taking risks. Alternatively, if the disputes can be resolved and the benefits of enhanced production and distribution spread around to all the parties, the region could witness significant economic growth.

The Status of the Caspian Sea and the Bosporus

Before examining the specific problems of energy extraction and distribution, territorial questions concerning the legal status of the Caspian Sea and the Bosporus Straits have to be addressed as oil traffic from the Black Sea has to pass through the latter to reach the Mediterranean and worldwide markets.

According to the 1921 Treaty of Moscow, which was reaffirmed in 1935, the inland Caspian Sea belonged to Russia and Persia and was referred to as a "Soviet and Iranian sea."[42] Today Russia maintains that the sea should legally be treated as joint property shared by all the bordering states—including the newly independent republics, Russia, and Iran—which would effectively delegitimize efforts by any one state to unilaterally exploit the sea's resources within its own demarcation lines, which would be drawn to equidistant lines between bordering states to effectively divide up the entire sea.

President Yeltsin has argued that, "The Caspian Sea is a special inner sea and it cannot be divided. . . . This concerns oil and gas as well." Russia's position is supported by the fact that the Caspian Sea had historically been treated as a unity (in that it lacked any formal delimitation) by Persia (Iran) and the Soviet Union (Russia). In the singular relevant precedent concerning joint sovereignty over enclosed or semienclosed bodies of water—the case of the Gulf of Fonseca, which formerly belonged to a single state (Spain) but is now bordered by El Salvador, Honduras, and Nicaragua—the International Court of Justice saw no advantage in disrupting the unity of the body of water after the emergence of successor coastal states.[43] Both Iran and Russia have a common interest in the "inner sea" concept since it is believed that neither has many resources in its particular sectors of the sea.[44]

A Russian memorandum to the UN on October 5, 1994, emphasized that Soviet-era treaties involving joint utilization remained in force as the existing legal regime for the Caspian and that the Law of the Sea (which would permit boundary delimitation) could not apply since the Caspian has no natural connection to other seas.[45] In late June 1995 similarly worded diplomatic notes were sent to Azerbaijan, Kazakstan, and Turkmenistan, declaring that Russia would not guarantee the safety of Caspian natural resource projects launched before the sea's legal status was agreed upon by the littoral states, ostensibly referring to an $8 billion 1994 agreement signed by Azerbaijan and a Western-led international consortium. In the memorandum and the notes Moscow declared that, "The Russian Federation reserves the right to take the necessary measures to restore order and to eliminate the results of such unilateral moves disregarding the Caspian Sea's judicial nature."[46] In an attempt to appear conciliatory and break the deadlock, in November 1996 Russia proposed a compromise whereby each Caspian state be given exclusive jurisdiction over oil fields lying within 45 miles of a zone extended out from national shorelines. To date no other Caspian state has agreed to this proposal.[47]

Disputes over the Caspian Sea are paralleled by an age-old dispute between Turkey and Russia over the Turkish straits, which continue to comprise a critical export route for energy resources from the former Soviet republics. The 1936 Montreux Treaty enjoins Turkey from restricting Russian merchant shipping through the Bosporus. However, it appears that Turkey is prepared to mobilize environmentalists and world opinion to limit Russian oil transiting the waterway and will promote its own alternatives for bringing Caspian oil to the Mediterranean via pipelines that bypass the Black Sea.[48]

How Much Energy?

According to the 1994 IEA forecast, oil production in Central Asia and the Caucasus, particularly in Kazakstan and Azerbaijan, could reach 1.5 mb/d by the year 2000, provided the problem of transportation can be solved.[49] The massive Tengiz field in Kazakstan, for example, is served by only one pipeline through Russia—a pipeline that is old and poorly maintained—with a limited carrying capacity. The field currently produces 130,000 b/d and is expected to increase that to 160,000 b/d in 1996.

Kazakstan's Statistical Committee reports that 190 explored oil and gas fields hold 15.33 billion barrels of oil, 5.69 billon barrels of condensate, and 83.35 trillion cubic feet (tcf) of free and associated gas, with prospective resources estimated at around 33 billion barrels of crude oil, 13 billion barrels of condensate, and 205 tcf of natural gas.[50] However, estimates of hydrocarbon reserves and the value of this potential vary widely. The U.S. Department of Energy has broadly estimated that 30 to 60 billion barrels of oil lie underground in Kazakstan,[51] and Chevron estimates 6–9 billion barrels, worth at least $114 billion, in the Tengiz field alone.[52] Even richer in potential are the shoreline and north Caspian fields currently under exploration; it has been estimated that up to 183 billion barrels lie along the shoreline and 36 billion barrels can be found in the Kazak north Caspian alone.[53]

Two major inland agreements involving vast sums of foreign capital for the extraction of oil from Kazakstan have already been signed. Chevron's investment in the Tengiz field—the largest oil deal signed thus far in the FSU—involves a total pledge of $20 billion over the forty-year life of the project.[54] The other major agreement, signed by British Gas (42.5 percent), Italy's Agip SpA (42.5 percent), and Russian Gazprom (15 percent), is for the development of the huge Karachaganak gas-condensate field in the Ural Mountains of western Kazakstan, officially estimated to hold 53.8 tcf of gas, 6.9 billion barrels of gas condensate, and 1.5 billion barrels of crude.[55] Offshore, geophysical, geological, and ecological studies began in September 1994 in the Kazak sector of the Caspian shelf, estimated to contain 25 billion barrels of oil and 70 tcf of gas recoverable resources, under an agreement between the state company Kazakstancaspiishelf and an international consortium involving BP-Statoil, British Gas, Agip, Mobil, Shell, and Total. Negotiations will begin on exploration and development contracts after the studies are completed, with output scheduled to begin in 2000.[56]

Azerbaijan's explored recoverable reserves are estimated at 8.8 billion barrels, with 80 percent being offshore. Thus far twenty-seven oil and gas fields have been discovered in the Azeri sector of the Caspian Sea, fourteen of which have been tapped or are under development.[57] In September 1994 the Azeri parliament ratified a production-sharing contract involving a $7.8 billion program between the State Oil Company of the Azerbaijan Republic and the Azerbaijan International Operating Company (AIOC), a consortium planning for the development of three Caspian Sea fields, Azeri, Guneshli, and Chirag, which are estimated to contain 4 billion barrels of oil. Consortium members now involved in the British-led extraction deal include Amoco-Caspian Sea (17.01 percent), British Petroleum (17.13 percent), Russia's Lukoil (10 percent), McDermott-Azerbaijan (2.5 percent), Pennzoil (9.8 percent), Ramco Energy (2 percent), Unocal (11 percent), Saudi Arabia's Delta Oil (0.5 percent), Turkish Petroleum (6.75 percent), and Exxon (5 percent).[58]

The location of these three major fields—offshore in the Caspian—has made the Azeri export situation somewhat more complicated than that faced by Kazakstan; not only does Azerbaijan face environmental problems in the Caspian similar to those frustrating the Kazaks, but Russia has contested Azerbaijan's authority to grant exploration rights in what the Azeris perceive to be their section of the sea.

Turkmenistan is another former Soviet republic having rich hydrocarbon potential, with natural gas reserves second only to Russia within the FSU.[59] Turkmenistan has roughly 120 oil and gas fields[60] with a combined 98.9 tcf of natural gas—the most of any Central Asian republic and one of the biggest concentrations in the world.[61] (Iran is second to Russia in gas reserves, having 700 tcf.)

On the eve of independence Turkmenistan's first president, Saparmurat Niyazov, declared his nation the "Kuwait of Central Asia" and promised his people imminent prosperity, which they have yet to experience. The Turkmen economy has not improved for several reasons: Russian restrictions on the flow of gas through its pipelines; the slow pace of alternative pipeline planning and resource development; the expenditure of limited cash earnings on question-able "pet" projects; and Niyazov's reluctance to institute important market reforms.[62] Few major energy deals have gone through and Turkmenistan is having difficulty attracting Western companies because of the republic's repu-tation for reneging on contracts.

Thus while the region looks extremely promising in terms of actual oil and gas reserves, numerous, substantial obstacles must be overcome in order for resource development to proceed. In Kazakstan, for example, those companies that have made large investments must also deal with difficulties presented by the physical environment before they can begin the complicated problem of distribution.

Some 1,400 wells that were drilled onshore nearly thirty years ago were submerged when the level of the Caspian Sea rose unexpectedly. It has risen by eight feet since 1978 and has advanced up to forty-three miles in some areas. In June 1995 a strip of land around the shore with an area of approximately 3,500 square miles was flooded, threatening 700 oil wells with serious damage. Flooding of 127 oil wells in the Tengiz field alone caused annual oil losses of approximately 1 million barrels valued at $7.5 million.[63] The development of oil reserves in the north Caspian Sea will also be hampered by ice that forms in winter, the high hydrogen sulfide content of associated gas, temperature ex-tremes (temperatures can soar to 130 degrees in the summer and drop to 30 below during the winter), and deep mud.[64] Aside from generally inhospitable conditions, there is also a lack of local power sources; other than railroad cars filled with huge electric generators, there is no local source of electric power. (Several of the barely explored but potentially rich Russian fields pose equally difficult conditions for technicians and crews.)

Obstacles to the further development of gas and oil fields in the Caspian Basin involve environmental concerns as well. The region is home to several rare and valuable species of wildlife, including caviar-producing sturgeon, 600,000 seals, and over 300 species of rare birds.[65] As a result some exploration has been put on hold while the consortium companies commission environmental impact studies to determine the possible consequences of further development in certain areas. Environmental reports caution that an enclosed sea such as the Caspian is particularly vulnerable from an ecological standpoint to oil spills and other related sources of pollution.

Pipeline Politics

To understand the highly complex and changing saga concerning oil and gas pipelines from the Caspian Basin to market, several basic facts must be understood. First, when Azerbaijan, Kazakstan, and Turkmenistan were part of the Soviet Union, oil and gas were delivered to various destinations, including refineries, by an internal network of pipelines. While this network still exists, much of it is antiquated and its structure was geared for the Soviet Union's energy needs. New pipelines are needed to cope with the increase in oil and gas production that is anticipated if investment plans for the three countries are implemented. Second, in planning for increased oil production it is necessary to distinguish between "early oil," which will come on-line and be available for export in 1997–98, and "late oil," which probably will not be available before the turn of the century, after substantial additional drilling and development are complete.

Quotas and tariffs imposed by the Russians on the flow of oil and gas through leaky and overburdened pipelines into that country—remnants from the days of the Soviet Union—have accelerated the search for alternative outlets.[66] Many companies, both Eastern and Western, are eager to construct new pipelines, but their zeal is matched by that of countries battling to put the pipelines, and thus control of the oil, in their own territory. As a result most pipeline construction has yet to begin, and many of the final routes remain undetermined as the chief players war among themselves. Map 21 provides an impressionistic interpretation of the various options under consideration.

Ideally, for the world's major energy consumers, the near future will a see a harmonious settlement of the legal access disputes with a sufficient redundancy being built into the pipeline distribution system so that if one transport route is interrupted for any reason, other routes can still keep the energy flowing. In theory the United States should welcome pipelines that run through Georgia, Turkey, Russia, and Iran to ensure this redundancy. However, given the currently poisonous relations between the United States and Iran, official U.S. policy is to punish Iran by discouraging financial institutions from supporting a pipeline through it.

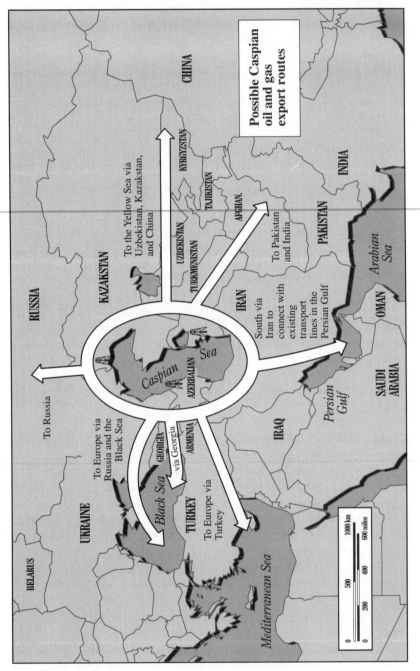

Possible Caspian oil and gas export routes

To the Yellow Sea via Uzbekistan, Kazakstan, and China

To Pakistan and India

South via Iran to connect with existing transport lines in the Persian Gulf

To Russia

To Europe via Russia and the Black Sea

via Georgia

To Europe via Turkey

Map 21

U.S. support for pipelines through Russia is lukewarm because Moscow's precise role in the future exploitation of Caspian oil resources is uncertain. Undoubtedly Russia, which is well aware of the power and control to be gained from ownership of oil and gas pipelines, will continue to protect its own interests. Until recently the riches of the Caspian Sea and its surrounding territories were exploited to Russia's great advantage. For one thing the Russians discovered the 200-square-mile Tengiz field—capable of producing 750,000 b/d—in the mid-1960s and had undertaken some development by the mid-1980s.[67] But today that field, among others, sits across internationally recognized borders—a difficult pill for Moscow to swallow.

Russia's current monopoly over pipelines out of Central Asia is a "consolation prize" for Moscow. To the extent that Russia can continue to charge for the transport of oil and gas through its territory and reap the concomitant benefits of enhanced economic and strategic leverage, it may remain willing to accept foreign ownership of these resources. In order to maintain this position, Moscow is not taking the pipeline scramble lightly. Competition with NATO-allied Turkey for influence in the region has been particularly stiff, and Russia is not going to stand by and watch while Turkish influence grows into Western hegemony in its own backyard.

Whether motivated by practical economic concerns related to the nonpayment of duties and tariffs or by problems of limited pipeline capacity, Russia's pipeline monopoly has allowed Moscow to slow natural gas exports from Turkmenistan as well as oil flows out of Kazakstan's Tengiz field. Whether or not this implies that Russia will strive for hegemonic control over the resources of its small neighbors at the expense of Western energy consumers remains to be seen. Nonetheless, the United States is anxious to avoid any situation in which one country, Russia or another, can control the energy riches of Caspian countries and manipulate the world energy consumers who will soon rely on these resources. Tables 4-2 through 4-4 display a number of the many proposals that have been made in recent years to export the oil and gas. The lists are constantly changing as political and economic considerations change.

Underlying the question of where to lay new pipelines are the political and military obstacles inherent in every proposed route out of the FSU. The Chechen capital of Grozny was traditionally an oil center—first by virtue of its reserves and later as a refining site—and several Russian proposals included it in transport routes from Baku—and even Kazakstan—to the Black Sea coast. With the Russian-Chechen battle for control of Grozny, however, investors became wary. Private companies interested in transporting the oil were concerned that a pipeline through Chechen territory would be constantly in danger of terrorist attacks. A second option would involve laying a pipeline through Georgia to the Georgian Black Sea port of Poti, but builders and investors have raised similar concerns. Domestic upheavals are a threat in regard to both options. Moreover, Georgia has been reluctant to support a single-line option

Table 4-2. Existing and Proposed Oil Pipelines

Organization	Location	Description
Former Soviet government/ Transneft	Russian Federation– Turkmenistan– Kazakstan–Azerbaijan– Georgia	Existing pipeline built under the control of the former Soviet Union. Composed of five major sections: (1) 1,900-mile pipeline linking Omsk (Russian Federation)–Pavlodar (Kazakstan)–Chimkent (Kazakstan)– Chardzhou (Turkmenistan); (2) 2,200-mile pipeline linking Novorossisk (Russian Federation)–Grozny (Russian Federation)– Atyrau (Kazakstan)–Tengiz (Kazakstan)– Aktau (Kazakstan); (3) 900-mile pipeline linking Novorossisk–Grozny–Baku (Azerbaijan); (4) 600-mile pipeline linking Batumi (Georgia)–Baku; (5) 600-mile pipeline linking Atyrau–Samara (Russian Federation)
CPC (Kazakstan, Russian Federation, Oman)	Kazakstan–Russian Federation (Black Sea)	Proposed rehabilitation of existing 470-mile pipeline from Tengiz, which travels north of the Caspian Sea to the Black Sea port of Novorossisk and can carry 110–125 mb/y. Would cost approximately $750 million and take an estimated one year to complete.
	Kazakstan–Russian Federation (Black Sea)	Proposed construction of new 470-mile pipeline from Tengiz to a new Black Sea port north of Novorossisk, which would travel north of the Caspian Sea, cost approximately $1.5 billion, and carry 367–403 mb/y.
	Kazakstan–Georgia	Proposed 1,200-mile pipeline from Tengiz, which would travel through Turkmenistan, across the Caspian Sea, Azerbaijan, and Georgia to the Black Sea.
CPC/Japanese firm	Kazakstan–China	Possible pipeline from Tengiz through China to transport Central Asian oil and gas to markets on the Pacific rim. Futuristic.
Kazakstan Pipeline Consortium	Kazakstan–Iran	Proposed 1,400-mile pipeline from Tengiz, through Turkmenistan and Iran, to Kharg Island in the Gulf.
Kazakturkbunay	Kazakstan–Turkey	Pipeline to be built between Kazakstan and Turkey to deliver Kazak oil under a new thirty-six-year petroleum agreement between the two governments.

Table 4-2. Existing and Proposed Oil Pipelines (*continued*)

Organization	Location	Description
AIOC	Azerbaijan–Russian Federation, and Azerbaijan–Georgia	Existing pipelines from Baku to Novorossisk will be refurbished and, for a second evacuation route, new pipelines will be built between Baku and Supsa, on the Georgian Black Sea coast. Other routes for the export of long-term output have not yet been determined.
Botas-Socar	Azerbaijan–Iran–Turkey	Proposed 660-mile pipeline from Baku, through Iran, to Ceyhan, on Turkey's Mediterranean coast.
UNOCAL and Delta Oil of Saudi Arabia	Turkmenistan–Afghanistan–Pakistan	Proposed 1,100-mile pipeline with a carrying capacity of 1 mb/d from Chardzhou, Turkmenistan, through Afghanistan, to a Pakistani export terminal at Gwadar, on the Indian Ocean. Estimated cost is $4 billion.
Turkmenistan–Iran	Turkmenistan–Iran	Possible pipeline from Turkmen oil fields to Iran.

Sources: Map of Caspian Basin and Black Sea, *Petroleum Economist*, 1995, and additional sources.

through its territory and Turkey for fear of antagonizing its powerful northern neighbor. A Georgian negotiator in talks with Turkey in late July 1995 asserted that, "Although Georgia's relations with Turkey are brilliant, the republic still favors balancing interests in the region and builds its policy on this principle."[68]

The options for a Turkish-controlled pipeline also harbor multiple potential pitfalls. Southeastern Turkey has become a hotbed of Kurdish insurgency, and of Turkish invasions across the Iraqi border in attempts to quell it. Although the Clinton administration has ignored the parallels between this situation and that in Chechnya—in terms of minority repression and egregious human rights violations—pipelines through Kurdish-dominated areas could also be perceived as symbolic of Western support for Ankara's campaign against one-fifth of Turkey's population. Growing Islamic radical movements in Turkey also worry investors.

There is the additional problem of transporting the oil to Turkey for further transport to refining and export facilities. Several proposed routes pass through Azerbaijan, which continues its confrontation with neighboring Armenia in Nagorno-Karabakh and the enclave of Nakhichevan. The alternative route to Turkey—through Iran—is even less desirable for an investment consortium led by American companies, and without U.S. financial backing the whole project might well fall apart. Large sums are needed for investment, and involving Iran would probably mean giving up necessary U.S. support: The U.S. government prohibits investment of U.S. dollars in Iranian infrastructure in light of Presi-

Table 4-3. Other Proposed Oil Pipelines (Hypothetical)

Organization	Location	Description
CPC (Kazakstan, Russian Federation, Oman)	Kazakstan–Azerbaijan–Iran	Proposed 500-mile pipeline from the Aktau oil fields, across the Caspian Sea and Azerbaijan, to Iran. The lack of U.S. support will make financing difficult.
	Kazakstan, Tajikistan	Proposed 1,030-mile pipeline from Tengiz to Tajikistan through Uzbekistan and Kyrgyzstan. Tajikistan is not stable enough to ensure uninterrupted oil delivery.
	Kazakstan–Russian Federation–Georgia–Turkey	Proposed 1,000-mile pipeline from Tengiz, which would travel north of the Caspian Sea through the Russian Federation, Georgia, and Turkey to the Mediterranean Sea. Moscow will probably not permit Tengiz oil to go to market via Turkey. Kazak oil transported across the Russian Federation will likely end up at a Russian Black Sea port en route to Europe via the Bosporus or Burgas.
Tefken Holding A.S. (Turkey)	Kazakstan–Azerbaijan–Iran–Turkey	Proposed 1,200-mile pipeline from Tengiz across the Caspian Sea, Azerbaijan, and Iran to the Mediterranean coast of Turkey via an existing Turkey-Iran pipeline. Capacity would be 293 mb/y, including 183 mb/y from Azerbaijan. Moscow will probably not permit Tengiz oil to go to market via Turkey.

Sources: Map of Caspian Basin and Black Sea, *Petroleum Economist,* 1995, and additional sources.

dent Clinton's decision in 1995 to impose a comprehensive U.S. boycott of Iran. In sum no option is totally promising, but given the value of the oil and gas in question there is reason to hope that in time a solution will be found that is acceptable to all the major parties.

KAZAKSTAN. The value of Kazakstan's hydrocarbon potential has put Turkey and Russia at odds over the placement of one or possibly two separate pipelines out of the republic, and Kazakstan is finding itself in murky political waters in the search for an economically and politically safe solution. Chevron's involvement in the Tengiz field predates Kazakstan's birth as an independent state, and the original agreement was signed with Moscow, which is interested in maintaining a central role in the development of Tengiz resources. The Kazaks signed an agreement with Russia in 1992 to transport Tengiz oil to a new terminal near the Black Sea port of Novorossisk.[69] At the time Russia had envisioned using the Turkish-controlled Bosporus Straits to export the oil by

Table 4-4. Existing and Proposed Gas Pipelines

Organization	Location	Description
Gazprom	Russian Federation–Central Asian Republics	Existing 18,600 miles of pipelines owned by Russian state gas concern.
Botas, Turkmenistan	Turkmenistan–Azerbaijan–Armenia–Turkey	Proposed 3,000-mile, $8.5 billion twin pipelines from Ashkhabad to Turkey via the Caspian Sea, Azerbaijan, and Armenia with a capacity of 1.4 tcf/y.
Enron (Botas-Gama Guris)	Turkmenistan–Azerbaijan–Georgia–Turkey	Proposed $16 billion pipeline from Turkmenistan, under the Caspian Sea to Baku, and through the Caucasus Mountains to Turkey with a capacity of 1.4 tcf/y.
Iran, Russian Federation, Turkey, Turkmenistan	Turkmenistan–Iran–Turkey	The four countries signed an agreement to build a pipeline with a capacity of 989 bcf/y to carry Turkmen gas to Turkey and Europe. The pipeline has an estimated cost of approximately $8 billion. Iran has agreed to finance 50 percent of the cost of the project. U.S opposition, which has increased the difficulty of finding financing, was countered by endorsements from Germany, Britain, and Austria, whose firms may participate in the project. In the meantime, Ashkhabad will deliver 282 bcf of natural gas to Iran for re-export during the next two years.
Tefken Holding A.S., Iran	Turkmenistan–Iran–Turkey	Proposed 3,100-mile pipeline from Turkmenistan, through Iran, and to Turkey. Cost estimates exceed $10 billion.
Delta Oil, Unocal of Saudi Arabia, Turkmenistan	Turkmenistan–Afghanistan–Pakistan	Proposed 800-mile pipeline from southeastern Turkmenistan to Gwadar (Pakistan) with a carrying capacity of 2 bcf/d. The deal gives Unocal the right to buy guaranteed Turkmen gas at the Afghan border, transport it to Pakistan, and market it locally. Cost of the pipeline is estimated at $2.5–3 billion.

Table 4-4. Existing and Proposed Gas Pipelines (*continued*)

Organization	Location	Description
Taplines International[a]	Turkmenistan– Afghanistan– Pakistan	Proposed 1,200-mile high-pressure pipeline through western Afghanistan to Karachi, Pakistan. Bridas, the Argentine industrial group that is one of the largest foreign investors in Turkmenistan, signed an accord with Afghanistan on behalf of Tap in early March 1995.
Exxon, Mitsubishi, China's state oil company	Turkmenistan– Uzbekistan– Kazakstan–China (–Japan)	Proposed 4,200-mile gas pipeline from Turkmenistan to the Yellow Sea—possibly including an underwater spur to Japan—which would pass through Uzbekistan, Kazakstan, and China, all of which have already given their consent. Estimated cost $10 billion.

Sources: *Petroleum Economist,* 1995, and various newspaper and journal sources.
a. Consortium including Turkmen, Afghani, and Pakistani interests.

tanker from the Black Sea to the Middle East and to European markets. Turkey, however, has instituted a new regime for the passage of ships through its Straits that strictly limits the number and type of ships that can travel through the Bosporus with ecologically unsound freight—a move motivated by very real fears of environmental hazards to Istanbul, with its population of twelve million.

Eager to make up for the pipeline transport revenue it lost when Iraqi oil was embargoed, Ankara has proposed an alternate route that would involve lucrative pipeline export contracts for Turkey—an extension of the existing pipeline between Midyal and Ceyhan (previously used to ship Iraqi oil) so that oil could flow across the Caspian to Baku, across Azerbaijan into Georgia, and then across Turkey to the Mediterranean. Russia, in turn, has countered this proposal with a new alternative that would bypass Turkey altogether. In mid-April 1995, under the auspices of the Conference on Black Sea Economic Cooperation, the foreign ministers of Russia, Greece, and Bulgaria discussed the possibility of a trans-Balkan pipeline that would run from Burgas (Bulgaria) to Alexandroupolis (Greece), connecting the shores of the Black and Aegean seas while skirting the Black Sea straits on the way to Europe (see maps 22 and 23). Russia, Bulgaria, and Greece have discussed an agreement to build the $1 billion, 174-mile pipeline with a potential shipment capacity of 600,000 b/d of crude, but as of spring 1996 problems of financing remained unresolved.

The question of pipeline financing has further complicated matters in Kazakstan. Chevron, which is funding the exploration of the Tengiz fields, was

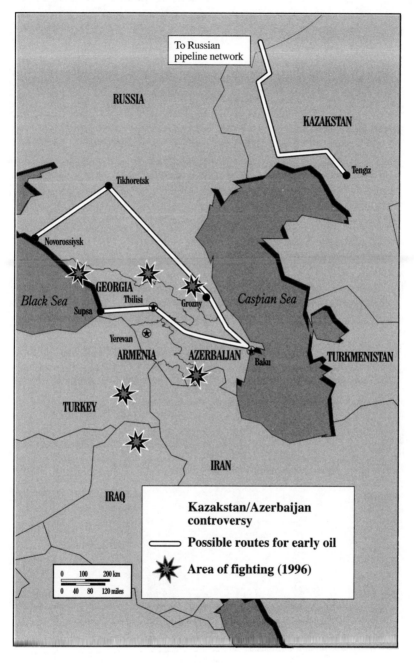

Map 22

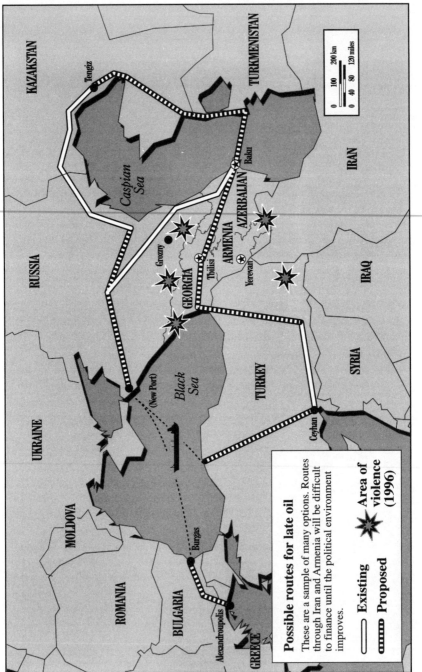

Possible routes for late oil

These are a sample of many options. Routes through Iran and Armenia will be difficult to finance until the political environment improves.

Existing

Proposed

✸ Area of violence (1996)

Map 23

invited to join with Russia, Kazakstan, and the Oman Oil Company in the Caspian Pipeline Consortium (CPC), which holds exclusive rights to build and operate the pipeline from Tengiz to Russia's Black Sea coast. However, since Chevron and the CPC have been slow to agree on the terms of their venture, Chevron was forced to slash 1995 funding for the project to 10 percent of the amount originally planned.[70] Without the new pipeline production levels— maintained at 55,000–65,000 b/d rather than the projected 130,000 b/d—were constrained by Russia's refusal to permit increased shipments of oil through its limited-capacity, existing pipeline system. The consortium has opened up to include Lukoil, Mobil, Atlantic Richfield, British Gas PLC, Agip, Transneft of Russia, and Kazakstan's Munaigas.[71] The involvement of any or all of these companies will provide the Kazak government a much-needed infusion of cash and might jump-start the stalemated process of developing an export mechanism for the oil.[72]

Notwithstanding the Bosporus complications, a Tengiz-Novorossisk route that crosses west from Komsomolsk is geographically and financially most logical and feasible. In a region filled with pockets of unrest, this option avoids major conflict centers such as Chechnya, Nagorno-Karabakh, Abkhazia, Turkish Kurdistan, and proximate areas where a couple of pounds of plastic explosives are as easy to find as a motivation for sabotage. This route also avoids Georgia, which could be a weak link because of its potentially unstable central government.

The northbound Tengiz-Novorossisk route, which passes just south of Astrakhan before heading westward to Tikhoretsk, is also logical from a strictly geographical perspective because it follows the quickest path westward without crossing the Caspian itself—a more ecologically hazardous alternative. In addition much of the pipeline along this route is already in place and can be used while a new line is being constructed. Acknowledging Russia's ability to make or break the deal, Chevron has professed a strong preference for this alternative partly because it takes Moscow's interests into consideration.[73] However, since any solution involving Novorossisk as an export point would have to resolve the issue of Turkish limitations on Bosporus shipping, this option will probably require the building of another pipeline bypassing the Bosporus, tacking on more costs. Furthermore, the facilities at Novorossisk are seriously inadequate for the quantities of oil that it planned to export in coming years, as its maximum handling capacity is only 235–257 million barrels annually, far below expected Caspian production.[74] The current Burgas–Alexandroupolis proposal has brought two more players and their respective wishes and concerns into the picture.

Turkey has given various estimates for several proposed pipeline routes, arguing that a pipeline from Tengiz to the Ceyhan terminal on Turkey's Mediterranean coast, which would bypass both Georgia and Grozny, could be built for $9 billion.[75] One version of such a pipeline would pass through two undesirable areas (Armenia, and Kurdish-dominated southeastern Turkey) but would prevent complete Russian control of a major source of oil, which is an issue for the West. A second alternative through Georgia, bypassing both Grozny and Kurdish areas, could also be built but would involve mountain passes in Turkey and would therefore require expensive pumping stations, increasing costs to $2.5–$3.2 billion.

In essence the options are virtually endless. Each alternative, however, presents its own political hazards and economic downsides. The United States, a vital player because of its capacity to back pipeline construction financially, currently favors a two-pipeline option for the export of Caspian oil, including one through Turkey and one north into Russia. Since it appears likely that for political and geographical reasons Kazak oil will be transported through Russia, the United States has been pushing for the transport of Azeri oil through pipelines that would traverse Turkey, thereby spreading influence and profits among the regional powers. However, because of stalled negotiations between Chevron and the CPC, major decisions regarding the export of Kazak oil will probably not be made until the outcome of Azeri pipeline deliberations is clear, which could make the transport of Kazak oil through the Azeri pipeline arrangement more likely; should the CPC fail to move forward on the development of a new pipeline for Tengiz, Chevron will not ignore the possibility of tapping into the Azeri route after transporting Tengiz oil across the Caspian to Baku. Opinions may still change several times, however, before final decisions are made. In the meantime Kazakstan and the oil companies, anxious to develop their extensive fields, wait, frustrated, for a solution to their transportation dilemma.

AZERBAIJAN. Russia's belligerent position with regard to the legal status of the Caspian Sea is undoubtedly related to the bickering surrounding the selection of a transport route for Azeri oil; Moscow is clearly using the legal status dispute as a bargaining chip in the effort to secure a pipeline through Russia. As with Kazakstan's energy potential, Azeri oil can be transported through Armenia, Georgia, Turkey, Iran, and Russia in a variety of ways. In this case, however, the preferences of the Azeri government seem to lie with Turkey owing both to long-standing cultural ties between the two nations and to the fact that Azerbaijan is particularly anxious to avoid Russian control of its resources.

In June 1995 as the race to find an oil route was heating up, Vafa Gulizade, foreign affairs advisor to Azeri president Heidar Aliyev, stated at a news conference that "We are Turks, we are one nation divided into two states."[76] Earlier, in October 1994, President Aliyev, after meeting with President Suleyman Demirel, declared his country's recognition of Turkey's indisputable right to construct a strategic pipeline running through its territory.[77] And in August 1995 Azeri foreign minister Gasan Gasanov declared that Azerbaijan's promise to Turkey that Caspian oil will reach Europe via Turkey will be kept.[78]

Indeed, many of the new Central Asian republics have cultural ties to Turkey, leading to speculation among experts that a new axis could form around Turkey, a potential replacement for Russia as the central player. The fledgling states might see Turkish control of their resources as a welcome alternative to continued Russian control. Azerbaijan, the sole Commonwealth of Independent States (CIS) country completely free of Russian forces, has been faced with increasing pressure from Moscow to join the CIS collective security arrangements, and has asked the Turkish president and prime minister to seek Washington's assistance in efforts to support Baku's resistance to Russian demands for military facilities at Gyandzha and Gabala, "joint protection" of the "CIS external border," joint naval patrols of Azerbaijan's sector of the Caspian Sea, and Azeri inclusion in the CIS collective air defense system. Georgia, through which the early oil would flow should the Turkish option be selected, has been less successful in resisting such demands, particularly because Tbilisi is depending on Moscow for the settlement of disputes in the self-declared republic of Abkhazia and in South Ossetia. There is already a Russian naval presence in Poti, and Moscow has announced its intention to redeploy border troops in the Batumi area.

Nonetheless, fearing that a complete rejection of the Russian pipeline option might lead to retribution from Moscow in terms of the Caspian Sea's legal status and Azerbaijan's right to exploit its resources, Azeri officials continue to emphasize that no final decision has yet been made. A protocol outlining the possibility of exporting early Azeri oil through Russian territory, signed by the president of Socar (State Oil Company of Azerbaijan Republic) and Russia's minister of energy and fuel, was approved by the Azeri government and by Russian vice-premier Oleg Davydov in June 1995. Meanwhile a duplicate Georgian-Azerbaijani document has also been prepared and is due to be signed in the near future.[79]

The eleven-company international consortium tentatively agreed in mid-September 1995 to use two pipeline routes for the early oil, one through Groznyin Chechnya and into Russia (an 870-mile route, which will require $50million of investment to cover the upgrading of certain sections and the construction of approximately 17 miles of pipeline), and the other through Georgia (a 525-mile route, of which 90 miles have yet to be constructed,[80] at a projected cost of just under $200 million).[81] Fearing that deliveries of early oil solely through Russia might lead to a permanent arrangement along those lines, Turkey has agreed to finance the construction of a 75-mile pipeline segment between Baku and Tbilisi and has agreed to purchase the early oil that would reach Supsa, near Georgia's Black Sea coast, for domestic consumption. Ankara has offered to buy this oil in order to avoid permitting its passage through the Bosporus, a move which could later weaken Turkey's environmental argument against the pipeline through Russia. The most likely routes for the early oil are shown on map 22.

The route for the main long-term pipeline had not been decided on as of the end of 1996.[82] In the meantime, although the consortium appears to be moving in the direction of a long-term dual-pipeline option, officials close to the AIOC discussions have implied that the early oil will move primarily, if not solely, through Russia's pipelines for basic commercial reasons, including the timeline for pipeline completion;[83] upgrading of the existing line from Baku to the Russian border will be completed before the Georgian line is built. Nevertheless, AIOC has announced that when the pipeline through Georgia and the loading facilities are built the consortium will begin to send some of the 80,000 b/d of Chirag oil to Turkey.[84] The most-discussed routes for late oil are shown on map 23.

Even the current Azeri regime, which seems relatively stable, faces serious opposition from Russia because of the division of Caspian Sea resources. Nevertheless, the Azeri parliament ratified the contract with the Western consortium by an overwhelming majority and, in spite of some belligerent statements from Russian officials, is hoping to smooth matters over. The speaker of the Azeri parliament, Rasul Kuliyev, declared after the ratification of the contract that Russia remains Azerbaijan's economic partner and that his country is ready to grant most favorable trading status to Russian business. Trying to "console" its other neighbor, Iran, Azerbaijan declared that it intends to give 5 percent of its share of the project to the Iranian national oil company,[85] a move that prompted an immediate backlash from the United States. Fearing the loss of American financial support, Azerbaijan retracted its offer and began private negotiations with Iran for the development of a separate field. A semigovernmental Iranian company has been alloted a 10 percent share.

TURKMENISTAN. Currently, the primary recipients of Turkmen gas are the Transcaucasus and the Ukraine, including Georgia and Armenia, which are experiencing economic difficulties and have been unable to pay in full for their gas purchases. These countries receive Turkmen gas through the Central Asia–Russia–Europe pipeline system, which was used effectively before the breakup of the Soviet Union, but has not helped to promote Turkmen economic stability, owing to Turkmenistan's creditor status with regard to gas provided along this route; of all its customers only Armenia has a decent payment record. Russia, which controls the pipelines, prefers to sell its own gas to Europe, leaving Turkmenistan to deal with the delinquent buyers. Consequently, Ashkhabad has been searching for southern-border pipeline alternatives to ensure that its gas will get out to the international energy carriers market to be purchased by buyers who will pay for it.[86] Options for exporting Turkmenistan gas are shown on map 24.

Turkmenistan and Pakistan signed a memorandum early in 1995 expressing their intention to build a pipeline to export Turkmen gas to Pakistan via Afghanistan and on June 19, 1995, reached an agreement in principle in Islamabad. The importance of these developments is highlighted by the fact that at the time Pakistani prime minister Benazir Bhutto personally took charge of the implementation of this project, which could bring 1.6 billion cubic feet of gas a day from Turkmenistan to Pakistan through a pipeline to be built by Unocal. In October 1995 Unocal and the Saudi-owned Delta Oil Company signed a protocol of intentions with the government of Turkmenistan to build the pipeline, which would be linked with a new marine terminal at Gwadar in Pakistan.

In 1995 the Exxon Corporation, the Mitsubishi Corporation of Japan, and China's state oil company announced their intention to study the feasibility of constructing a 4,200-mile gas pipeline from Turkmenistan to the Yellow Sea—possibly including an underwater spur to Japan—which would pass through Uzbekistan, Kazakstan, and China, all of which have already given their consent. This extravagant project—with an estimated cost of $10 billion if it ever begins—will not be completed until early in the next century.[87]

Another pipeline, through Iran, is also under discussion. Talks between Iran and Turkmenistan were convened in Tehran in June 1995, and Iran has agreed to finance 50 percent of the cost of the project involving a 1,600-mile, $6 billion pipeline into Turkey.[88] U.S. opposition to the project, which has increased the difficulty of finding financing, was countered by endorsements from Germany, Britain, and Austria, whose firms may participate in the project. However, new U.S. laws, specifically the Iran-Libya Sanctions Act, could make this a risky proposition for the Europeans. In the meantime Ashkhabad will deliver 283 billion cubic feet of natural gas to Iran during the next two years.[89] Turkmen president Saparmurat Niyazov said that this contract "has become the first move toward implementing the project of transporting natural gas from Turkmenistan to Europe via Iran."[90] Financing remains problematic, however, and could prove to be an insurmountable obstacle to the project's implementation.[91] In the final analysis the geographically counterintuitive option of a pipeline to Japan may prove more feasible simply because of U.S. support. However, in order to avert a domestic upheaval in Turkmenistan during the costly and time-consuming pipeline construction, Ashkhabad may require large-scale credits from Japan or the United States.

Indeed, one of the most intriguing problems concerns the role of Iran as an alternative route for oil and natural gas from Azerbaijan, Kazakstan, and Turkmenistan. For the new republics, pipelines to the Persian Gulf via Iran have some clear benefits. First, the route is feasible from both an economic and geographic point of view. Second, it would reduce their dependence on Russia and its potential control of their economies. In the short run the U.S. decision to hinder efforts by Iran and its partners to raise money for such activities serves American interests, but over the longer term it is not clear that the benefits outweigh the costs. Denying the Iranians a role in the export of the oil and gas from the CIS increases Russia's power and its hegemony over an area that could eventually become an important supplement to Persian Gulf oil.

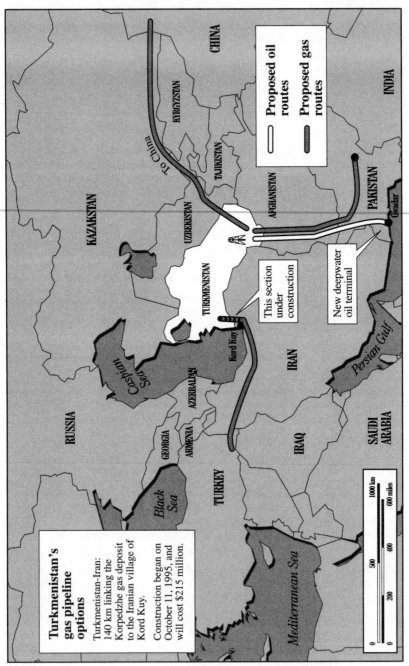

Map 24

Turkmenistan's gas pipeline options

Turkmenistan-Iran: 140 km linking the Korpedzhe gas deposit to the Iranian village of Kord Kuy.

Construction began on October 11, 1995, and will cost $215 million.

Proposed oil routes

Proposed gas routes

This section under construction

New deepwater oil terminal

The complex situation in the FSU changes from day to day, and along with it the current plans of the oil-rich states and the companies developing their fields. At present it is impossible to know for certain the details of when and how the fields will begin to produce large quantities of oil and gas, or where and when the pipelines to transport the oil will finally be built. Given the size and promise of the FSU's resources, it is safe to assume that, barring a total descent into political chaos, the region will continue to make slow but steady progress toward becoming a major source of energy. In the meantime, however —and it could be many years before the region is stable enough to allow for such a major development—the Persian Gulf remains the most important energy exporting region in the world.

Part Three

MILITARY OPERATIONS
AND PLANNING

Part 2 outlined the primary reasons for the fact that the greater Middle East remains a vital strategic prize and why there is need to be concerned about future military conflict in the region. Part 3 examines the relationship between geography and military planning and operations in the region. Specifically, chapter 5 reviews recent greater Middle East wars and highlights the continued importance of key geographic factors and how they vary within the region. Chapter 6 focuses on the specific lessons of the Gulf War and how the Revolution in Military Affairs, including new developments in technology, has led to a reassessment of the importance of certain geographic factors that influenced the parameters of military conflict in the past. Chapter 7 considers the effect of the region's geography upon the ability of external powers, especially the United States, to deploy and sustain a military presence in the area in the face of possible military threats and challenges. Chapter 8 focuses on the specific dangers of the spread of weapons of mass destruction to the greater Middle East and the impact that ballistic missiles and long-range strike aircraft have on regional strategy.

FIVE

Geography and Conventional Warfare: Lessons of Recent Wars in the Greater Middle East

SO LONG AS the greater Middle East remains an arena for major conflict, the impact of the region's geography on potential military operations is important. However, before examining future contingencies, it is necessary to draw lessons from the recent past.

Multiple geographic factors, many of which are aspects of physical geography, influence the nature and conduct of conventional war and low-intensity conflict (LIC), and this has been clearly demonstrated in countless confrontations in the greater Middle East. This chapter examines the various major geographic problems that have an impact on warfare, illustrating them with reference to the many recent instances of military confrontation. First, we outline the basics of military geography—terrain (including cover and strategic depth) and weather—and discuss how these factors influence wars of attrition and maneuver as well as the advantages and disadvantages they impose on the conduct of insurgency or counterinsurgency warfare. We then review these military-geographic principles and lessons as they apply to recent wars in the Middle East region. The specific lessons of the 1991 Gulf War and the Revolution in Military Affairs (RMA) will be examined in detail in chapter 6.

Although very different military encounters, the 1980–88 Iran-Iraq war and the 1991 Gulf War both demonstrated the crucial, though understated, role of military geography. The same could be said about the 1967, 1973, and 1982 Arab-Israeli wars, the 1965 and 1971 India-Pakistan wars, the 1962 war between India and China, and the long Afghanistan war, as well as about the numerous smaller military conflicts in Yemen, Western Sahara, and Kurdistan and along the Libya-Chad border. In all of these contests, the relationship between the state of military technology and the traditional concerns of military geography has been crucial. This is true both at the tactical level, where factors such as terrain and weather are important and at the operational and strategic levels, where matters such as strategic depth and the relationship of borders to certain major terrain features are germane as are the technological and geographic determinants of the historical variations between wars of maneuver and those of attrition.[1] At the level of grand strategy, one can further speculate on how geography may have influenced how wars start and why they end, and how geography may have influenced war aims, both before and during conflicts.[2]

For the most part here we are examining broad strategic factors related to geography. On a tactical level, in the most general sense, we are considering the factors of terrain and weather, for example, the implications of flat desert terrain versus mountains covered with thick-forested canopies. On a more specific tactical level, of course, one would need to consider, in a comparative sense, the elements of military terrain analysis taught by the U.S. Army under the acronym OCOKA: observation and fields of fire, cover and concealment, obstacles, key terrain, and avenues of approach. These are germane, of course, at the level of small-unit tactics, though one can make broad generalizations for whole theaters, such as have to do with the open nature of the terrain in the Sinai Desert or the obstacles presented by the marshes of the Tigris-Euphrates estuary. Key terrain features (including artificial ones) can often be identified in certain battle areas, for example, the dunes near Abu Agheila that might have blocked an Israeli advance into Sinai in 1967, the Beaufort Castle in the 1982 war in Lebanon, or the Kara Marda Pass between Somalia and Ethiopia in 1977–78.

Lessons of Recent Conventional Wars in the Greater Middle East

A number of traditional geographic factors are crucial, in a comparative sense, to an understanding of the land operations of recent Middle Eastern wars. They vary in applicability along a spectrum from the tactical to the operational to the strategic levels of such warfare. Among these factors are the size of theaters and depth of battle lines; strategic depth; the ethnography of battle areas or borderlands; patterns of settlement and road networks; mountains and rivers; the existence of two battlefronts for a centrally located power; and weather and seasons.

Size of Theater and Length of Forward Line of Troops

The sizes of the theaters of the various conventional wars with which we are dealing vary immensely. On one end was the large-scale Iran-Iraq war between two very big countries, fought mostly all along their mutual frontiers but also involving long-range bombardments of cities and oil installations as well as some naval action. In what was a mix of a conventional and unconventional war, the conflict in Afghanistan was fought all over an immense territory roughly the size of Texas. At the other end of the spectrum, the 1982 war in Lebanon was fought in a surprisingly small area, and that between Ethiopia and Somalia along only a limited portion of their very lengthy shared frontier.

The lengths of the battle lines or "frontal widths" vary greatly and do not necessarily reflect the wars' magnitude or intensity, in terms of the sizes of the forces involved and the human and material losses.

At one extreme were the Iraq-Iran and India-Pakistan wars, the former with a frontier of some 1,300 miles and the latter with a frontier about 730 miles long. (In 1971 much of the fighting in the latter conflict was along India's three-sided frontier with East Bengal, now Bangladesh, which has some 800 additional miles of frontier.) The fighting in the Iran-Iraq war (see map 25) was heavily concentrated for eight years on the extreme southern end of the line in the Shatt-al-Arab estuary between Abadan-Khorramshahr and Basra and to the north in the marshlands west of Susangird and Ahwaz.[3] But there were other areas of frequent fighting—west of Dehloran, near Mehran only some seventy miles from Baghdad, near Khanaqin and Quasr-e-Shirin, also some seventy miles from Iraq's capital, and in Kurdish-populated areas around Panjwin in the mountains east of Kirkuk and Mosul. In between these areas, there was little fighting, and certainly no major offensives by either side, because of difficult terrain: rugged mountains in the north and marshlands in the south along the Tigris and Shatt-al-Arab. Hence the frequent comparison between this conflict and World War I, with its endless and (literally) continuous trenches and fortifications from the English Channel to the Swiss border, is only partially valid.

The India-Pakistan border is also nearly 800 miles long, but the fighting in 1965 and 1971 was concentrated along only a small portion of that front: in the areas north and south of the Lahore-Amritsar axis, particularly around Sialkot and Chhamb and in the Rann of Kutch marshlands east of Karachi and Hyderabad.[4] There was only limited fighting along the huge stretch of frontier between the Thar and Great Indian deserts, involving an Indian advance, and in the high mountains between Jammu and Kashmir. Presumably that limitation was related to the remoteness of other areas as well as the fact that both sides concentrated forces at the approaches to the key cities of Lahore and Amritsar.

On the other hand the Lebanon war of 1982 was fought over a much narrower front. The Lebanon-Israel border is only some forty to fifty miles long, east to west, and the moving battlefront as Israeli forces struck northward toward the Beirut-Damascus axis to Syria measured only forty miles at its widest point (thirty miles from Beirut to the Syrian border along the route to

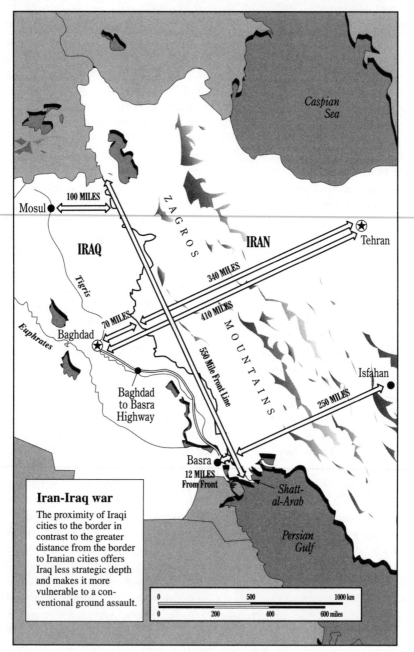

Caspian Sea

100 MILES

Mosul●

IRAQ

ZAGROS

IRAN

Tehran ✪

340 MILES

Tigris

410 MILES

70 MILES

Euphrates

Baghdad

MOUNTAINS

Isfahan ●

550 Mile Front Line

250 MILES

Baghdad to Basra Highway

Basra ●

12 MILES From Front

Shatt-al-Arab

Persian Gulf

Iran-Iraq war

The proximity of Iraqi cities to the border in contrast to the greater distance from the border to Iranian cities offers Iraq less strategic depth and makes it more vulnerable to a conventional ground assault.

0	500	1000 km
0	200	400	600 miles

Map 25

Damascus). Hence the combat here was channeled into a rather constricted frontal width, even though the opposing forces were two large, modern armies.[5]

Between these extremes there was the Sinai battlefield of 1967 and 1973. The Israeli frontier with Egyptian Sinai runs about 120 miles, with the larger distance from El Arish to the southern tip of Sinai measuring some 200 miles. The 1967 war was fought along a much shorter line in the northern Sinai, perhaps some eighty miles long between El Arish and An Nakhl,[6] and in 1973 the fighting was mostly along the Suez Canal between Port Said and Port Suez, also along a front about eighty miles long.[7] The comparison with the Iran-Iraq war is clear: here was a much more condensed line of battle.

Strategic Depth

One important aspect of military geography concerns strategic depth and the distances between the front lines or battle sectors and the combatants' industrial core areas, capital cities, heartlands, or other key cities or centers of military production. How vulnerable are these assets to a quick, preemptive attack or a grinding, methodical offensive? Conversely, to what extent can a country withdraw within its own territory, absorb its opponent's initial offensive thrust, and eventually take advantage of the fact that an opponent's offensive becomes more difficult as it extends further from its source of power? Stated another way, the issue is the trade-off between space and time. These are traditional staples of the military literature, illustrated, for instance, by Germany's failure to knock out the Soviet Union in 1941–42.

How do these traditional assumptions apply to recent wars in the greater Middle East? How have they influenced attempts at preemptive attacks intended to result in dramatic, conclusive victories? How important are the asymmetries involved in specific pairings of adversaries that have fought recent wars?

A good illustration of these issues was provided by the early stages of the war between Iran and Iraq in 1980. Iraq—then considered the weaker power in the Iran-Iraq pairing on the basis of population, gross national product, or both —launched an offensive, with the goal of seizing and holding the Khuzestan oil region, with its largely Arab population, in the hope that the eventual truce would make that a "permanent" outcome. Initially the strategy met with some success. But there was little chance of a quick, decisive Iraqi victory because of Iran's vast strategic depth, further buttressed by the formidable barrier of a lava plain and the Zagros Mountains between the border and the major cities of the Iranian heartland. Teheran is some 460 miles from the border and Isfahan approximately 310 miles. The smaller cities of Kermanshah, Abadan, and Ahwaz are closer, but still not so easy to overrun with a quick offensive. Only the area around Khorramshahr was easily subject to a quick "seize-and-hold" operation, with a more ambitious and longer-reach offensive rendered difficult

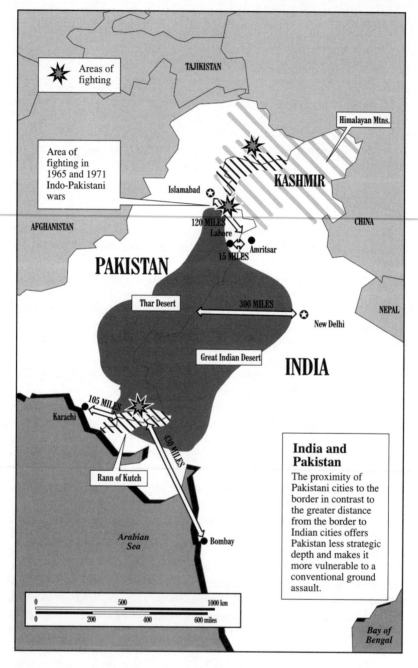

Map 26

by Iraq's lack of a real offensive punch, which would have required both better coordination of combined-arms tactics and better logistics.

On the other hand throughout the war, although Iraq remained highly vulnerable to Iranian offensives across the Shatt-al-Arab, these never succeeded despite enormous and costly efforts. Basra (a major city and one heavily shelled during the war) is only about ten miles from the frontier. Baghdad is only some sixty-five to seventy miles from the Iranian frontier further north. The major northern cities in the oil-rich Kurdish area, Kirkuk and Mosul, are also only some 70 and 100 miles, respectively, from the frontier. Further, the main roads connecting these cities run parallel to and close to the frontier, providing other points of vulnerability, so that in theory Iraq is subject to knockout by a quick offensive. These factors are shown on map 25.

The current India-Pakistan military balance illustrates the impact of asymmetries in strategic depth, as illustrated on map 26. Pakistan is highly vulnerable to a quick, preemptive attack.[8] As is the case with Iraq but even more so, its main cities lie near the border, with critical road and rail communications running close to and parallel to the frontier. Karachi is 100 miles from the border, Hyderabad 80, Islamabad and Rawalpindi 50, and Lahore only 20. (Yet in 1965 and 1971 Pakistani forces managed to defend the border areas against superior forces in short wars.) By contrast, New Delhi is more than 200 miles from the frontier, Bombay more than 400, with Amritsar, more vulnerable, only 20 miles away, and Ahmadabad farther at 120. These distances were critical factors in view of earlier, well-publicized Pakistani ambitions to imitate Israel in conducting a lightning preemptive strike toward Delhi, albeit over terrain that is relatively favorable for mechanized forces.

The Arab-Israel conflict has also illustrated the importance of strategic depth, although perhaps in some surprising ways. This aspect is illustrated on map 27. Before 1967 it was common to speak of Israel's extreme lack of depth along its borders with Egypt, Jordan, and Syria—from the West Bank a Jordanian advance of only nine miles north of Tel Aviv could have literally cut Israel in half. Syria was in near proximity to the Galilee settlements, and Egypt was poised to strike quickly at Eilat, Beersheva, and Ashdod, and indeed at all of Israel. But Israel's preemptive assault in 1967 was able to take advantage of interior lines, as has been so often described, allowing the small state to act like a "coiled spring."

The denouement gave Israel an additional 120 miles of strategic depth across the northern Sinai, then widely thought to be a strategic plus as a margin of badly needed safety. But Israel's setbacks in the early phases of the 1973 war along the Suez Canal proved again that the advantages of strategic depth are at least partially offset by the vulnerability resulting from extended lines of communication.[9] On the other hand the 1967 capture of the Golan Heights proved critical for Israel in 1973. Then, and again in 1982, it was of enormous concern to Syria that Israeli forward positions on the Golan were only some

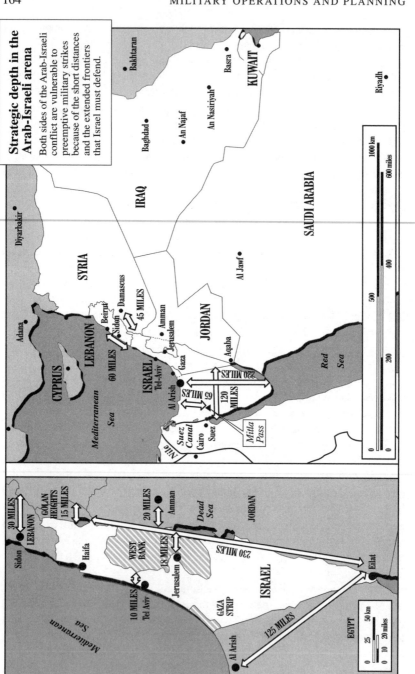

Strategic depth in the Arab-Israeli arena

Both sides of the Arab-Israeli conflict are vulnerable to preemptive military strikes because of the short distances and the extended frontiers that Israel must defend.

Map 27

thirty miles from its capital in Damascus. But paradoxically, despite its having acquired greater depth, Israel was also more vulnerable on the Golan in 1973, because of its extended supply lines and the proximity of the Golan to Syria's core area around Damascus.

Generally both Israel and Syria are in a precarious situation of shallow defensive depth vis-à-vis one another, which is why the Golan has remained such a contentious issue. Amman, meanwhile, is only some twenty miles from the Jordan Valley, and Jerusalem is almost as close to the Jordanian frontier on the Jordan River, albeit with a steeper, more imposing defensive terrain rampart. Eilat and Aqaba, the two key ports for Israel and Jordan on the Gulf of Aqaba, are contiguous mutual hostages. In 1982 the fact that Beirut is only fifty-five to sixty miles from Israel's northern frontier rendered it highly vulnerable to a quick armored strike, abetted somewhat by amphibious leapfrogging operations by Israeli naval forces along the Mediterranean coast. Generally speaking distances are very short in the core Middle Eastern zone of conflict, producing fast moving wars with quick outcomes.

Ethnography of Battle Areas or Borderlands

Several of the recent wars in the Middle East have involved one or more multiethnic (or multireligious or multilingual) societies. These aspects can have profound military implications, particularly if areas along a border or in regions involved in major fighting contain either populations related by blood or otherwise linked to the opposing side or dissident or disaffected groups ripe for rebellion or secessionist movements.

Iraq, for example, counted on support from the Arab populations of Iran's oil-rich Khuzestan Province. It hoped to end up in control of that region and to justify that control on a consensual "popular" basis. Iran, meanwhile, banked on support from disaffected Kurds along the northern end of the Iran-Iraq frontier and from the very large Shia population groups within Iraq, particularly in the south along the border with Iran, who were thought to be vulnerable to the enticements of the Ayatollah Khomeini's Islamic fundamentalism. (The Shias remained loyal to Iraq during its war with Iran but revolted in Basra and elsewhere after Saddam Hussein's defeat by the allied forces in 1991.) Somalia, in attacking Ethiopia in 1977, counted on extensive support (and received it) from the Somali population of the desert-like Ogaden Province, some of it nomadic. Somalia's war plans (involving an attack on a far more numerous foe) banked on Addis Ababa's strength being sapped by ongoing, full-scale insurgencies in the Eritrea and Tigre provinces, even though these were neither directly linked to nor supported by Somalia.

There are some other relevant recent examples. The Shia population in southern Lebanon first welcomed Israel's forces in 1982, but subsequently became its bitter foes, in part because of the impact of the 1979 Iranian revolution and the Israeli occupation. Turkey has a large Kurdish population in an area near its eastern border, which is the site of a massive, sustained counterinsurgency that has spilled over into Iraq. But Turkey's neighbors, Iran and Iraq, also have large Kurdish problems on their northern frontiers, and so far this has not induced conflict between them and Turkey.

Patterns of Settlement: Road Networks

Most wars in the greater Middle East have been fought in regions of low population density but varying topography and in areas not served by road or rail networks even remotely comparable to the extensive infrastructures of Central Europe or North America.[10] The latter factor has been cited as one reason for the relative absence of long-distance, offensive-maneuver warfare. The former factor has also meant that there has been little in the way of house-to-house urban warfare, as was seen, for example, in Stalingrad, Warsaw, and Berlin. The exceptions were Israel's operation in Lebanon in Tyre, Sidon, and Beirut in 1982 and Iraq's fighting in Khorramshahr and Abadan in 1980, in Kuwait City in 1990, and against its Shia population in Karbala and Basra in 1991. But the historical absence of urban warfare in the region may be changing, perhaps mostly with respect to LIC, as illustrated by the events in Grozny, capital of Chechnya.

As we noted above most of the wars in the Middle East have been fought away from settled areas. The images are familiar: the stark deserts of Sinai, the rocky plateaus and "tels" of the Golan, the Bekaa Valley (sparsely settled farmland), the empty wastes of the Ogaden, and the marshlands and foothills along the Iran-Iraq border (the Shatt-al-Arab area and some of the smaller towns along the border farther north being exceptions).

Consideration of some of these wars, fought as they were in places where there was a dearth of good roads, highlighted the need to control roads and road junctions and how that need is related to critical terrain features. Iraq's inability to capture the important junctions at Ahwaz and Dizful has often been noted as a key failure at the outset of the Iran-Iraq war. Israel's offensive in 1967 focused attention on several key road junctions or roads in central Sinai, at Abu Agheila, Bir Lahfan, El Arish, Bir Gifgafa, and Bir El Thamada; Somalia's assault on Ethiopia was critically dependent on several roads and junctions: specifically the road between Jijiga and Harar over the mountainous Kara Marda Pass and the key rail junction at Dire Dawa. In 1982 Israel made use of several roads leading north along the coast and overlooking the Bekaa Valley but generally was constrained by the fact that there were only those few narrow roads. During the Afghanistan conflict Soviet control over the main road leading south from the Soviet-Afghanistan border, particularly the tunnel through the Salang Pass in the Hindu Kush Mountains north of Kabul, was critical for its logistics. These factors are underscored in many cases by the combination of great distances, the mechanization of armies, logistical requirements for supplying ammunition and other materiel, and the vulnerability of road transport to attack from the air. Few if any of the wars alluded to here have included long-distance power projection by dismounted infantry or mechanized forces.

On the other hand if a war lasts long enough new road networks can be constructed, as was done by Iraq during its eight-year war with Iran. During this period it greatly expanded the road network necessary for the movement of tank transporters that was required for a mobile defense. Now, in theory, modern highways linking Iraq and Jordan would allow much more rapid movement of armored forces in either direction between Iraq and Israel, albeit movement subject to interdiction from the air.

Mountains and Rivers: Parallel to or Across the Line of Advance

Mountainous terrain, riverine terrain, and large rivers are critical features of military geography. There have long been generalizations about mountains and rivers providing barriers that favor the defending forces, and these factors have been important in the wars in the Middle East and South Asia. But the exact location of these barriers and their relation to potential axes of advance for offensive operations warrant closer scrutiny.

In Iran and Ethiopia, for instance, significant mountain ranges more or less parallel to the borders (in any event straddling possible avenues of advance toward capital cities or otherwise important core areas) have been critical terrain features. Near the Iran-Iraq border, but back some distance on the Iranian side, the Zagros Mountains form a defensive glacis protecting Iran, at least along northern sections of the border, from Iraqi attack. But the mountains also make an Iranian offensive against Iraq more difficult. Similarly, a Somali assault toward Addis Ababa and the Rift Valley must surmount two mountain ranges that are nearly perpendicular to an otherwise preferred line of advance (the mountains run southwest to northeast).

Mountains and rivers are equally important in the Israeli-Arab context. The mountains of Lebanon, running north-south, ran parallel to the Israeli line of advance in 1982 (but across the grain of any Syrian advance toward Beirut). But along both sides of the Jordan Valley, north-south mountains block an advance toward Jerusalem or toward the Jordanian heartland, while in Sinai, the north-south mountain line in the west (crossed by the strategic Mitla and Gidi passes) acted first as a barrier to Israeli forces in 1967, then potentially as a barrier to a further Egyptian advance into central Sinai in 1973.

One must keep in mind that there are mountains and then there are mountains! Those in western Sinai are only 1,500 to 2,000 feet high, really just big hills. Those along the Ethiopia-Somalia border are much higher and more rugged, measuring 9,000 to 11,000 feet, indeed almost precluding fighting by large units in the contested areas around Harar and Jijiga. (Mountains in front of Addis Ababa reach 13,000 to 14,000 feet.) Some of the Caucasus Mountains also reach these heights, and those of Kashmir are higher still, around 15,000 feet. The spine along the center of Lebanon's north-south mountain range has an elevation of 4,000 to 6,000 feet. The Zagros Mountains along the Iran-Iraq border reach 7,000 feet, with the altitude of most of the area being over 3,000 feet. However, these numbers must be taken with a measure of caution, as height above sea level is not an indicator of ruggedness or of the steepness of inclines. In addition mountains of any given height may differ in regard to tree cover, extent of rockiness, and type of soil, all of which have implications for tractability for vehicles, cover for defensive forces, and visibility from the air.

Several major rivers have figured prominently in recent conventional wars. The Shatt-al-Arab, from the confluence of the Tigris and Euphrates rivers at Al-Qurna southeast to Basra, Khorramshahr, Abadan, Khorsabad, Al Faw, and hence the Persian Gulf, largely defined the major battle areas of the Iran-Iraq war, forming a major barrier to advances in both directions. Its tributary, the Karun River, coming first from the east and then from the north, was also a major barrier to be crossed by the forces of both sides. The seasons and flood levels were crucial at various times. Minor rivers further north did not figure in the fighting, nor did the Tigris from Baghdad to Al-Qurna. In the event of an Iranian march west the latter could have been a major obstacle in the way of Iran's cutting of the critical Baghdad-to-Basra highway.

Israel's advance north into Lebanon in 1982 required crossing several east-west river barriers—the Litani, Zahrani, Awali, and Damour rivers. None proved to be a major obstacle, in the absence of organized linear PLO defensive lines along the rivers. After the war, the rivers provided some obvious natural lines of demarcation for the several stages of the Israeli withdrawal. The small northward flowing streams in the Bekaa Valley did not figure importantly in the fighting.

India's multipronged attack against Pakistan's East Bengal in 1971 involved the crossing of numerous north-south rivers leading to the mouth of the Ganges. Indeed, its main thrust from north of Calcutta to Jessore and on to Dacca required several river crossings even before the huge one created by the confluence of the Brahmaputra and Ganges rivers.[11] However, India's forces, being well trained and well equipped with bridging equipment, amphibious vehicles, ferries, and helicopters, were not much hindered. No such problems existed for India in the tougher fighting in the west in 1965 and 1971, but if it were ever to attempt to invade Karachi, a crossing of the major Indus River would be required.

In the Afghanistan war, the Oxus River defined the border between Afghanistan and the old USSR. The Soviets had to span it with pontoon bridges in order to bring in supplies and supposedly (though often violated) it provided a "red line" beyond which the Soviets would not countenance incursions by Islamic forces into the USSR proper.

Mountains and rivers running perpendicular to the main line of advance of large forces are the two features which come to mind first in considering defensive terrain barriers. But there are others. The swamps and marshlands of the southern Iran-Iraq border area provided good defensive terrain for both sides, particularly during rainy seasons. Large desert areas along the India-Pakistan border are so forbidding as to have precluded more than limited fighting over large stretches of the border in both 1965 and 1971. The desert has also made for formidable barriers to offensive action in Sinai. In 1956 and 1967 even the rapid Israeli advances were largely restricted to a small number of roads, though in 1967 Israeli tank forces surprised the Egyptians with flanking moves across the sands near Abu Agheila and along some dry wadis leading south and west from that key junction. The deserts in northern Chad similarly channeled combat in the Libya-Chad war onto a few roads and otherwise passable areas.

The Military Geography of Two-Front Conventional War

In some conventional wars there may be vast differences in the terrain along an extended battle area, sometimes amounting to entirely different theaters in regard to terrain and weather. For example, we have noted the wholly different environments provided by the northern and southern regions of the Iran-Iraq border and those of the various border regions between India and Pakistan. In the former case Iran shifted its focus back and forth between its southern and northern fronts in attempting to mount successful offensives, sometimes in relation to the changing seasons. But these are all examples of a single, albeit sometimes lengthy, forward line of battle, usually defined largely by an extended border or mutual frontiers or the areas adjacent to (or lines parallel to) those borders.

Some wars, however, have involved countries having to fight along two fronts, usually against two allied enemies. Examples of this type of conflict include the Arab-Israeli wars of 1967 and 1973 and the India-Pakistan war of 1971. More recently there was extensive speculation about the possibility that Iraq would have to fight a two-front war against the Gulf coalition in the south and Turkey in the north. As it turned out Turkey abstained, but did tie down some Iraqi forces.

In 1967 Israel in planning a preemptive war strategy hinged on its (precariously) advantageous use of interior lines had to decide whether its initial emphasis would be on the Egyptian or the Syrian-Jordanian front. Geographical issues momentarily aside, Israel's concentrated power was outnumbered but qualitatively superior and it had to go for a quick, decisive victory before its more numerous foes could mobilize for a lengthy two-front war of attrition.

The issue then was whether to attack the stronger of two opponents or the weaker first. The interior power struck first at the stronger opponent (in this case Egypt), perhaps in defiance of what would appear to be a more logical, practical alternative, that is, knocking out a weaker foe and *then* concentrating all one's forces on the stronger one. (If the stronger side is successfully knocked out first, defeating the weaker side next becomes more or less automatic.)[12]

Israel in 1967 chose to deal with Egypt first, which allowed it to take maximum advantage of its mobile, fast-moving warfare capability ("armored pace") in a larger theater, albeit one of tough desert terrain, dry wadis, and few roads. But Israel was ready for desert warfare, or so it was to prove in a quick four-day victory. On the Golan Heights, en route to Damascus, it faced a very narrow frontal zone that featured rugged mountain and plateau terrain that could easily accommodate fixed fortifications in depth and was not at all suitable for mobile, armored operations. After its victory over Egypt, Israel defeated the Syrian army on the Golan, but eschewed a pursuit to Damascus, both because of external diplomatic pressures and the realization that such a pursuit would entail extensive casualties.

In the 1973 war Israel was initially on the defensive, but then took the offensive on both fronts. Here too, in both phases, it had to decide on an allocation of emphasis. It concentrated its early defensive efforts against Syria because of the danger to its settlements in the Galilee if the Syrian army moved down from the Golan Heights.[13] The Sinai, on the other hand, provided immense spatial depth for retreat and regrouping. However, when Israel moved onto the offensive—and despite some back-and-forth uncertainty about which front to focus on—it again chose to make its major effort against Egypt, for many of the same reasons as in 1967, that is, the extent to which the large, open desert theater allowed for mobile offensive warfare, in contrast to the axis from the Golan Heights toward Damascus.

India in 1971 fighting an inherently weaker foe on two fronts from a position on interior lines (in this case the same foe on both fronts) chose to conduct an offensive on the weaker side first. To a great extent this decision appears to have been based on political grounds (independence for Bangladesh and the subsequent division and weakening of Pakistan) and on the fact that Pakistan's military was far stronger in the Punjab than in Bengal; the latter was not only difficult to resupply during the war but Pakistan also faced a strong internal insurgency there from the Mukti Bahini forces aligned with India. The Indian army fought through very difficult riverine terrain in Bengal but nevertheless was able to mount a fast-moving offensive, while in the Punjab it faced much tougher going against better entrenched forces.

The Weather and the Seasons

Terrain conditions correlate closely with climate and seasonal weather patterns. However some additional points need to be noted concerning the latter, again as it pertains to recent conventional wars in the broadly construed Middle Eastern region.

Three important categories of weather conditions are relevant: (1) the overall prevailing or dominant climate patterns of a region, (2) the seasonal variations in climate, and (3) the vagaries of day-to-day changes in the weather. Regarding the relevance for military operations, the critical factors are the impact on (a) ground terrain (particularly regarding mobility for tracked and wheeled vehicles); (b) concealment, as it pertains to surveillance from both the air and the ground; and (c) temperature, particularly the effect of extremes of heat and cold on combat troops and their equipment.

This is a large subject and we can do little here but provide some relevant illustrations from recent wars. In connection with broad, general climatic considerations, the gross disparities among, for example, desert, tropical jungle, and northern forest conditions obviously have military implications in regard to cover, visibility, and maneuverability, and these in turn are affected by seasonal and day-to-day variations. Some wars—for example the Iran-Iraq and Arab-Israeli conflicts—were fought in different climates and with correspondingly different terrains, even where the distances involved—say between Sinai and the Golan—were less than 1,000 miles.

In some recent wars wide variations in seasonal weather patterns had a big impact on the planning as well as the execution of military operations. Indeed in the Iran-Iraq war, the pace and scope of military operations varied tremendously according to the seasons. Offensive operations—such as the several efforts at a "final push" by the Iranian Pasdaran hordes in 1984–88—were timed for the dry seasons in the Shatt-al-Arab estuary area, which allowed for better use of armor as they were less hindered by flooding of the rivers and marshland areas. Armored operations are hampered by the winter rains from November through February, and then by the spring flooding caused by thawing of snow in the mountains. On the other hand during the summer months temperatures in the southern and central sectors can reach over 100°F, while those in enclosed vehicles go up to 130°F, and high temperatures and lack of water can greatly hamper the effectiveness of infantry operations on open terrain. Operations further north along the Iran-Iraq border were not affected by heat and sand and, indeed, in those areas winter conditions in the northern mountains were seen as a hindrance to mobile operations.[14]

A few further examples: The Israelis in 1967 attacked at a time of extreme heat in the Sinai, but used the fact that they had better trained troops and better equipment to their advantage (the cramped and unventilated Soviet tanks used by the Egyptians were unsuited for the climate—being equipped with heaters rather than air conditioners and not having sand filters).[15] In contrast the 1973 war, fought in October, meant wet and cold weather on the Golan Heights, a far cry from desert conditions and also posing the problem of accompanying cloud cover, which could nullify air operations and the ability to interdict ground forces with tactical aviation. This was a particular problem for Israel, which relies so heavily on its air force in the early days of a war. If Ethiopia had wished to capitalize on its victory over Somalia in early 1978 by moving into Somalia proper, it would soon have faced that country's rainy season, which would have made it difficult to move mechanized forces in an area with just a few desert roads. India in attacking East Bengal in December 1971 was able to surmount riverine terrain outside the seasons of heavy rains.[16] In the Afghanistan war, the Soviets took advantage of seasonal floods that made it difficult for the Mujahideen to escape in relatively flat regions. But they also had a tough time in the winter, sometimes losing whole units to snowslides.

On a more "micro" level, one can always point to examples in which the day-to-day vagaries of the weather had a big impact on specific military operations, albeit the difficulty of predicting to whose advantage. A sudden idiosyncratic sandstorm helped wreck the U.S. heliborne hostage rescue attempt in Iran in 1980. The cloudy weather over Iraq in January 1991 at the end of the first week of the air assault phase of the Gulf War hampered U.S. aircraft operations for several critical days.

At times the weather can also affect the efficiency of certain types of weapons. According to one source, in one of the Iraqi chemical weapons attacks against Iran, "weather and water effects robbed Iraqi nerve agents, cyanide agents, and mustard gas of much of their impact."[17] Wetness was also said to have adversely affected Iraq's use of artillery and airpower, whereby "shells and fragmentation bombs lost much of their effectiveness because the soft or marshy terrain absorbed much of the force of the explosion and offered a relatively high degree of shelter."[18] One study on the Afghanistan war relates that the Mujahideen had trouble operating British-origin Blowpipe surface-to-air missiles (SAMs) because of problems caused by the low temperatures.[19]

It is obvious from the foregoing that it is not always easy to predict which side will benefit most from "good" or "bad" weather or from the early or delayed advents of summer or winter weather, even if one can expect that such answers would be provided by asymmetries in training and equipment. As with football, both sides have to play (or fight) with bad weather.

Some Key Questions

Based on the preceding analysis several questions relating to geography and warfare in the greater Middle East now need to be addressed.

—How does geography affect wars of maneuver and attrition and the possibilities for quick, decisive outcomes, resulting either from preemptive opening strikes or, generally, from long-distance offensive operations?

—What is the impact of geography on logistics lines running from core areas to the forward line of troops (FLOT)?

—Does different geography favor certain kinds of key weapons systems, for example, attack aircraft, armor, and heavy artillery?

—To what degree does the quality of military personnel and equipment moderate the effects of weather and terrain?

—What are the implications of this relationship for training, in order to secure the advantages accruing to better trained forces?

—To what extent can one or another side in a conflict benefit from greater acclimatization to terrain and weather from the perspective of the military equivalent of a "home field" advantage?

—How did military geography affect initial perceptions about predicted war outcomes: specifically regarding the perceptions of aggressors planning offensive operations, preemptive strikes, and seize-and-hold operations?

—What is the relationship between military geography and arms acquisitions; that is, how suitable are the weapons acquired from preferred suppliers for the specific terrain and weather involved, whether or not there were reasonably available alternatives?

The impact of terrain and weather on the extent to which conventional wars are fought as wars of maneuver or of attrition leads to very vexing questions but ones that are particularly important in some theaters, especially in the Arab-Israeli conflict, where the relationship between the duration of war and casualties is of enormous concern to Israel. In the immediate aftermath of the 1973 Arab-Israeli war, some analysts rushed to contrast the wars of 1967 and 1973 as having been, respectively, wars of maneuver and of attrition (though the Arab plan of 1973 had elements of both), with the fundamental difference said to have been the result of new precision-guided munitions (PGMs) in antitank and surface-to-air missiles that blunted the effectiveness of Israel's old offensive tactics.[20] But later and with the aid of hindsight, many of these analysts saw less of a fundamental shift in 1973. Israel's ultimately successful offensive operations in the Suez area reaffirmed the dominance of maneuver warfare.

Geography is only one of several factors determining whether wars will be dominated by maneuver or attrition, though probably a major one. Geography may modify, but will not eliminate, fundamental imbalances that may allow one side to roll over another, by either frontal or encircling offensives. That is, it may serve merely to raise or lower the ratios of offensive to defensive forces required for successful offensive operations. The state of military technology and the ability of combatants to use that technology effectively are also major factors. Probably most important, it is ultimately the extent or absence of military balance between contending forces that will determine whether a war is fought as one of attrition or maneuver.

Along these lines, the record of the recent wars in the greater Middle East presents a mixed and confusing picture. Israel was able to fight a war of maneuver in 1967 and again in 1973 despite tough, sandy desert terrain because that terrain with its lack of vegetation and cover provided an advantage for its tactical airpower. On the other hand the terrain made it tougher on the Golan-Damascus axis in 1973. Iran and Iraq, however, fighting with many of the same weapons, fought a long war of attrition from 1980 to 1988, despite terrain that at least on the southern front outside of the rainy seasons was deemed suitable for mobile, mechanized operations. The reasons for the disparity include: the nature of the opposing armies; cultural proclivities; an inability to sustain long-range combined arms operations, especially the use of tactical airpower; overly hierarchical command structures that did not allow for much initiative by low-level commanders; and generally the failure of Iraq to defeat Iran by imposing its will through tactical decisions.

In 1977 Somalia was able to fight a war of maneuver initially, despite relatively difficult terrain in the area of the Kara Marda Pass (so was Ethiopia in 1978 after it and its Cuban and Soviet friends had seized the advantage). In 1971 India, facing very difficult riverine terrain, was able to fight an offensive war against a well-trained and competent army, albeit one that was sorely outnumbered and suffering a marked logistical disadvantage. Again, geography can be only one of many contributing factors to an analysis of the reasons for relative degrees of maneuver and attrition warfare. Generally speaking it is clear that competent and well-led armies, for example Indian and Israeli, can find ways to transcend tough terrain and weather. But less effective armies using the same weapons can find factors of military geography highly daunting.

One key issue regarding the relationship between logistics and geography is whether terrain and weather permit interdiction from the air of supplies moving forward to the front, and this depends on the proximity of civilian road and rail networks. Here the importance of cover is obvious. Israel, it is assumed, would easily be able to interdict Iraqi supply or tank convoys moving across Jordan toward the West Bank because the roads are in open, desert terrain. (This also assumes, of course, control of the air.)[21] In some wars, however, the absence of usable airpower on either or both sides has precluded interdiction of supply routes even in the case of relatively open terrain or where roads were largely subject to surveillance in wooded, mountainous terrain. Iran and Iraq from 1980 to 1988 and Somalia and Ethiopia in 1977–78 both fit this description.[22] Perhaps only Israel and India among the nations in the region have a capability that reflects the strategy of the earlier U.S. Air-Land Battle Doctrine, that is, a heavy reliance on long-range interdiction of enemy logistics and reserve forces, as was so well exemplified in the Gulf War of 1991.

The problem of training armed forces for fighting in specific terrain is an interesting one. Some greater Middle East protagonists have, of course, fought on familiar, home ground and in familiar weather. The Israeli and (some) Arab armies are trained for desert warfare and U.S. forces trained in the Mojave Desert needed only additional acclimatization in the Saudi desert in 1990, but, of course, not all deserts are alike. Earlier, Egyptian forces, drawn mostly from the Nile Delta, seemed poorly trained for combat in the Sinai in comparison to the Israeli soldiers, who had trained on the nearby Negev Desert. Many of these countries have special units for specific terrain, for instance, the Indian and Pakistani mountain troops. The Soviets did use some specialized mountain troops in Afghanistan but they continued to be at a disadvantage facing the Afghan guerrillas, who were accustomed to fighting in their own rugged mountains and defiles, even during the cold winters.[23] The Soviets tried to compensate by training mountain warfare units beforehand in the Bulgarian mountains.[24]

Without access to classified files regarding military planning, it is difficult to say just how, in some cases, misapprehension of factors of military geography affected planning for such actions as offensive preemptive attacks and preventive wars, in regard to broad strategic considerations or narrower operational or tactical ones. Syria may have underestimated the difficulties of overcoming the rugged "tels" of the Golan in 1973 as Israel was mobilizing its forces while fighting a delaying action. Iraq surely seems to have overestimated its ability to conduct offensive operations across the Euphrates-Karun estuary and the nearby marshes in 1980. Somalia may have underestimated the barriers presented by the mountains shielding Harar and Dire Dawa in hoping for a quick seize-and-hold operation in 1977.

The fit of different countries' arms acquisitions to the military geography of their conflicts is an interesting topic. Most combatant less-developed countries, of course, are completely or largely dependent for their arms on big power or second-tier suppliers. Israel and India are to some degree exceptions, through rarely when it comes to the so-called "big-ticket" items, for example, high-performanceaircraft (and their electronics and weapons), tanks, and helicopters. Usually—though perhaps less so today in an era of fading bipolar, ideological conflict—less-developed countries do not have an open choice of arms supplier(s) as these choices tend to be dictated by political alignments.

American, Soviet, and Western European weapons systems have been optimized for full-scale warfare in Central Europe. They often need to be modified (or substituted for) if the big powers themselves are contemplating combat in the deserts or mountains of the Middle East or South Asia.

However, for some regional powers, this particularity can present problems. The Egyptians complained bitterly about the lack of sand filters on Soviet tanks after the 1973 war as well as about the cramped crew space that exacerbated crew fatigue in the desert climate, and about the incapacity of the Soviet tank guns to achieve the depression needed for firing behind sand dunes. The Israelis have had to modify U.S.-supplied aircraft extensively for Middle East conditions, and have been able to build their own cheaper models in part because they have no need for the all-weather capability that would be taken for granted in the European context. Numerous recent articles have detailed the need for the United States to improve its maintenance of weapons on the

Saudi-Kuwaiti border, for instance, through more frequent attention to sand filters for tanks and aircraft engines. Iran's Kilo-class submarines, originally developed by the Soviets for northern waters, could not hold an electric charge in the warmer waters of the Persian Gulf. There has even been a revival of interest in the innovations introduced by Rommel's army in North Africa during World War II, for instance, using trucks with giant blowers to cause dust storms that simulate the approach of a formation of tanks.[25]

But aside from the specifics of the "fit" of certain major power–supplied weapons to given third world clients there have been more generic problems regarding the utility of certain classes of weapons for some specific combat conditions in recent wars in the greater Middle East.

In Afghanistan, for instance, Soviet tanks and armored personnel carriers were not particularly useful for most of the off-road combat in the mountains. Similarly, most medium and heavy artillery was not especially useful, both because of its weight and bulk but also because it could not be elevated high enough to fire at guerrillas operating from mountain crests.[26] The latter drawback also curtailed the usefulness of multiple rocket launchers. Indeed mortars and light mountain guns both proved more effective for fighting in the mountains, and antiaircraft guns were used instead of field artillery because they could provide adequate elevation. Helicopters proved to be better than artillery in mountain fighting, but were vulnerable to Afghan guerrillas, who were able to fire down upon them from the mountain tops.

Israel learned a number of useful lessons about some of its own and its opponents' weapons while fighting in Lebanon in 1982, in terrain and climate vastly different from that of the Sinai Desert. Its artillery proved less useful in the mountains than in open terrain.[27] Its armored personnel carriers proved to be not particularly effective in rough or urban terrain, being vulnerable to ambushes by infantry armed with antiarmor weapons. (Syria's Soviet-supplied armored fighting vehicles were similarly hampered.) But its heavy Merkava tank proved very effective—with its front-mounted engine and wide tracks—for both mountain and urban warfare, the primary milieus in its Lebanese war. Similarly, Israel learned that it needed more lethal light infantry weapons for fighting in mountains and cities. Both Israel and Syria made effective use of attack helicopters in rugged terrain, taking advantage of terrain-masking. (The same was true for Iran and Iraq, both of which had problems operating helicopters in relatively open terrain.) Israel also learned about the drawbacks of not having sufficient infantry forces trained for mountain warfare.

In the Iraq-Iran war it was apparent early on that Iraq's easy access to the world arms market gave it an advantage in armor, artillery, and airpower, which in turn enabled it to take advantage of good weather. By way of compensation Iran had to make more effective use of night attacks, put a greater emphasis on mountain warfare in its allocation of effort among areas where offensives were conducted, and also make effective use of both natural and artificial water barriers.[28] On the other hand in inclement weather on the southern front, Iran appears to have been unable to maneuver and resupply in the mud, owing to the limitations of its mechanized forces and road networks. Further, although Iraq fielded an advantage in armor and artillery, in the early phases of the war it appeared to be rather weak when it came to infantry operations across the Karun River and in the urban fighting in Khorramshahr and Abadan.

In the long Iran-Iraq war, fought for the most part along quite stationary front lines reminiscent of World War I, both sides made extensive use of "alterations" of the terrain in the form of artificial barriers and the like.[29] The Iranians used deliberate flooding and water barriers to force Iraq to build new roads. The Iraqis did likewise, using dikes, dams, flooding, and huge earthen barriers and artificial ridges. These techniques were both the cause and effect of the stationary nature of the fighting. Earlier, in 1973, the Egyptians had built elevated sand ridges along the Suez Canal to provide better observation and firing positions at the outset of the war.[30] The Israelis, meanwhile, were unable to implement a plan for flooding the canal with burning napalm in the face of a cross-canal attack. Both Syria and Israel have built formidable networks of artificial barriers on the Golan Heights, which include such devices as earthworks, obstacles, and tank ditches.

Although many of the countries in the region have appeared to be greatly constrained by a lack of fit between the specific requirements imposed by terrain and weather on the conduct of fighting and logistics, and the weapons and other equipment acquired from major power suppliers, there have been some notable exceptions. Israel moved quickly in Lebanon in 1982 as it had earlier on the desert; it used three different types of tanks on different axes, depending on the terrain. India carried out a highly successful mobile campaign in Bangladesh in 1971 in the face of difficult riverine terrain that would normally be daunting to mechanized forces. It made particularly efficacious use of helicopters, ferries, and pontoon bridges to overcome what otherwise might have been overwhelmingly difficult terrain impediments to a fast-moving mechanized offensive.

Airpower and Geography

In the aftermath of the Gulf War, numerous analysts asserted, not without some rebuttal, that that war had been the first in which airpower had been altogether decisive, leaving the ground phase of Desert Storm to appear almost as a mop-up operation. It was routinely said, perhaps unarguably, that desert conditions in western and southern Iraq and Kuwait, magnified the superiority of Western technology over Iraq's Soviet-supplied technology. Billy Mitchell, Giulio Douhet, Alexander de Seversky, Curtis LeMay, and Arthur "Bomber" Harris all appeared finally vindicated, almost a half-century beyond the Strategic Bombing Survey that had concluded with skepticism about the decisiveness of strategic bombing in World War II, and twenty years after the failure of the U.S. Air Force to bomb North Vietnam and the Vietcong into submission, much less back into the Stone Age, as some American generals would have had it.

In recent decades, airpower has played a mixed role in several conflicts in the greater Middle East with the extent of that role being largely determined by a combination of geographical and technological factors. With respect to the former, our standard generalizations hold water: airpower tended to be most efficacious in flat and uncovered terrain and in good weather but effective only when it was in the hands of technically proficient and well-armed air forces that could establish air control.

Israel's air force (IAF) was a crucial if not dominant factor in the wars of 1956, 1967, 1973, and 1982. (It was not a major factor in 1948.) In 1956 it was virtually unopposed, as was the Israeli army in Sinai. In 1967 there were the heralded preemptive strikes that quickly destroyed the bulk of the Egyptian, Syrian, and Jordanian air forces, abetted by clear weather, good intelligence, and the lack of proper dispersal and revetments for the several Arab air forces. Following the initial strikes, the IAF provided critical support for Israel's armored forces in their quick *blitzkrieg,* nowhere better evidenced than by the total destruction of the Egyptian mechanized forces at An Nakhl in the southern Sinai sector. Israeli airpower was a vital counterdeterrent to Egyptian artillery in the 1969–70 War of Attrition, conducting numerous deep strikes against such targets as Egyptian industries and bridges and even tangling with planes piloted by Soviet flyers. In 1973 the IAF was initially thwarted by newly introduced Soviet SAM systems, but after initial setbacks and heavy losses regained its usual dominance, which in turn allowed for victory on the ground. In 1982 the IAF, using new weapons and tactics (RPVs [remotely piloted vehicles], antiradiation missiles versus SAMs, battle management, electronic countermeasure–electronic counter-countermeasure [ECM-ECCM] aircraft, and down-looking radars) completely obliterated Syrian SAM installations in the Bekaa Valley and chalked up a very lopsided margin of kills versus the Syrian air force, something like 92:2.[31] In between major wars one could also point to some sensational IAF operations such as the Entebbe raid in 1976 and the destruction of Iraq's Osirak reactor in 1981.

In the India-Pakistan wars, airpower was a relatively less decisive factor although there was significant activity. Generally speaking airpower was used primarily for the interdiction of the rival's airpower and to some degree for other strategic purposes—much less for tactical support by comparison with the Israeli case.[32] In 1965 Pakistan launched a preemptive strike and caught some Indian air bases by surprise; afterward there was considerable air-to-air combat in what was largely a land war. In 1971 Pakistan preempted again, but India had anticipated that and dispersed many of its aircraft to rear bases, illustrating in this case the advantage of strategic depth. Pakistan struck only some forward air bases, scoring no kills, and did not use its newly acquired Mirages for deeper strikes against available targets in New Delhi and Lucknow. India responded by moving planes up to forward bases and then attacking Pakistani airfields at Chander, Sargodha, Masrur, Risalwala, Chaklala, and Murid as well as radar stations at Sakesar and Bidan. Pakistan had no such equivalent strategic depth and was highly vulnerable to strikes by Canberra bombers. After that, the Indian air force played a moderate tactical role in the Chhamb offensive and a major one out in the desert in stalling a Pakistani armored attack on Longewala in the Rajasthan sector. The attrited Pakistani air force was unable to contest the Indian navy's bold foray off Karachi. On a strategic level the Indian air force did succeed in attacks against the Karachi port complex and against oil and gas storage facilities. Air reconnaissance over the Indian Ocean was also vital. The Indian air force averaged 500 sorties a day.[33] But as in 1965 land warfare was dominant and that was dictated by the combination of mostly covered terrain and limited tactical bombing capacity. However, the role of the Indian air force in the Rajasthan Desert mirrored Israel's in similar terrain conditions.[34]

The Iran-Iraq war saw a much more limited role for airpower even though much of the terrain (and the weather, depending upon the seasons and the sector of the shared frontier) was suitable for the use of airpower. Thus, according to Anthony Cordesman, "air power had an important impact upon the war, but it never had the major strategic or tactical impact that the number and quality of the weapons on each side should have permitted"; further, that "a variety of factors sharply limited the ability of fixed-wing aircraft to provide effective close air support, although helicopters were somewhat more successful."[35] Cordesman does not list geography among many factors he cites as explanatory: target mix, widespread proliferation of area and point defense antiaircraft weapons, problems in technology transfer, and problems in command-and-control and battle management.[36]

Both sides had some limited success against strategic economic targets. Iraq, for instance, using Soviet-supplied Tu-22 bombers, did extensive damage to Iranian automobile and steel plants near Teheran.[37] In some phases of the war Iraq caused a significant drop in Iranian oil exports by strikes against tankers, oil facilities, and Kharg Island. Iran had some success in attacking Iraqi oil facilities and power plants at the outset of the war, but Iraq had a virtual monopoly on strategic bombing after 1982. Both sides, despite the relatively open terrain, were unable to use fixed-wing aircraft effectively for close air support of their land forces. Iran, armed with Maverick missiles, could not use them against Iraqi tanks in the face of such factors as difficult terrain, poor visibility, and dust. Both were incapable of nighttime support operations.

On the other hand, mostly on the Iraqi side, helicopters were useful for close support operations. The Iraqis did learn to fly low, use pop-up tactics, and take advantage of terrain-masking. Both sides learned to use helicopters for moving across water barriers, for small tactical moves in mountain areas, and to fly around terrain barriers. But, unlike in Afghanistan, neither side learned, on a significant scale, "to use helicopters to seek out and kill targets that jet fighters cannot find and to operate in mountainous and built-up terrain where helicopters can locate and maneuver around light AA (antiaircraft) guns and man-portable missiles that fighters cannot spot and thus overfly."[38]

The Soviet experience in Afghanistan presents a mixed picture regarding the impact of geography on the use of airpower. On the one hand as measured by equipment, training, and the ability of pilots, the Soviet air force would presumably rank below the United States and (qualitatively at least) below Israel but, on the other hand, well above Iran and Iraq and somewhat above India and Pakistan. There was no air-to-air combat in Afghanistan nor—because the Soviets' foes were mostly bands of guerrillas armed with light weapons—was there much need for strategic warfare or concern with industrial infrastructure, fuel storage sites, and airfields. However, there was some terror bombing of Afghan villages.

The war saw extensive efforts by the Soviets to use both fixed-wing and rotary-wing aircraft for counterinsurgency measures, sometimes in support of large-scale and large-unit Soviet ground operations. Those operations were conducted in a variety of terrain and weather conditions, mostly in rugged mountains with limited cover but also on plains and in valleys and under desert conditions, and in weather ranging from hot dry summers to winter snows.[39] In the latter stages of the war, the Soviets had to contend with the U.S.-supplied Stinger SAM, at the time the world's most up-to-date man-portable air defense weapon.

The Soviets made extensive use of several different types of fixed-wing aircraft for close air support: Tu-16 and Tu-22 bombers, Su-17, MiG-21, MiG-23, MiG-27, Su-24, and Su-25 fighters and fighter-bombers.[40] Their targets were mostly hardened mountain locations that controlled the heights and narrow ravines as well as villages and the creeks leading to them. The Soviets used some chemical weapons, napalm, and cluster bombs as well as conventional bombs, many of which apparently did not explode and, according to Cordesman, "had little lethality in the rough ground and unique terrain conditions in Afghanistan."[41] Soviet ability to locate and hit targets in rough terrain apparently improved by 1984–86, particularly with the use of Su-25s and related An-12 transports used as spotter planes and with the use of more advanced targeting and delivery technology, such as laser range finders, low-light television, infrared sensors, computers, and radar. Laser-guided bombs came to be effective against tunnels, cave mouths, and hard-to-reach strong points by about 1986. The Soviets also launched high-altitude attacks on area targets, including extensive use of "carpet bombing." Long-range navigation systems and electro-optical delivery systems were apparently somewhat successful at night and in poor weather. But overall airpower was not finally decisive in this war as it would be in Desert Storm because of the guerrilla nature of the war, notwithstanding the massive civilian casualties that resulted from the bombing.

The Soviets made extensive use of helicopters in combined operations in mountainous terrain, adopting, if only slowly, the American pop-up and terrain-hugging tactics.[42] They were lacking then in adequate night target acquisition capability and night vision aids such as forward-looking infrared equipment. Further, the high mountainous terrain in the contested areas of Afghanistan and the fact that Soviet helicopters are less well armored on top than on the sides or bottoms made them vulnerable to Mujahideen fire from high points on the mountains. The Soviets in turn countered with newer tactics, diving from high altitudes and using terrain-hugging flight patterns. By the end of the conflict, improved versions of Soviet Hind helicopters allowed for more rapid turns and flight at nap-of-the-earth altitudes. The Soviets also made extensive use of heavy-lift helicopters to position troops on mountain tops in advance of large formations, tactics referred to as "cresting the heights."[43] Generally, the Soviets had always used helicopters in conjunction with infantry operations in the mountains and now employed them in lieu of artillery where the latter could not be transported into the mountains.

Overall, the lesson of these wars (and before the Gulf War offered still newer evidence) is that by 1990 the steady development of new technologies associated with tactical and strategic bombing had to some extent transcended traditional obstacles of terrain and weather, but only for the most advanced air forces; it was much less the case in low-intensity conflict or conventional conflict fought in difficult terrain and weather.

On a more general plane and somewhat by way of summary, the issues dealt with regarding geography and air operations include: the impact of radar ground clutter on air-to-air operations (aircraft may use ground clutter to hide and then pop up to surprise an enemy); terrain-masking and its impact on air defense operations; overland versus overwater operations; terrain and helicopter operations (terrain-masking is often vital here).[44] These issues are for the most part tactical or operational, and we have noted their importance in the several contexts of the Arab-Israel, India-Pakistan, Iran-Iraq, and Afghanistan wars. Israel, for instance, was able to transcend these factors in its successful aerial assault on Syrian SAM installations in 1982 in the Bekaa Valley. Both Israel and Syria in 1982 and mostly Iraq in the Iraq-Iran war were able successfully to use terrain-hugging and pop-up tactics with helicopters in attacking armor and troop concentrations. Nowhere in these varied contexts was there an expanse of open and uncovered terrain equal to what the United States and Iraq would contend over in 1990–91.

Additionally, terrain features in combination with the state of radar, air defense, and battle-management technology may affect long-distance interdiction strikes. Israeli aircraft flying to and from Negev Desert bases en route to the Iraqi Osirak nuclear reactor were apparently able to fly low routes that used terrain-masking, allowing them to avoid radar detection by Saudi Arabia, Jordan, or both. Presumably Saudi Arabia's later acquisition of airborne warning and control systems (AWACs) would render a repeat of such an operation much more difficult, assuming, of course, that the Saudis would be willing to pass along such information to Iraq.

On a broader and more strategic level one might point to some additional geographic issues in relation to airpower. Strategic depth versus aircraft ranges and tanker refueling capabilities of opponents is an obvious one. We have noted India's ability in 1971 to disperse its aircraft to the rear, out of range of the Pakistani air force, and Pakistan's inability to do the same. Israel was easily able to reach all of Egypt's, Syria's, and Jordan's airfields in 1967, and those in western Iraq, even without refueling, and again in 1973 with respect to Egypt and Syria (the distances were shorter than those faced by U.S. planes based in Saudi Arabia en route to Iraq in 1990–91). Iraq, of course, also flew many aircraft to Iranian sanctuaries of sorts (Pakistan might one day do the same). Israel is similarly vulnerable but its preemptive strategy and stronger defenses, based mostly on airpower, have not yet brought this to the test. (There has been much speculation about the possibilities of using chemical weapons on Israeli airfields.) This vulnerability is one of the main reasons for its reluctance to relinquish high ground in the West Bank and the Golan that provides for early warning installations that cannot easily be replaced by airborne systems.

There is also the question of (air) force to (air) space ratios and their impact on air defense operations, leading to the somewhat paradoxical situation that having smaller spaces to defend (such as Israel) may be easier than defending industrial and population centers dispersed over a large area. But that is complicated! It involves, among other factors, the possibilities for omnidirectional approach routes to a country's heartland, which depend on the capabilities of the opponents. Thus, the United States long thought in terms of omnidirectional bomber approaches to the USSR (including over the Middle East from Diego Garcia). That in turn compelled the Soviets to defend very long borders and immense spaces. On the other hand Iran and Iraq are large countries in a conflict situation where each can only attack the other head-on (neither has viable aerial refueling capabilities), which prescribes defenses based on forward air bases not far from their mutual frontiers.

Naval Warfare: Geographic Considerations

In none of the recent major wars in the greater Middle East were naval warfare or naval operations crucial in the sense that they were, for example, in the U.S.-Japan war in 1941–45 or Britain's Falklands campaign. As a matter of fact with the exception of India to some degree, all of the region's navies can be characterized as "white water" rather than "blue water"; that is, they are largely engaged in coastal defense and short-distance power projection operations. But in each of the recent regional wars there was a significant naval component, as exemplified by the massive U.S. logistical effort by sea in Desert Storm; the successful feint used by the United States in regard to a possible Marine amphibious landing in Kuwait that tricked Iraq into positioning its defensive forces badly; the Iran-Iraq tanker battles during the 1980–88 war in which each side attempted to interfere with the other's logistics in the Persian Gulf (Iran to interdict supplies coming *in* via Saudi Arabia and Kuwait, and Iraq to interdict Iran's oil exports *out*); India's successful blockade of both Chittagong and Karachi in 1971 and its well-executed naval assault on the latter; and Israel's destruction of significant parts of the Egyptian and Syrian navies in 1973 and its successful, leapfrogging amphibious landings along the Lebanese coast in 1982. Then too there was the sinking of the Israeli destroyer *Eilat* by Egyptian Styx missiles after the 1967 war, often cited as the beginning of PGM warfare. These examples and some hypothesized future naval questions can be set in the context of the sum of the tactical and strategic variables and factors that inform the military naval geography of the greater Middle East.

John Collins has described a variety of tactically related naval warfare components, of which some are related to past or anticipated naval operations in the greater Middle East and others are more germane, for example, to the prospect of all-out submarine and antisubmarine warfare (ASW) in a major power oceanic context. Among those factors are: sea water attributes such as salinity, density, stratification from surface to sea bottom, and permeability to light and sound; and sea surface behavior such as currents, tides, waves, and swell. Then there is marine topography, important here mostly in regard to mainland littorals and islands, especially beaches, their seaward approaches, and straits. The first set of these factors, in the context of greater Middle East naval operations, pertains mostly to ASW, as conducted by rival submarines, aircraft, or surface ships as well as to mine warfare. The second set of factors has relevance primarily for amphibious landings, either over a beach or in less challenging circumstances involving built-up areas, opposed or unopposed.[45]

Submarines and ASW have, so far, played a fairly limited role in the wars between the region's three major conflict pairings: Arab-Israel, Iran-Iraq, and India-Pakistan, as well as in the Gulf War. The Iran-Iraq war and the several Arab-Israeli wars have seen little if any action by submarines or ASW units. In between major wars, the Israelis have apparently used submarines in some long-distance clandestine operations, for instance, the raid on Tunis against the PLO leadership there. There has been a certain amount of talk from time to time about the Israelis deploying long-range, nuclear-armed missiles on new submarines acquired from Germany, which could afford them some invulnerability to a preemptive first strike.[46] The waters of the eastern Mediterranean are, for these purposes, fairly shallow just west of Israel, but up to 3,500 meters deep between Alexandria and the southern Turkish coast near Antalya. During the 1973 sealifts on behalf of Israel and the Arabs by the United States and USSR, respectively, neither side dared to use submarines to interdict a supply ship.

On the other hand submarines did figure to some degree in the naval phase of the 1971 India-Pakistan war in the relatively shallow waters of the Arabian Sea and Bay of Bengal. India feared Pakistan's Midget and Chariot submarines, one of which sank an Indian ASW frigate, and a Pakistani submarine in the Bay of Bengal that was to keep the Indian carrier *Vikrant* further south near the Andaman Islands, was sunk in Vizag harbor.[47] One article commenting upon India's ASW problems in 1971 noted that "the Indian Ocean's peculiar sea conditions and relatively shallow depth only subvert the Indian navy's efforts to neutralize them"; further, in this context, that "submarines cruising in shallow waters do not make for good long-range propagation of sound, and the long-range acoustic path is predictably obstructed by seamounts and sounds emanating from other natural sources."[48] That article further discusses the possibilities for Pakistani (or Indian) submarines operating in particularly impermeable near-surface water layers, from which long-distance sound propagation is attenuated, placing ASW forces at a serious disadvantage. Both Iran and Israel have quiet diesel submarines that can operate in such a milieu.

The recent Iranian purchase of a few Soviet diesel submarines has raised the question for the United States as to whether such vessels operating out of Bandar Abbas near the Strait of Hormuz could be effective against U.S. forces in the Persian Gulf. While the shallowness of the water would appear to render ASW detection difficult, the limited operating area for submarines in the face of the vaunted U.S. ASW capabilities would appear to render them highly vulnerable.

Generally speaking, submarines are considered to be well suited for littoral warfare of the type anticipated in the greater Middle East. Israeli Admiral Yedidia Ya'ari, looking to the future of submarines "designed to confront ASW with firepower rather than improving their capability to hide from it," claims that the submarines' optimal operating niche is just below the surface "with sensors just above it and weapons systems ready to engage ASW patrols, air or seaborne." In this position, "nearly hidden by the coastal radar's sea-clutter," it is said to have "the best signature management a surface combatant can ever hope to get."[49]

There have been some instances of amphibious operations during these wars and contingency planning for much more. Israel, as noted earlier, used leapfrog operations reminiscent of U.S. operations in New Guinea and Italy during World War II in an attempt to bypass and trap the PLO forces fleeing north toward Beirut in the 1982 war. The tides and beaches appear not to have been impediments. In 1971 India, concerned that Pakistani troops were escaping from East Bengal via Cox's Bazar, landed a force there from a merchant ship, an action called by one commentator "the first 'combined operations' landing ever made by the Indian armed forces."[50]

The possibilities for amphibious warfare in the Persian Gulf have been widely discussed and analyzed within the U.S. defense planning establishment. The Saudi Arabian Persian Gulf coast (and those of other Gulf Cooperation Council states) would appear not to offer serious obstacles to a landing. Iran at any rate lacks the capability for such operations. The U.S. Marines have conducted numerous amphibious exercises on that coast, as the United States might conceivably need to conduct a "forced entry" amphibious landing either in the same area or on the Iranian side. In neither case would such a landing be considered subject to an overly challenging set of geographic barriers; the same would obtain for Saudi Arabia's Red Sea coast. The mountainous topography of the coastline outside the Strait of Hormuz along Iran's Makran coast extending into Baluchistan would appear to be more of a challenge.

Generally speaking, however, the geographical constraints on amphibious warfare are gradually being removed by technological change, specifically, the use of the helicopter. The United States, at least, can now land significant forces over the beaches by helicopter, which is the reason that the U.S. Marines have wanted so much to acquire the V-22 Osprey. Other nations have less of this kind of capability, but developments toward an RMA, to the extent that they entail reduced requirements for large masses of troops, may result in some other nations becoming more capable of circumventing the need for large-scale amphibious invasions in order to establish a beachhead against a rival.

The possibility of amphibious operations must be considered not only in terms of the feasibility of landings along various coastal areas but also with a view to the vulnerability of some countries to amphibious assault owing to the proximity of key assets to coastal regions. This is another twist in the strategic depth problem. Iran's Makran coast is a long way from principal Iranian cities and resource areas. On the other hand, at least hypothetically speaking, Pakistan, Israel, Syria, Egypt, Libya, and Algeria all have key assets near potential amphibious landing areas. This is also true of Turkey to some degree as evidenced by the history of the Gallipoli operation during World War I.

On this more hypothetical note, there are other long coastlines that might figure in scenarios for amphibious operations. The Black Sea coast could become a military arena if a resurgent Russia were to try to regain the USSR's former holdings in the Caucasus, or if Russia and Turkey should come to loggerheads again as they were for so long in the nineteenth century. Then, too, there are possibilities that U.S. military operations could be conducted against Libya or Algeria from the Mediterranean—both of which would be nearly defenseless against a large-scale amphibious operation—particularly if either of those countries were to escalate terrorist activity or threaten the use of weapons of mass destruction against Europe or the United States.

On a more macro or strategic level, some further generalizations can be made about the naval military environments of the region's existing and potential conflict pairings. Generally this involves such factors as distance, contiguity, the spatial relationships between contending forces, the lengths of coastlines, the particularities of specific situations involving choke points, and the configurations of the key regional bodies of water, that is, the Mediterranean, Red, and Arabian seas and the Persian Gulf.

Israel has a contiguous coastline with Egypt and Lebanon-Syria. Iran and Iraq have coastlines that face one another across the narrow Persian Gulf. India and Pakistan have a contiguous or continuous coastline along the Arabian Sea (they had another in the east up to 1971). Those are entirely different than the geographic relationships between, for example, the United States and Russia, the United States and China, or the United Kingdom and Argentina. Within the former group there are possibilities for quick preemptive naval strikes at the outset of a conflict in which one or another navy can be caught in or near port and immediately destroyed (airpower used to attack ships is also, ipso facto, close by). Israel rapidly knocked out a good part of the Syrian navy in 1973; India did likewise to Pakistan in 1971, particularly with its daring and effective naval attack on Karachi harbor. This factor of contiguity also means that in all of these regional conflict pairings, land-based aircraft will be a vital adjunct to naval warfare, as it was, for instance, in the Iran-Iraq war.

Some nations have particularly narrow and vulnerable access to the sea and so are easily bottled up and blockaded. Iraq's hopeless position at Basra-Umm Qasr resulted in its navy being almost wholly inoperative during the Iran-Iraq war. Pakistan also labors under the disadvantage of a short coastline, and in 1971 the not-much-larger Indian navy was easily able to block its access to Chittagong in the east and to Karachi in the west, though in the latter case the coastline is actually several hundred miles long and there is another decent port further west at Gwadar in Baluchistan that was coveted by both the USSR and the United States at various times during the cold war. India by contrast has an immense 3,750-mile coastline with numerous excellent ports, and is thus in a situation of very favorable maritime strategic depth. Iran could position some naval units outside the Persian Gulf on a rather remote coastline but would be forced to operate largely within the Persian Gulf, where its naval units would be highly vulnerable to air and naval attacks from superior U.S. forces. Other than India the naval powers in the region must operate from a small number of good harbors: Israel from Haifa, Eilat, and Ashkelon (only the first being a good natural harbor); Egypt from Port Said and Alexandria; Syria from Tartus and Latakia. These navies are all highly vulnerable if an opponent achieves air superiority.

Israel not only has a short coastline, but also has to take into account the vulnerability of Eilat to closure of the Strait of Tiran, as was threatened by Egypt in 1967. However, with its long-range patrol boats and submarines, it also has the capability and seaborne window for relatively long-range operations in the Mediterranean and Red seas and maybe even further afield. One is reminded that the Israeli navy operated down to the Straits of Bab El Mandeb, to provide an ECM capability for the Entebbe raid, sailed to Tunis in an antiterrorist operation, moved stolen patrol boats out of Cherbourg harbor and got them home, and used its maritime capability in the "Plumbat Affair" to hijack uranium on the high seas.

The role of the region's various choke points, the Straits of Bab el Mandeb and Hormuz, the Suez Canal, and Gibraltar, will be discussed later in the context of power projection, for example, U.S. power projection into the region. Suez has figured prominently in the wars of 1956, 1967, and 1973, and the Strait of Hormuz at least at times seemed vulnerable to Iranian actions during the tanker battles during the Iran-Iraq war. The narrowness of the Persian Gulf was an obvious factor in the latter instance, rendering shipping to and from Kuwait vulnerable, and ultimately leading to the U.S.-led tanker reflagging operation near the close of that war.

These factors are also pertinent to the possibilities of mine warfare in this regional context. This was a minimal if not nonexistent feature in the Arab-Israel and India-Pakistan conflicts, perhaps owing to geographical considerations. But the Iran-Iraq war was different. The West feared an Iranian effort to mine the Persian Gulf from the outset of that war and was particularly concerned that there would be an attempt to close the mouth of the Gulf. But that would not have been easy. As Cordesman pointed out, the deep-water channel between the island of Jazye Larak in Iran and the rocks off Didamar in the south is nearly twenty miles wide and with an average depth of around 190 feet and a maximum depth of around 280 feet. Strong currents course through the Strait of Hormuz, and pressure and acoustic mines cannot be used below 125 to 160 feet, though huge magnetic contact mines could be moored to the bottom. It is said that 2,000 to 3,000 such mines would be needed to close the Straits, but that the West at any rate, could deal with it.[51] Mines could also be used elsewhere in the Gulf, as was experienced by the tanker *Bridgeton* and several Soviet ships.[52] Iran used old Soviet mines near Kuwait that could be used at a depth of 350 feet, and it had, variously, contact, magnetic, acoustic, bow wave, pressure, temperature, and remote-controlled mines.

Later, in planning for the possibility of war with Iran, the United States had to concern itself with the possibility of mine warfare, along with the threats of Silkworm missiles, missile-armed patrol boats, and submarines. The combination could make a traverse of the Gulf very precarious, given the narrowness of the Strait of Hormuz, the need to pass by the Iranian-occupied islets of Abu Musa and the Tunbs, and another point near the Shah Allum shoals that is close to Iranian waters. Generally fears about mine warfare have been an important impetus to the United States having upgraded its mine warfare capabilities, long a neglected area of the U.S. Navy's force deployments. Whether mine warfare could become a focus of defense activity, say, of Egypt, Turkey (the Straits), Israel, or Syria, or for Pakistan in protecting Karachi or menacing Bombay is merely a point of speculation.

With one exception none of the region's military powers has made extensive use of outlying, perimeter bases as part of a broader power projection strategy. India, which has often been discussed in this context in recent years, has access to its Andaman and Nicobar Islands in the Bay of Bengal, and the Lakshadweep Islands off its southwest coast (perhaps also to Maldives, an independent republic). With an expanding navy and the proclamation of 200-mile economic exclusion zones (within which it has valuable off-shore mineral and oil deposits), it has an extended peripheral reach.[53] Iran has been rumored to be seeking some naval access in Sudan, an ideological bedfellow, that would allow it to flank Saudi Arabia at sea, for whatever that is worth in the light of its naval capabilities. As noted above, Israel has apparently had some access to Ethiopian islands in the Dahlak Archipelago at the southern end of the Red Sea for operations involving patrol boats near the Bab el Mandeb choke point, which controls access from Eilat to the Indian Ocean (Israeli–South African ties were germane here). The possible naval implications of growing ties between Israel and Turkey, if any, remain to be determined.

Low-Intensity Warfare: Geographic Considerations

In recent years the bulk of the fighting in the greater Middle East region has been on the low-intensity end of the conflict scale. Indeed since the end of the Iran-Iraq war and the war in Afghanistan (a mixture of guerilla and conventional warfare), only the Gulf War has been fought on the major, conventional warfare level. By contrast recent years have witnessed a surprising number of LICs, while a number of others are temporarily dormant but potentially explosive.

The long Western Sahara conflict pitting Morocco against the Polisario guerrillas is now dormant, but the underlying causes of the conflict (Moroccan control versus an independent Western Sahara) remain unresolved. In Algeria the government battles Islamic insurgents in the cities, while an impending regional conflict with the Berbers in the southern mountains seems a real possibility. The Egyptian government, too, battles Islamic insurgents. The long border war between Libya and Chad (linked to a civil war in the latter) is now dormant but latently dangerous. In Sudan the long bloody war between the northern Islamic government and the non-Islamic separatists in the south winds on unabated, while in Somalia war among the various warlords has returned more or less to what it was before the UN-led intervention, but is now exacerbated by the specter of regional separation.

Israel continues to battle the Hizbollah, Hamas, and the Islamic Jihad along the Lebanese border, in the West Bank, and in Israel proper. Lebanon's former multisided war is, for the moment, relatively inactive. Yemen, too, is quiet after a civil war in which forces from the former North Yemen invaded the former South Yemen. The Kurdish insurgency continues in Turkey and Iraq, and also on a minor scale in Iran and Syria. Additionally, there are remnants of a Shia insurgency in Iraq in the marshlands of the south. In the Caucasus in addition to the recent rumbles in Chechnya, Georgia (Abkhazia and South Ossetia) and Nagorno-Karabakh (Azerbaijan versus Armenia), several other flashpoints within Russia appear ready to flare up. In Central Asia there is still some fighting in Tajikistan's Fergana Valley along a dividing line between remnant communist government forces and Islamic insurgents. Several Islamic forces continue to vie for control in Afghanistan, and they each control large chunks of territory. In Kashmir Islamic guerrillas continue to harass the Indian army, while in Sri Lanka a long bloody war between the Hindu insurgents (the Tamil Eelam Tigers) and a Sinhalese-based government continues on at a deadly level. In short there is no shortage of LICs and of nasty fighting in this large, bloodstained region.

One can outline a continuum of LICs, running along a spectrum from interstate border warfare (Chad versus Libya) to fairly large-scale civil warfare (Sudan and Yemen) through various levels of guerrilla warfare and insurgency and on to brief border skirmishes, coups, and terrorism. That continuum involves various mixes of the standard criteria for gauging levels of conflict— severity(casualties), magnitude (number of combatants involved), and duration. There is a major division according to whether the warfare is linear (a moving, visible "front") or nonlinear (more dispersed and hit-and-run in character, without large-unit combat formations).

There is one other simple scheme for classifying LICs that has been widely applied as a basis for comparison in the recent issues of the *SIPRI Yearbook* and it is interesting from a military-geographic standpoint, having to do with the simple dichotomy between wars over the control of government and those over control of territory.[54] The former category involves conflicts normally fought within the established and accepted boundaries of one country over political control of that country. As such those conflicts are not normally centered on ethnic, linguistic, or racial identities and associated territorial bailiwicks or enclaves, but are rather along an ideological divide, construed broadly to incorporate religious divides. There have been wars over territory classed as LICs that have involved either interstate combat or territorial problems within a more or less established state. In the first, as distinguished from conventional wars, were those wars fought at a relatively unsophisticated level, or at a modest level of magnitude and severity, as, for example the Libya-Chad war. One could also point to minor border skirmishes that were well short of large-scale combat such as the fighting in recent years along the Mali–Upper

Volta border or that between Mauritania and Senegal. However, most LICs over territory have involved intrastate ethnic-tribal-religious conflicts, sometimes where there is an issue of autonomy or separation from a state. Such wars will normally feature regional demarcations between contending forces; visible physical differences between them including easily demarcated identities; and perhaps even a moving "front," that is, they will be somewhat linear in nature.

Numerous examples of wars over government and wars over territory can be seen in the current and recent history of the greater Middle East. Conflicts over control of the central government are being fought in Afghanistan (various Islamic factions, though some are also regionally based), Somalia (various warlord factions), Tajikistan (former communists versus Islamic insurgents), Algeria and Egypt (Islamic fundamentalists versus incumbent and more secular regimes), and Yemen. Those involving primarily territorial concerns, that is, ethnic conflicts, can be seen in such disparate locales as Chechnya, Georgia (central regime versus Abkhazian insurgents), Sudan (northern Muslims versus southerners), India (various regional insurgencies in Punjab, Assam, Nagaland), Turkey (Kurds), Iraq (Kurds), Israel (Palestinians), and Azerbaijan (Armenian insurgency in Nagorno-Karabakh). There are some conflicts in which both central government control and territorial issues are involved. In Algeria, Islamic fundamentalists are conducting urban guerilla warfare in Algiers with the aim of overthrowing the government while the Berbers in the south seem on the verge of a split-away movement. Iraq's government faces Shia and other struggles against central control, but also a territorial-based Kurdish insurgency. The now latent but many-sided Lebanese conflict has had elements of both, but perhaps with a predominant territorial cast, centered on rifts between Muslims and Christians, Shia and Sunni Muslims, Palestinians, and Druzes.

It is important to note that some of the above distinctions affect the geographic basis upon which low-intensity conflicts are fought. Furthermore, generally speaking, there are major differences between the geographic considerations of conventional wars and those of unconventional wars, though there are also some areas of commonality as they relate to issues of military geography.[55] The dualism of attrition and maneuver is relevant to both, albeit in somewhat different ways. Similarly relevant are matters related to offensive initiative and its opposite on the defensive side in the context of the momentum or "tides" of battle.

The primary criteria for analysis of the importance of military geography (terrain and weather) as applied to low-intensity warfare are as follows:

—The size of the theater or country involved as measured by square miles; in turn, the relationship of this factor to the ratio of forces between the insurgent and counterinsurgent sides.

—Cover, that is, vegetation and visibility from the air as it applies to such aspects as hiding, reconnaissance, the opportunities for ambush, and impediments to the use of high-technology weapons.

—Urbanization, for example, the proportion of population in urban areas and the importance of large urban areas overall relative to the countryside.

—Strategic depth, that is, the spatial or distance relationships of urban or built-up areas to remote redoubt areas (such as mountains, forest, and swamps), affecting the ability to strike or respond in *both* directions. If we relate to these criteria, it is clear that the recent wars in the greater Middle East have involved a broad mix of, for example, size of theater, population density, terrain, and weather.

In regard to the size of theater, for instance, the range is immense. Sudan, Afghanistan, Chad, and Western Sahara have vast expanses with areas between 100,000 and 900,000 square miles. While the fighting in Sudan has actually taken place over only a small part of that nation's territory in the south, the area is still vast relative to such smaller combat venues as Israel's West Bank or Lebanon. As another illustration, the previous phase of the war in Afghanistan involving the USSR was fought over a wide expanse of territory, but the current battles among the various Islamic factions seem more concentrated in and around Kabul, reflecting the nature of the conflict as one over control of the government. At the other extreme, Lebanon and the West Bank have areas of less than 10,000 square miles, and Chechnya, the battle area of Georgia-involving Abkhazian insurgents, and Nagorno-Karabakh are all also small. In more of a medium-area range would be the conflict in Kashmir. Large areas can make counterinsurgency (COIN) operations difficult, as evidenced by the problems involved in government-based operations against insurgents in Sudan and Kurdistan. But as Israeli attempts at suppressing the Intifada suggest, COIN operations can also be difficult in smaller areas, a matter sometimes but not always related to terrain.

Population density is another important factor.[56] Where classical COIN operations are involved, a denser population means greater problems of control and suppression. Most successful COIN operations have been carried out in large countries with low population densities—Sudan, Western Sahara, Chad —but here still another variable intrudes, that is, the factor of terrain—cover (vegetation), ruggedness, weather, to which we now turn.

Recent low-intensity warfare in the Middle East region has been conducted in a variety of terrains, soil, cover, and weather. In Yemen, (a war over territory) forces from the north attacked the remnant Marxist forces around Aden across mountains, along wadis, and over the desert, finally capturing the city, which is some twenty-five miles from the mountains. This war was fought with fairly sophisticated weaponry, including attack aircraft that could be effective in a theater altogether lacking in cover. This war could have been classified as bordering on a limited conventional war, but the short duration and limited number of combatants kept it in the LIC category.

In the Kurdish regions of eastern Turkey, an insurgency has been going on in rugged but covered mountainous terrain that is good for hiding, which makes effective COIN operations difficult, just as it does in northern Iraq. In the Caucasus, the locus of the fighting in the several recent or ongoing conflicts has been in rugged terrain, in mountains with fairly extensive cover that are favorable for guerilla warfare. Such terrain characterizes Nagorno-Karabakh, where the war has been fought mostly by infantry but with some small mechanized forces in the rugged hills west of the lowland plains stretching eastward into the core of Azerbaijan. There the Armenians, backed by Russian legionnaires, have been successful so far, in part because of the availability of good Soviet weapons. In the Chechnya conflict, after the fall of Grozny, the follow-on fighting has been concentrated in wooded, mountainous terrain in the Caucasus foothills, where the Russians face conditions similar to those in Afghanistan, albeit on a smaller scale.

In Kashmir a guerrilla war has been fought in valleys ringed by high mountains covered with glaciers, in and around Srinagar. Along the Chad-Libya border, light mechanized forces have fought in the open desert along the few passable roads, while in Sudan an interminable conflict has been fought in semiarid hard scrabble and savannah-like terrain. These conflicts have all been over territory so they have had somewhat of a linear character, with clear demarcations between areas controlled by contending forces.

Some LIC locales have involved desert or semidesert conditions, though within that rubric the terrain can vary markedly. Much of Afghanistan is without extensive cover, but the terrain changes, from the deserts of the southwest to the rugged mountains of the northeast. Western Sahara is real desert, but with broken and uneven terrain that does allow for some cover and concealment. Chad is mostly desert in the zones of fighting in the north below the Libyan border and the Aozou Strip, but the northern area also comprises the rugged albeit bald Tibesti Mountains, which rise up to over 11,000 feet.

Some of the region's LICs have featured urban warfare, usually at the level of terrorism or of low-level insurgency. Such examples have been evidenced in Beirut (Lebanon), Cairo (Egypt), Algiers (Algeria), and in Israeli cities, though in all these cases terrorism or low-level guerrilla warfare has also been conducted in smaller towns and in the countryside. If Islamic insurgencies should succeed in Algeria or Egypt or both, they would no doubt have to culminate with large-scale uprisings in the capital cities.

There are few easy generalizations to be drawn from these diverse cases. If there is one it concerns the difficulties of achieving decisive victories in LICs fought on rugged covered terrain, as has been underscored by the lengthy stalemates in the Caucasus, Afghanistan, and Lebanon. By contrast, in more open, less covered terrain, there have been more conclusive outcomes in Western Sahara, northern Chad, and Yemen, where there are few places for the weaker side to hide.

SIX

The Gulf War and the Revolution in Military Affairs

THIS CHAPTER deals with two related topics discussed in chapters 5 and 7. The first section focuses on the Gulf War and its implications for conventional war, the role of geography, and power projection. The second section deals with the Revolution in Military Affairs (RMA). The RMA is already having profound implications for the battlefield, as was demonstrated during the Gulf War. The new military technology and the new approach to warfare, as embodied in the RMA, arguably is changing the relevance of geographic factors, but in ways not yet clear.

Lessons from the Gulf War: A Revolution in Military Affairs?

The aftermath of the Gulf War has included an explosion of post-hoc analyses focused on "lessons learned." Books have been published ranging from memoirs by participants, to more or less descriptive and chronological coverages, to more analytical works devoted variously to the war's military and political-diplomatic lessons, even to its environmental consequences.[1] Both the Department of Defense and the U.S. Congress issued their own official "lessons learned" reports.[2] The Los Alamos National Laboratory produced a series of papers interpreting the lessons of the war from the perspectives of other nations, large and small, comprising most if not all of the world's serious military powers.[3] An Israeli think tank produced a volume that looked at various lessons emerging from the war from an Israeli perspective.[4] One volume edited by American scholars on the left examined a host of contextual

issues surrounding the war: environmental consequences, the politics of television, arms transfers, the role of international law, and the lost alternative of
reliance on sanctions, among others.[5]

The lessons learned literature from Desert Storm has become intertwined
with that on the Military Technical Revolution (MTR) and RMA. Primarily that
comes down to relative arguments, pro and con, over whether Desert Storm
inaugurated and exemplified such a revolution, versus more cautious analyses
that see the U.S. weapons systems used there as representing just the leading
edge, the mere glimmerings, of a much more fundamental revolution that will
be revealed and gradually consummated in the decades to come. Those arguments often revolve about the consequences of the unique geographical venue
in which Desert Storm was fought, that is, about whether the new military
technologies could *only* have been so effective in that environment.

In any case, MTR-RMA "advocates" (those who perceive us now to be at the
bare beginnings of a monumental revolution) also project enormous future
changes, even over the next fifteen years, in such principal criteria as data
transfer rates (a key indicator of the development toward information warfare
capability) and the number of soldiers needed to cover ten square kilometers
(battlefield density). In the 1991 Gulf War, computers worked at 192,000 words
a minute, and 23.4 soldiers were required to "cover" ten square kilometers. The
respectively comparable figures projected for a future war in 2010 are 1.5 trillion words a minute by computer and 2.4 soldiers to cover the ten square
kilometers. Those are, of course, massive projected changes.[6]

Terrain and Weather in the Gulf War

What can one say about the war's lessons that relates to the geographical
focus of this work? At the combined tactical-technical and operational levels,
the primary geographical lessons of the Gulf War are obvious and virtually
unarguable. The Pentagon's lessons learned report states that "we fought in a
unique desert environment, challenging in many ways, but presenting advantages too," and that "enemy forces were fielded for the most part in terrain
ideally suited to armor and air power and largely free of noncombatants."
Further, "the desert also allowed the U.S. armored forces to engage enemy
forces at very long range before our forces could be targeted, an advantage that
might have counted for less in a more mountainous or built-up environment."
The combination of relatively flat and open desert terrain, the lack of cover, the
normally clear (day and night) weather, and the nature of the desert terrain,
which allowed easy movement for armored forces, provided a huge advantage
for the technologically superior side; specifically in connection with complete
control of the air and a monopoly on space surveillance.[7] Summing it up, the
Pentagon's report states: "That assessment acknowledges that the desert climate was well suited to precision air strikes, that the terrain exposed enemy

vehicles to an unusual degree, that Saddam Hussein chose to establish a static defense, and that harsh desert conditions imposed constant logistical demands that made Iraqi forces more vulnerable to air interdiction."[8]

It was a perfect place for the United States to demonstrate its new MTR technology; indeed, to the extent that it could hardly be replicated. With the lack of cover one side was able to see almost everything with its satellites and control of the air; the other was almost completely blinded. In terms of information warfare there was a completely lopsided asymmetry, which was primarily a function of the prevailing terrain (topography and cover) and weather. The result was the first war to have been won largely by airpower, with the ground phase of the war appearing in retrospect to have been little more than a mopping-up operation.

However, in comparing the U.S. experience in Desert Storm with that in Vietnam, Luttwak has noted that aerial bombardment has "always been much more successful in the Middle East than in temperate zones with their abundance of clouds, mist, and fog, let alone the tropics with their all-concealing vegetation."[9] Airpower was, of course, crucial in the Arab-Israeli wars from 1956 onward. In 1956, 1967, and (after initial failures) 1973 the Israelis had taken control of the air and then were able to win decisively on the basis of unchallenged tactical airpower and armored "shock" backed by air superiority. The Gulf War represented an extension of this history because of the new capabilities of stealth aircraft, high-precision bombing (mostly laser-guided bombs) and the systematic destruction of air defense systems allowed by the newer MTR sensors, electronic countermeasure systems and precision strike systems.

The synergism produced by the nature of the terrain and weather in the Gulf War theater and some of the newer technologies was remarkable. Some examples will suffice from a broad menu. The Joint Surveillance Target Attack Radar System (JSTARS) aircraft rushed hurriedly into battle were easily able to pinpoint and track Iraqi vehicles all over the battlefield with an ease that would not have been achievable in more covered terrain. The flat and unobstructed desert also magnified the advantage of the U.S. Army's superior stand-off weapons, such as its tank guns and antitank missiles.

Hence, according to one source, "with an average range of up to 6,000 meters, the Hellfire [antitank missile] allowed the helicopters to attack Iraqi tanks outside the range of most antiaircraft weapons. . . . The desert is a perfect place to do this, for there are few places for the tanks to hide. In more built-up, or forested, areas, it's common for the helicopters to be shot down by enemy armored vehicles that are hidden from view."[10]

Similarly, the terrain magnified the superiority of the U.S. M1A1 tanks that exceeded all expectations in Desert Storm. Their thermal sights, unhindered by cover or other obstructions, could see targets as "hot spots" at a distance of over three miles, penetrating through oil fire smoke, even if they could not positively identify targets at about one mile's distance.[11]

Actually, there is a bit of a paradox involved here. As we have noted the flat desert and normally clear weather would, normally, advantage the side that dominates the air and has better and more accurate stand-off weapons. But ironically, as was noted in the Pentagon's report, the unusually bad weather that prevailed during the war might, in some ways, have added to the U.S. advantage, as did the haze caused by oil fires and sandstorms. The United States had the thermal-ranging and night-sighting equipment to deal with these conditions, and the Iraqis did not. Hence "the combination of austere terrain and desert weather coupled with extended periods of reduced visibility let U.S. forces exploit the advantages of long-range weapons and all-weather, day-sight systems."[12]

RMA Technology and Terrain

Some of the new U.S. aviation technology was favored by the flat, open terrain. The F-16s and F-15s used dual low-altitude navigation and targeting infrared for night system (LANTIRN) pods, originally developed as a cheaper way to obtain more high-performance bombers without building all the expensive electronics into fighter aircraft that would spend some of their time operating as fighters and some as bombers. Similar pods built into the U.S. F-111s and F-117As use terrain-viewing radar "that allows the aircraft to fly low and fast at night or in bad weather," avoiding radar and antiaircraft weapons.[13] In this open terrain the pilot could see the target at three to nine miles (twenty to sixty seconds' flying time), "paint" the target (magnified by up to sixteen times) with a weapon-system laser, and then hit it with a Maverick missile or laser-guided bomb.

The widely heralded U.S. ability at "tank plinking" during the Gulf War was also abetted by the terrain and weather. Here aircraft such as the F-111 (whose pilots apparently discovered this possibility) were able to use their thermal sensors after dark to find tanks and other vehicles whose metallic bodies cooled more slowly than the surrounding sand, presenting a thermal spot for the sensors.[14] Numerous dug-in Iraqi tanks were destroyed this way, unsuspectingly and at a distance. Such a feat would have been far more difficult in covered, nondesert terrain.

The U.S. forces also had a big advantage on the desert because of the widespread distribution of the handheld global positioning system (GPS) satellite navigation device. While navigating the treacherous terrain of the near featureless desert is very difficult for soldiers equipped only with maps and compasses, the GPS provides precise location information (to within eighty-two feet or less). The United States did not have all of its planned twenty-four navigation satellites in orbit by the time of Desert Storm, but it had enough to provide almost total coverage. Not only did the GPS allow units and individuals to navigate the desert and know where they were, but it was a big advantage in calling in artillery and air strikes, in reconnaissance, in marking the location of minefields, and coordinating the movement of ground units. One postmortem on the war characterized the GPS as "the one unsung techno-hero of the Gulf War." As it happens, the United States had some 4,500 GPS receivers in the war zone by February 1991, with 5,500 more on order. The U.S. military plans to acquire up to 30,000 in the near future, some of which also will be used to replace terrain countour mapping (TERCOM) in guiding cruise missiles.[15]

According to one source, the following are the key considerations involved in outlining the operational and tactical problems of desert warfare: lack of water, general lack of resources, harsh conditions for troops and equipment, few civilians, few built-up areas, few natural obstacles, and lack of cover and concealment. Following from that, success in desert warfare is said to depend on the following time-tested techniques: control the water, "get there first with the most" (that is, stress mobility and agility), information superiority, and experience in desert warfare.[16]

Regarding water (and food), for example, the Iraqis had an enormous problem. There was none of either to be found out in the desert, so it had to be transported from Iraq proper. As noted by one source, "when you are out in the desert, what you don't have with you, you don't have,"[17] a situation that existed in stark contrast, for example, with the experience in Vietnam, where the Vietcong could easily provision itself with food and water off the land. But the total coalition air dominance, the small number of available roads, the ease of aerial interdiction of key bridges and railroads in Iraq, and the fact that transport vehicles could be picked up by JSTARS and other sensors and then destroyed all rendered supply of water and food very hazardous.

The fact there were few civilians out in the desert meant that the Iraqis could not rely on coalition sensibilities about "collateral damage" (civilian casualties), as was the case for bombing in and around Baghdad. The absence of built-up areas eliminated one of the major causes of delays and casualties, though one main reason that the coalition did not move on to invest Baghdad was the fear of having to absorb heavy casualties in an urban environment. In addition, for the most part in Kuwait and southern Iraq there were few if any natural obstacles, except the unimposing Al Mut'la ridge north of Kuwait City, which might have been used by an army stronger than Kuwait's to slow down the Iraqi advance at the war's outset, and which the Iraqis were later unable to exploit to stop the advance of the U.S. Marines.[18]

Technology, Training, Strategy, and Tactics on the Desert

It might have been assumed at first that the harsh desert conditions (sand, sandstorms, poisonous insects, big fluctuations in temperature between day and night) might have seemed to favor the Iraqis, who supposedly were on familiar ground. But that was not really the case. As stated by one source, "the Iraqi army had never maneuvered much in the desert, and this contributed to a fatal ignorance of the key problems of desert warfare."[19] The Iran-Iraq war had been fought in marshes, salt plains, and mountains, but not at all in the desert. Few Iraqis live or have spent time on the desert. By contrast, the same source averred that "American forces had more experience with desert fighting than just about any other army in the world except perhaps Israel."[20] Indeed, many of the U.S. troops had received extensive training in the desert at the National Training Center at Fort Irwin, California (Army) and at the Marines' desert training center at 29 Palms in the Mojave Desert. Both army and marine forces and associated air units had conducted extensive, large-scale, live-fire armored maneuvers at these centers, under conditions very similar to those in Kuwait and southern Iraq, with temperatures up to 110°F.[21] Ironically the Iraqi army had had no such equivalent experience.[22]

There are varying kinds of desert terrain, ranging from soft sand dunes to hard scrabble lava-type surfaces, and these can be very different in terms of the mobility of vehicles (tracked or wheeled) and soldiers on foot.[23] In some parts of the greater Middle East, soft and shifting sand dunes virtually preclude rapid movement of large-scale armored forces. This is the case, for instance, in much of the northern Sinai and the Egyptian and Libyan deserts, most of the Sahara, and much of the desert in Saudi Arabia. Apparently the U.S. military had concluded just after the Iraqi takeover of Kuwait that further armored movements along the coast toward the oil fields would be easy and hence a threat to the Saudis, but that a thrust inland toward Riyadh would have encountered nearly insurmountable soft sand dunes.[24]

The U.S. Army worked hard at ascertaining the trafficability of the Iraqi desert terrain west of Wadi Al-Batin, where the "left hook" or "Hail Mary" was to be performed. Specifically that involved the area out to where the Iraqis had built a road from As-Salman to the Saudi border and a possible area of operations from the vicinity of As-Samawah to the east along Highway 8. Hence according to the Pentagon report:

> Throughout, trafficability issues played a role in planning. There was concern as to whether wheeled vehicles could negotiate the terrain north of the Saudi-Iraqi border. A secondary concern was cross-country mobility for large trucks west of the Kuwait-Iraq border. A trafficability test was conducted by XVIII Airborne Corps in the area east of Wadi Al-Batin and south of the Kuwait-Saudi border. The terrain in this location most closely resembled that west of the Wadi Al-Batin and north of the intended line of departure. Tracked and wheeled vehicles were driven cross-country to confirm the terrain could accommodate them.[25]

Elsewhere it was reported that special operations forces (SOF) were used to analyze soil conditions to determine whether it would permit passage of heavy armored vehicles. Low-light cameras and probing equipment were used near the border to determine trafficability, and soil core samples were taken behind the lines.[26] In particular, there was a focus on the problem of off-road mobility for heavy wheeled trucks, that is, heavy-equipment transporters (HETs) and ammunition and fuel carriers.

The planning for the operation determined that the terrain in the area designated for the "left hook" invasion was suitable for a rapid, mobile armored advance, illustrated in map 28. There was no problem for the tanks, and some armored units, featuring the M1A1 tanks, were able to advance 124 to 230 miles in 100 hours. The Third Armored Division was able to move 300 tanks at night across 124 miles of desert without any breakdowns, after the tanks had been moved into position by HETs. Elsewhere 60,000 marines were moved considerable distances along a single dirt road. The Bradley AFVs (armored fighting vehicles) and the carriers for the multiple launch rocket systems did very well in the desert.[27]

These armored movements were not without problems. Some of the U.S. Army's new cross-country vehicles performed well, as, for instance, the heavy expanded mobility tactical truck (HEMTT), which was able to keep going even after the rain had turned some desert areas into quagmires. On the other hand, there were some problems with off-road mobility for some logistics vehicles, including the HETs, tracked ammunition carriers, and the line haul tractor. Under some circumstances the armored maneuver forces could have outrun their fuel and other resources.[28]

The weather during the Gulf War actually turned out to be somewhat of a surprise, as it was considerably more inclement than the norm. According to the Pentagon report, coalition planners had assumed the standard 13 percent cloud cover typical for the region at that time of year, but in fact the cloud cover persisted 39 percent of the time, which was the worst in fourteen years.[29] The weather caused cancellation or diversion of many planned aerial sorties and forced others to operate at lower altitudes and to use attack profiles that made them more vulnerable to Iraqi air defenses, a problem compounded by the haze created by the Iraqis' deliberate torching of hundreds of Kuwaiti oil wells. The early morning of G-Day, the day that the ground operations began, was marked by very adverse weather, with blowing sand and rain, making visibility very poor. Earlier, at the beginning of the air war, the sky had been overcast at 3,000 feet with visibility at three miles with fog.[30] As we noted this may have been an advantage for the technologically superior coalition forces, particularly with regard to tank gunnery exchanges, but one area in which the coalition was disadvantaged by poor weather was in the Scud-hunting operations, as Iraq was able to take advantage of the variations in cloud cover to use its mobile Scud launchers against Israel and Saudi Arabia.

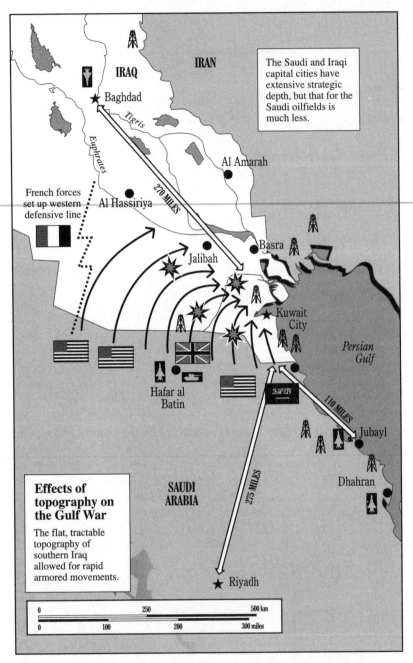

The Saudi and Iraqi capital cities have extensive strategic depth, but that for the Saudi oilfields is much less.

IRAQ

IRAN

Baghdad

Tigris

Euphrates

Al Amarah

French forces set up western defensive line

Al Hassiriya

270 MILES

Basra

Jalibah

Kuwait City

Persian Gulf

Hafar al Batin

110 MILES

Jubayl

275 MILES

Dhahran

Effects of topography on the Gulf War

The flat, tractable topography of southern Iraq allowed for rapid armored movements.

SAUDI ARABIA

Riyadh

0   250   500 km

0   100   200   300 miles

Map 28

The Seasons and Desert Warfare

Of course the strategy and tactics of the Gulf War were heavily affected by seasonal variations in the weather. Indeed, the United States and its allies were loathe to attempt a major offensive operation in the extreme summer heat, and ended up initiating operations in midwinter with the expectation that the war would be over before the most extreme summer conditions would come around again.

At the outbreak of the crisis it was believed that U.S. forces would be badly handicapped by summer conditions in Saudi Arabia. It would be too hot for infantry combat, all the more so if gas masks had to be worn. It was also claimed that the extreme heat would disadvantage a high-tech military force by affecting the performance of aircraft and tanks. In addition, in the summer the logistics of water supply for the troops would be unusually difficult.

Regarding the difficulties that would have obtained if the war had been fought during the summer season, the following appeared in the *Washington Post Weekly* (but the problem of sand was for all seasons):

Metal on military vehicles becomes so oven-hot that troops are forced to wear thick gloves to protect their hands. Helmets turn into Dutch ovens. On the day the 82nd Airborne Division was ordered into the desert, one member of the outfit barbered 1,000 of his fellows to the quick in the hope that bare scalps would fare better in the hot helmets than full heads of hair.

The sun pounds the space-age skin of the Air Force F-15 fighter jets, expanding the metal and creating fuel leaks, according to crew chiefs who service the aircraft.

The heat also forced the Marine Corps to change all the oil in all its tanks to a different grade. Marine officials say they were surprised to discover their preposi-tioned ships—which are stocked with tanks, other weapons, and supplies at far-flung bases in readiness for conflict under various conditions—came equipped with oil unsuited for desert operations.

For some equipment, the abrasive effect of the desert sand and salt is even more damaging than the heat. The velvety sand is creating serious maintenance problems for virtually every U.S. military unit here. The grit seeps, for instance, into the belts that feed the bullets in M-60 machine-guns. "It causes it to jam," says Pfc. Tim Barnett. "It doesn't work. . . . Sand gets into all the moveable parts."

Most combat troops here say that if they're not on the job or eating or sleeping, they're cleaning weapons. One Marine recently sat outside his camouflaged desert bunker cleaning the barrel of his machine gun with a toothbrush. Barnett says the troops clean the weapons daily, usually after dark when the winds have calmed.

Because of the omnipresent sand, the Army flushes the engines in its Apache attack helicopters daily and cleans the air filters on its tank and truck engines far more frequently than when the equipment was used back home in Texas and Georgia.[31]

It is clear that if such desert conditions cause those kinds of problems for a first-rate superpower military force, they would cause far greater problems for a force from a developing country that would presumably fare much worse when it came to discipline and equipment maintenance. Indeed, that had long been a great advantage for the Israelis in wars with the Arabs.

Also not subject to seasonal variations was the oft-reported disorientation of troops because of the vast empty desert theater, which had prompted hurried acquisition of the handheld GPS devices. Also the infrared scanners used by aircraft at night probably did not achieve their full effectiveness, since they were subject to interference from temperature variations.[32]

Generally speaking, however, the Gulf War did highlight the relationships among planning, training, and the vagaries of possible combat in unexpected places. Numerous analyses pointed out that by 1990, the elite units of the U.S. Army and Marine Corps were mostly prepared for low-intensity combat in jungles or temperate forest climes, if not also for high-intensity conventional combat in Europe. Even those troops trained in California's Mojave Desert were said to be wholly unaccustomed, psychologically speaking, to the vastness of the empty spaces of the Saudi desert.[33]

Hence, at the end of 1990, the U.S. press was full of speculations about when would be the most propitious time for an all-out offensive against Iraq's occupation of Kuwait. Most military analysts pointed to January because temperatures then would be most reasonable and there would be some three months for operations under favorable conditions before the overwhelming heat returned. Other analyses also called attention to the variations in moonlight as critical to military operations—infrared notwithstanding, a full moon would provide some additional visibility for nighttime operations. As it turned out, of course, the coalition did not have to worry about long-term climatic conditions, and, ironically, the ground war began under unusually rainy and muddy conditions.[34]

Strategic Depth and Frontal Width

Some of the other major issues of military geography analyzed with application to recent third world wars can also be discussed in connection with the Gulf War. Of particular interest is the question of strategic depth, which in this case is actually a bit complex. On the one hand, Iraq was faced with only a sixty-mile distance from the Saudi border to Kuwait City, seemingly the coalition's main war objective, at least as applied to its declaratory posture. Baghdad's some 280-mile distance from the western end of the coalition's deployments appeared to provide considerable depth, but as it turned out that expanse was left largely undefended as Iraq concentrated its forces to defend Kuwait and Basra. On the reverse side one could point out that important Saudi oil fields and loading points were only some 140–150 miles from the Kuwaiti border, that is, vulnerable to an Iraqi drive had there been one immediately after Kuwait was overrun in August 1990. Postmortems after the war also noted that if Iraq had moved south of Kuwait in August, it would have been much more difficult for coalition forces to move and assemble their (ultimately) massive, decisive military forces. Or would it have? The lessons of the earlier Iran-Iraq war and even those of the Iraqi failure in the brief attack on Al-Khafji might be read to support the thesis that Iraq could not easily sustain and support a long-range offensive even against weak opposition, particularly without control of the air. In addition, the distances involved were formidable. These factors are illustrated in map 28.

One could argue about whether Baghdad's distance from the central Kuwaiti area of the war played a role in the U.S. or coalition decision not to pursue a total victory. Apparently as the bulk of Iraq's army and its key Republican Guard formations were being destroyed near Basra and along the lower Euphrates (the extent of the destruction of the Republican Guard units was apparently vastly overstated by General Norman Schwarzkopf at the time the hostilities were concluded), Iraq had meager and weakened forces and no discernible line of defense between the advanced positions of U.S. and French screening forces near As-Salman and Baghdad itself. It appears that coalition armored forces, assuming adequate fuel and other logistics, could have reached Baghdad within about forty-eight hours. The primary *military* reason for not doing so concerned the fear of extensive coalition casualties in the event of urban street warfare in Baghdad, or so it was publicly stated. But there were also overriding political reasons, devoid of geographical content, that is, fear of the consequences of political dismemberment of Iraq and the unwillingness of the United States to govern Iraq, as well as British and Saudi pressures about not "piling on" a defeated Iraq. It is strange that an interesting and crucial geographic problem in relation to these issues is rarely discussed. One might say that destruction of Iraq's weapons of mass destruction facilities *should* have been the primary coalition war aim. Most of those facilities were outside Baghdad proper, many in its suburbs and near periphery. They could have been captured without risking urban warfare in Baghdad (one of the main justifications for halting the war because of the risk of high casualties)—an interesting matter at the juncture of geography and decisional cognition. In retrospect, the coalition strategy should probably have been one of putting a ring around Baghdad without entering it, and then capturing the nuclear, chemical, and biological facilities around it.

In one other sense Iraq was favored by its strategic depth. During October 1990, according to the Pentagon study, a plan was developed "to conduct an attack with ground forces against Scud fixed launcher complexes at H2 and H3 airfields in the extreme western part of Iraq" (these are designations of pumping stations along the now-defunct Iraqi pipeline that terminated at Haifa). This plan was later rejected "because of the extended LOC [lines of commmunication] required to support the operation and the risk and the demands of planned corps operations."[35]

The Gulf War represents an intermediate case in regard to the criterion of frontal width. It first appeared that coalition forces would attempt to assault Kuwait across a relatively narrow 150-mile front comprising the Saudi-Kuwait border. But the coalition widened the front by moving a concentration of forces westward, so as to enable an attack along a 300-mile stretch from the Persian Gulf westward almost to the Medina road, which runs north to An Najaf and Baghdad. As the western half of the front was virtually undefended, it allowed for a classic flanking and encirclement of the Iraqi army, in the manner of Cannae.

As previously noted human settlements (or their absence) and road networks (or their paucity) were critical features of the Gulf War. The sparse or nonexistent civilian populations in the main battle areas (outside of Kuwait City) removed, for the coalition, the problem of collateral damage as a factor in the conduct of the ground war and related aerial support operations. In addition, the fact that Iraq had to rely so heavily on one main highway from Baghdad to Basra for its logistics multiplied the effects of the coalition advantage in airpower. The otherwise open terrain did not allow for an equivalent to the Ho Chi Minh trail, and, as we noted, the lack of roads did not significantly slow down coalition armored forces because the specific nature of the desert terrain allowed for off-road mobility. Meanwhile there were no mountains or rivers or other major terrain barriers to serve as a defensive glacis or strong point for the Iraqi defense, as the Zagros Mountains, for instance, had served in the Iran-Iraq war.

Iraq did face, as a function of its location, the possible threat of a two-front war. Turkey was a participant in the opposing coalition and it allowed coalition aircraft to fly missions against Iraq from its Incirlik base. Hypothetically, then, in an expanded war, the coalition might have opened a second front in the Kurdish areas of northern Iraq.

The Strategic Level: Logistics and In-Theater Basing Access

Moving to the strategic level of analysis, one could outline several main points regarding the Gulf War, some of which are altogether obvious and have been covered elsewhere in this work. Among them: the importance of Saudi and other Gulf Cooperation Council (GCC) nations' ports and airfields to the coalition's logistical and combat activities; the importance of relatively nearby European air bases for the bombing effort against Iraq as well as for logistics; the whole question of the partial annihilation of the distance factor by modern technology; and the fact that there was no blockage of access for the coalition through several key straits and choke points leading into the region.

Paramount among the advantages held by the coalition—and one not assured in future conflicts—was the availability of modern, state-of-the-art airfields and port facilities in Saudi Arabia and other GCC states, which infrastructure was built mostly by the United States and paid for by the regional nations' oil wealth. This allowed for an elaborate complex of air bases from which combat missions could be flown against Iraq, and for sea and aerial ports of debarkation in connection with the logistical effort, that is, the movement of personnel and materiel. Some eleven major air bases in the region were available and ready for combat missions.[36]

The U.S. logistics effort also benefited from (and could not have been conducted without) access to state-of-the-art air bases in Saudi Arabia and neighboring Gulf states. The two initial primary aerial ports of entry were at Dhahran and Riyadh, and later King Fahd, King Khalid, and others were used. Also important were the facilities provided for this purpose by Oman at the preposition bases of Thumrait and Masirah. All of these bases (the one at Masirah had long been used by U.S. P-3 Orion ocean surveillance aircraft) have long, modern runways, modern navigation aids, communications capabilities, and, according to the Pentagon, "more ramp space than most commercial airports." Early in the logistics operation KC-135 tankers operated out of Abu Dhabi, and they operated throughout at Masirah and Seeb in Oman, at Cairo West in Egypt, Incirlik in Turkey, and Riyadh and Jidda in Saudi Arabia.[37]

Similarly, the U.S. logistics effort was eased by modern seaport facilities. There were seven major ports capable of handling large quantities of materiel. Four were on the Persian Gulf coast, the other three on the Red Sea coast. According to the Pentagon, the two principal ports, at Al Jubayl and Al Dammam, "had heavy lift equipment, warehouses, outdoor hardstand storage and staging areas, and good road networks around the port facilities."[38] There were still other in-theater port facilities used to move prepositioned stocks and provide storage, in Bahrain, Oman, and the United Arab Emirates. These aerial and naval facilities, in addition to the vast network of available en route basing access in Europe, Asia, and North Africa, provided the United States with an almost ideal situation for fighting in the Persian Gulf area and supplying a large force there. Perhaps most important of all, these facilities were never seriously threatened by Iraq. The coalition had secure sea lines of commmunication and secure ports and forward air bases, and it fought in a "permissive environment" in which the conditions were almost ideal.

Finally, the U.S. logistics effort was greatly eased by access to local petroleum resources, which meant that the coalition forces did not have to bring their own fuel. The U.S. Navy did, however, lose access to Kuwait, which was an important source of aviation fuel, when it was overrun by Iraqi forces.

Air Operations and Geography in the Gulf War

Almost all of the postmortems on the Gulf War have concluded, not surprisingly, that it was won largely by airpower, which was enabled by a combination of advanced technology and terrain and weather that were deemed almost quintessentially suited to its use. A reading of the massive, multivolume *Gulf War Air Power Survey* (*GWAPS*) does not offer a general contradiction to those conclusions (since countered somewhat by a General Accounting Office [GAO] report that calls the role of precision-guided munitions [PGMs] overrated). But it does provide a wealth of additional material in relation to the geographic aspects of the air domain of that war.[39]

For instance, *GWAPS* does provide somewhat of a picture in regard to which of the U.S. Air Force's weapons were or were not usable in the face of inclement weather or nighttime. Among the air-to-ground missiles, for instance, the AGM-62B Walleye, an electro-optical weapon, is usable only in daytime and clear weather. In addition, there was the need for relatively clear visibility to deliver laser-guided bombs, day or night; weather was thus a constraining factor and had an adverse effect on F-117 stealth aircraft operations in particular. Planners learned to schedule F-117 sorties around the poor ceilings and visibility associated with frontal weather passages and to attack targets suitable for radar deliveries with F-111s, F-15Es, A-6s, and B-52s.[40]

Overall the United States did demonstrate an increasing ability in Desert Storm to bomb accurately at night and in adverse weather. That had come from two improvements in technology dating from the mid-1960s. One was ground-mapping and terrain-avoidance radars that made low-altitude penetration of radar-controlled, ground-based enemy defenses feasible. In Desert Storm A-6s and F-111s were able to penetrate at night, at altitudes generally below 1,000 feet so that they were masked by terrain enough of the time to defeat Iraqi radar. Secondly, there were night-viewing devices capable of discerning point targets, buildings, and vehicles that were used initially on side-firing gunships, such as the AC-130 first tested in combat in 1968. Then, too, there was the development of the "ability to navigate accurately and locate targets at night and through clouds, smoke, and haze with sufficient precision-delivered ordinance." The pivotal development was the joining of night-viewing devices such as forward-looking infrared sensors (FLIR) with designators for laser-guided bombs. This combination allowed accuracy in night bombing with genuine precision from higher altitudes, if the target could be observed on FLIR and the laser designator brought to bear. However, in inclement weather it was a difficult maneuver, and numerous sorties had to be canceled and the effect of many intended strikes was frustrated.[41]

Because of the dangers posed by Iraqi antiaircraft artillery, the U.S. Air Force decided early in the war to limit its vulnerability by conducting bombing raids from medium to high altitudes, that is, after knocking out the surface-to-air missiles (SAMs) that are most effective above medium altitudes. However, A-10 pilots using medium-altitude release tactics had difficulty hitting armor with MK-20 Rockeye bombs, and could have gotten better results from steeper dive angles, but bad weather frequently forced them to use shallower dive angles. It was reported that because of the previous emphasis on low-altitude training, and lack of medium-altitude releases, few pilots were exposed to some of the associated problems, such as extremely high crosswinds and high-G releases.[42]

Despite the extensive training conducted for years beforehand on deserts in California and Nevada, U.S. pilots found the desert conditions in Southwest Asia unfamiliar, with different weather and visibility considerations. The effects of heat made it necessary to change the way airplanes were flown, and some sensors could not be turned on while the planes were on the ground.[43]

Even some of the effects of bombing proved to be unique in the desert arena. It was learned that area bombing by B-52s and other aircraft that was intended to hit troop concentrations and to disorient and demoralize Iraqi troops did not have to be all that accurate because vibrations and sound travel great distances in the desert and that "horrified" men would quiver in fear as far-away units were hit.[44] Though never commented upon, this has undoubtedly been an additional benefit to Israel's air control in several wars, though more limited because of its absence of capability for such large-scale area attacks.

Despite the relatively open terrain in the relevant areas of southern and western Iraq, hunting down the Iraqi Scud missiles proved difficult. Iraq was able to use hiding places such as culverts, overpasses, and bridges. But *GWAPS* further speculates that one reason that more Scuds were fired at Saudi Arabia than at Israel (the latter presumed to be a preferred target) was that in the Tigris and Euphrates valleys, Iraq could use an area in which there were both villages and vegetation, which was easier to hide in than the open deserts of the western part of the country.[45]

The geographic lessons of the air segment of Desert Storm have largely to do with the location and distance factors associated with the U.S. and coalition air bases in Saudi Arabia in relation to targets in Iraq and Kuwait. As pointed out in *GWAPS,* Saudi Arabia is an enormous country, one that, if its map were superimposed on that of the United States, would run east to west from eastern North Carolina to South Dakota, and north to south from Minnesota to South Carolina. In addition, Saudi Arabia had constructed its network of air bases to confront diverse threats from Israel, Iran, and the Soviet Union as well as Iraq, so that the coalition air forces deployed to those bases found themselves at great distances from Iraq. To reach Iraq F-117s faced a journey of more than

665 nautical miles (nm); F-111s at Taif had a 525-nm trip to the Iraqi border; fighters at Al Dhafra and Al Minhad had flights of 528 nm, and even F-15s at Dhahran, Al Kharj, and Tabuk and the F-4s at Shaikh Isa all had trips of some 250 nm to reach Iraqi air space. For naval aircraft based in the Red Sea, there was a 600-mile trip to Iraq and one of 800 miles or more from the mouth of the Persian Gulf or outside it, a problem that was rectified when the Iraqi navy was neutralized to the point that U.S. carriers could move up into the Gulf, and thus relieve some of the need for tanker support. Thus coalition aircraft required extensive midair refueling to execute their missions, and thus was made possible by the formidable and unmatched tanker force fielded by the U.S. Air Force.[46]

GWAPS presents a fascinating wealth of detail on how a series of "tanker tracks" running across northern and central Saudi Arabia was established, which allowed for refueling at drop-off points just short of Iraqi territory. This required careful coordination as the tankers and attack aircraft often came out of different airfields. It also required coordinated refueling of complex strike "packages" typically combining F-16 fighter bombers, F-15 air superiority fighters, F-4G Wild Weasels for suppression of enemy air defenses, EF-111s for electronic warfare, and navy and marine EA-6Bs and F/A-18s, so that they could move across the frontier as a coordinated force that could jam radars, knock out SAM sites and airfields, fly combat air patrol against enemy aircraft, and attack targets.[47]

Not only aircraft sorties needed to be coordinated. *GWAPS* reports that coordination of Tomahawk cruise missile attacks with F-117 strikes against Baghdad was complicated by wind factors that affected the time of arrival of the missiles. In addition, it was noted that Iraq's huge area made the task of knocking out SAM networks much more difficult than it had been for the Israeli air force operating against a far smaller area in the Bekaa Valley.[48]

Above all bad weather adversely affected the U.S. strategic campaign against Iraq. Not only the F-117s but also other aircraft using laser-guided and infrared weapons were adversely affected to the extent that they averaged much lower than anticipated target hit rates. This is important to note in relation to possible future such campaigns—for instance, by the United States in Iran, by Israel in Syria, or by India in Pakistan—if they were attempted in rainy seasons or in periods of unexpected adverse weather such as the U.S. Air Force faced in early 1991. None of these nations apart from the United States has the elaborate and large-scale capability for strike packages such as those used in the Gulf War, and none has any remote equivalent to the U.S. tanker refueling capabilities. But then Israel is much closer to (a much smaller) Syria, and Indian bases are much closer to an (equivalent sized) Pakistan, in comparison to the distances involved in U.S. strike missions against Iraq.

The Revolution in Military Affairs and Its Impact on Geography

The military journals in recent years have evidenced an enormous interest in what for a while was commonly referred to as the MTR but what has now come to be called the RMA. Indeed, it is now widely assumed that the U.S. technological tour de force in Desert Storm represented the first large-scale illustration of such a revolution, though some analysts point to some of Israel's technological feats in the Syrian phase of the 1982 war in Lebanon as a mild precursor, along with some aspects of the Falklands War.[49] With regard to Desert Storm one recent article on the RMA referred generally to the technologies that were keys to its success: space systems, telecommunications systems, computer architectures, global information distribution networks, and navigation systems.[50]

As it happens the evolving MTR-RMA literature has acquired the character of a somewhat arcane set of arguments among specialists and devotees, many of whom are connected in one way or another with the Pentagon's Office of Net Assessment. Many of those arguments (couched in historical terms as well as with reference to the 1990s) have to do with what exactly characterizes such a revolution and whether, historically, there have been discrete and easily identifiable revolutions or, rather, slower evolutions cumulating to revolutions over a longer time span.

These arguments have, actually, been fought out over the characterization of the Gulf War experience. Many analysts have cited that war as evidence that a revolution had taken place, particularly because of the crucial nature of American airpower. Others, however, have taken a more evolutionary approach, both in pointing to the prefatory example of Israel's use of MTR technologies in 1982, but also in a more cautious and conservative interpretation of the Gulf War. Eliot Cohen, for instance, writing in *Foreign Affairs,* avers that "the cautious military historian (and even more cautious soldier) looks askance at prophets of radical change," and that "on closer inspection the apparently rapid rate of change in modern warfare may prove deceptive."[51] He and others have further noted that despite the attention lavished on "smart" bombs during the Gulf War, most of the bombs dropped during that war were 1960s–1970s vintage "iron" or "dumb" bombs (but on a unit basis the "smart" bombs were more effective).

As it happens a recent GAO report by the goes beyond that and concludes that the "smart" weapons used in the Gulf War were greatly overrated, and that on a cost-effective basis they did not actually outperform the "dumb" ones. Contrary to the aspirations of the U.S. military, the GAO questioned the wisdom of depending "increasingly on weapons that extend the state of the art of war at a cost of tens of billions of dollars."[52] The GAO report questioned whether the F-117 stealth fighters had actually outperformed older, cheaper aircraft on a cost basis. In addition, it was asserted that U.S. aircraft sensors—laser, electro-optical, and infrared systems—could not see clearly through clouds, rain, fog, smoke, or high humidity, creating doubts about whether the MTR weaponry really was transcending geography and climate, as so often claimed.

As noted by Admiral William Owens, there have been several general areas in which those who have heralded a real revolution have been subject to criticism. He puts them under the following categories: "opponents fight back," "relying on technology is an Achilles heel," "it applies only to the last war," "it ignores the fog and friction of war," and "if it ain't broke, don't fix it."[53] Sequentially the first two of these criticisms address the possibilities for asymmetric strategies (see later) and information warfare. The third, referring to a tendency to infer too much from the results of the Gulf War, is centered on geography and on the question of whether the unique geographical conditions of the Gulf War led to a general exaggeration of the revolutionary nature of MTR warfare.

Critical to the MTR-RMA literature is the proposition, argued about only in degree, that such revolutions are not defined solely by weapons developments, but by related or superimposed changes in organizational adaptation, training, integration of systems, and doctrinal innovations. The example of Nazi Germany's development of the air and armor *blitzkrieg* tactics used against France in 1940 is often cited (France having had at the time equivalent numbers of tanks and planes but insufficiently developed doctrinal and organizational architecture to go with it). In the context of the evolving MTR-RMA literature, there is frequent reference to the notion of "system of systems" (popularized by Admiral Owens) to make this point. FitzSimonds and Van Tol have stressed the synergistic effect of technological development, doctrinal (or operational) innovation, and organizational adaptation, and claim that "it is the increasing recognition of the importance of the doctrinal and organizational elements that has led to the term revolution in military affairs gaining currency over expressions such as military-technical revolution, which implied that technology was the predominant factor."[54] In the development of these themes, the distinction between MTR and RMA (the latter superseded the former in the literature) became blurred—but not entirely.

Some analysts strove for historical perspective in relating the new RMA to previous such revolutions and watersheds. Some, wedded to longer cycles, have seen the MTR as representing a break with a long period going back to Napoleon, at which point the era of massed armies and massive firepower was considered to heve been inaugurated, corresponding to the early phases of the industrial revolution.[55] Others pointed to more recent RMAs, using shorter cycles. Hence the German development of *blitzkrieg* warfare with the integration of tactical airpower and armor and the then contemporaneous U.S. and Japanese developments of carrier battle groups and amphibious warfare were cited as evidence of an RMA in the 1930s, while the development of nuclear weapons and of long-range aircraft and missiles was cited as evidence of another revolutionary break in the 1950s.[56] Almost all current commentators assume that the RMA inaugurated with Desert Storm is still in its initial phases and will be carried much further in parallel with the ongoing information revolution in the civilian economies of the leading nations.

What is involved in the MTR-RMA, and its basic terminology and imagery, has been elaborated upon in a large number of recent publications, perhaps most notably in a monograph, *The Military Technical Revolution,* published by a group at the Center for Strategic and International Studies, and the widely acclaimed book by the technorevolution gurus, Alvin and Heidi Toffler, *War and Anti-War.*[57] Indeed, the first public heralding of the MTR may have come from Soviet general Nikolai V. Ogarkov, in writing well before the collapse of the USSR about reconnaissance-strike complexes (elsewhere, later referred to as sensor-shooter systems) in recognition that the Soviets would have a difficult time matching U.S. technological capabilities in the new age of precision guided weapons and informatics. Soviet writers also argued that this period would be followed by one dominated by even more advanced technologies, for example, directed-energy weapons, robotics, and special materials. More recently, Secretary of Defense William Perry, in discussing the main lessons of the Gulf War, pointed to the crucial importance of the interrelated trilogy of intelligence sensors, defense suppression systems, and precision guidance subsystems for "smart" (now increasingly "brilliant") weapons, which developments were foreshadowed in the 1970s by advances in microelectronics and computers and by subsequent development of associated tactics.[58] Admiral Owens, popularizer of the now widely used term "system of systems" delineates (similarly) three general categories of requirements: intelligence, command and control, and precision force. The first of these really refers more broadly to intelligence, surveillance, and reconnaissance; the second, elsewhere called "fusion," pertains to advanced C^3I, "by which we translate the awareness of what is occurring in a broad geographical area into an *understanding* of what is taking place there" and communicate it to combat forces (target identification, mission assignment, and force allocation); and precision force means precision-guided weaponry.[59] Other analysts, with a slightly different version of a trilogy, saw the importance of sensors (reconnaisance, surveillance, and target acquisition), fusion (C^3I), and PGMs. Perry saw the genesis of these developments, tactically speaking, in the search by the outnumbered Western forces in Europe for an "offset strategy," which came to fruition with the Air-Land 2000 doctrine that hinged on effective use of intelligence sensors, stealth aircraft, and PGMs geared to interdicting Soviet rear and support echelons well behind the forward line of troops (FLOT). Perry emphasized that the U.S. advantage in these areas, illustrated by the Gulf War, was not so much in components as in total systems, training, and operational experience.

In the early RMA literature of the mid-1990s, a number of other, related themes are prominent. Two writers characterizing the expected nature of twenty-first century warfare claimed that its main characteristics will be lethality and dispersion, precision of fire, the irrelevance and vulnerability of massed forces, and invisibility.[60] Other observers have written of the impending blurring of the heretofore sacrosanct categories of the strategic, operational (theater), and tactical levels of warfare, as well as the blurring of the traditional service functions corresponding to the land, sea, and air realms.[61]

Some analysts claim that as a general consideration, information dominance over adversaries will become the major focus of operational art, further broken down into such categories as: comprehensive intelligence regarding an enemy's military, political, economic, and cultural targets while denying the same to him; disruption-manipulation of enemy C^3I systems and defense of one's own; space-based information usage and denial; sensor-to-shooter data fusion; flexible information-intelligence databases; and use of simulations to support operational decisionmaking.[62]

Generally speaking writers on RMA such as Mazaar, Libicki, Bernstein, and Kendall see a future battlefield on which the importance of today's weapons platforms (aircraft, tanks, warships) will be greatly diminished. The large and the complex would give way to the small and the many, that is, literally millions of sensors, emitters, microbats, and miniprojectiles.

Futuristic Technologies and the Soldierless Battlefield

Current MTR-RMA literature provides a bewildering array of possibilities for new weapons systems and related military technology, all of which would be meshed together in the much heralded system of systems, that is, new organizations and doctrines.[63] Naval forces would feature the "arsenal ships" with 500 or so missile launchers of all sorts, but possibly also converted Trident submarines to be used as stealthy troop transports or submersible "battleships" carrying PGMs, either in lieu of or supplementary to the arsenal ships. In the air and in space Boeing 747s armed with airborne lasers would target ballistic missiles aimed at U.S. forces; thirty years from now there may be satellites firing lasers to knock down missiles. PGMs of the next generation will be well advanced over those used in Desert Storm, among other features dramatically reducing the logistical requirements for long-range power projection. Some missiles would release numerous submunitions that could "hover like a hawk" over enemy forces, searching for tanks and other targets. Even the simple infantryman at the heart of ground forces is expected to go "hi-tech," with "smart helmets" incorporating "heads-up displays" of terrain and targets and live video capabilities to be fed back into the network, new personal weapons with advanced sensors that can see hidden targets, and new clothing and gear to deal with threats of chemical and biological warfare and to preempt fire by friendly forces.[64]

Some of the key futuristic technologies involve unmanned bombers that could travel as fast as Mach 18 and perform acrobatic maneuvers at up to eighteen times the force of gravity; robotic "exoskeletons" that could enable an infantryman to walk 100 miles a day and lift huge weights; "fotofighters" that can make themselves invulnerable by firing lasers in all directions; high-power microwaves to renew the power sources of unmanned aerial vehicles; satellite bombs that could be hurled toward earth at 4,000 miles an hour; memory-enhancing drugs; and space mines.

The overall trend is toward battles and battlefields without soldiers, or certainly much less of a concentration of them. Hence "some areas of the battlefield might be devoid of soldiers but filled with remote-controlled missile launchers, optical and acoustic sensors searching for targets, and mines that pop into the air and chase moving targets."[65]

Information would be collected, variously, by dispersed ground troops, newly powerful ship radars, large numbers of unmanned aerial vehicles, reconnaissance satellites able to pinpoint objects with better than one-meter resolution and sweep areas of more than 100,000 square miles, and a variety of remote sensors dropped onto the battlefield that could detect heat, movement, and truck exhaust fumes. All this information would be integrated to form a big picture of an ongoing war, at battlefield command posts but also back in the United States, and then instantly transmitted to troops throughout the battle theater so that all the parties would be looking at the same picture. Then with only small forces needed on the battleground itself, most of the fighting would be done through long-range strikes, often far to the rear with "massed effects," by combined use of surface missiles, air strikes with missiles and smart bombs, arsenal ships, advanced artillery guided by lasers, and field troops.

Simultaneity, Dispersion, Tempo, and "Battlespace"

Various writers who deal with the RMA claim that the key to winning on the fluid and multidimensional battlefield of the twenty-first century will be *simultaneity*—the simultaneous employment of overwhelming combat power throughout the breadth and depth of the operational area to paralyze the enemy. They note further that this will be linked to surprise and disruption of the opponent's decision cycle, with the objective of making the enemy incapable of responding to a rapid succession of initiatives devised to win quickly and decisively. This capability is said to allow a small, capable force to defeat a much more numerous foe.

One point agreed upon by all of the analysts of the RMA is that the tempo of warfare will be increased, day and night. Just how that would relate to matters of military geography is not entirely clear, but it would appear that that increase in tempo would apply more in flat and uncovered terrain, less so in rugged terrain and for that with cover.

We have noted some of the principal implications of these major transformations, that is, the blurring of lines between the strategic, theater, and tactical levels of warfare and the nonlinear battlefield.[66] Further, it is predicted that there will be a reduction of close-in combat; that is, armies will attempt to destroy each other from stand-off, distant positions, a follow-on to the aspiration of air-land battle advocates for a greater capacity for operations in depth.

Then, particularly because of the possibility of symmetrical developments of RMA capabilities, the probabilities for preemptive strikes will be increased, though that may also be true even in cases of greater asymmetry as, for example, in the Gulf War.[67] Hair-trigger situations may result, with both sides in a conflict fearing the specter of a sudden and simultaneous execution of "the net," which could be crippling to the side that hesitates.

It also seems that the advent of what Alvin Bernstein refers to as "pop-up and fire ant warfare" consigns the long-sacred military principles of mass and concentration of force to history's dustbin. Those principles have been virtual dogma for generations of military analysts going back to Jomini and Clausewitz, but more recently also including numerous versions of the U.S. Army's FM 100-5 on operations and tactics.[68] Now fires, not forces, will be massed, so concentration and mass may still apply, but in a different way.

In future conflicts, dispersion, mobility, and the parsimony of small-unit forces will, it is said, be much more advantageous than the traditional massing of a favorable correlation of force at a critical point and time on the battlefield, with the latter leading not to the traditional decisive victory but rather to a decisive vulnerability in reverse, as massed forces can now easily be destroyed en masse.[69]

In some recent literature on the RMA, a newer, mostly geographical concept has emerged called "battlespace," which is defined as the "physical volume in which military forces conduct combat operations."[70] In this context it is claimed that armies, navies, and air forces formerly fought largely within their own domains and against the same service on the opponent's side. Now the traditional demarcations of battlespace among services is said to have become blurred if not completely eliminated so that, as one writer claimed, "over the next few decades, we will see a dramatically increased overlap of networked sensors and weapons and a more complete extension of each service's battlespace into that of the others." This in turn will lead to the conduct of "parallel war," that is, simultaneous attacks on tactical, operational, and strategic targets. The new RMA, in this vernacular, is said to be distinguished from earlier concepts by the "structures and processes of battlespace reach and battlespace networks," adding to the new reality of "overlapping battlespace." This overlap would replace the current reality in which each service (at least in the United States) develops its own reconnaissance-strike complexes.[71]

The relevance of this discussion for the near future of possible warfare in the greater Middle East is worth speculating upon, near future here meaning perhaps five to ten years. It is interesting that some contemporary military writings, in dwelling upon future scenarios involving the United States pitted against an aggressive Iran or revived Iraq, are hinting of existing and impending development of capabilities that would go well beyond those that were in evidence in Desert Storm. One analyst, dealing with a "canonical scenario" for a renewed Iraqi invasion of Kuwait involving (as in the October 1994 feint) some 1,700 armored vehicles, sees a force of some 200 U.S. aircraft with precision antiarmor munitions being largely able to destroy such an invasion, particularly given the limited road network availed the attacking force."[72] Critical here would be fast-deployable minefields, with wide area mines, dropped behind the armor so as to trap it in a forward position or slow it down. Later we shall note the apparently near-term advent of the U.S. "robo-ship," which seems intended almost by itself to be able to deal with large-scale aggression, and the deployment of several such ships in or before a crisis will, presumably, eventually be possible.

There appears to be somewhat of a paradox in the RMA literature as it pertains to some of the geographic aspects of land warfare. On the one hand, military geography is often downgraded or considered less important in the sense that deep strikes and the easy destruction of massed forces on the battlefield seem to render traditional barriers such as rivers, heights, and other key terrain features irrelevant. But on the other hand, in those areas where there is an absence of cover and concealment and limited road networks, the new technology seems to make it easy to destroy tank and mechanized force convoys that are channeled into easily identifiable approach routes to the battlefield.

Several other major issues have been raised in the recent RMA literature, all of which have relevance to geographic considerations: the future of nonlethal weapons and of information warfare; the applicability of the RMA to low-intensity conflict or insurgency and counterinsurgency warfare doctrines and strategies or both; whether the RMA will any time soon be diffused to other nations at various levels of socioeconomic development; whether RMA technologies inherently favor the defensive side in warfare[73]; the implications of the capability of users of such technologies to conduct simultaneous strike operations over a wide area at great distances; and the continuing validity of order-of-battle data for measuring the comparative military strengths of contending regional rivals.

Nonlethal Warfare

Until recently thought of as an arcane and a politically suspect area of inquiry, the prospect of nonlethal warfare has now acquired a certain validity in serious national security discussions and, indeed, has been the subject of focused studies within the Pentagon.[74] To some degree its genesis lay in efforts to circumvent the problem of conducting lethal "CNN wars" without causing massive collateral damage (such as the civilian casualties shown on CNN from Baghdad after at least one U.S. bombing). But this subject has also been technologically driven, as recent years have seen the development of some truly interesting new technologies. Among the primary foci in this area have been microwave generators (to disorient foes), lasers, gels that can prevent the use of roads and bridges, and, most prominently, computer viruses designed to destroy an enemy's command and control systems.[75]

The RMA and Low-Intensity Conflict

There are some strong disagreements among analysts in regard to the applicability of the RMA to low-intensity conflict. Initially some observers saw these new technologies as far less promising in dealing with terrorism or low-level urban or jungle warfare than with traditional high-density warfare involving large conventional forces. But there are demurrers to this point. The recent study by the Center for Strategic and International Studies on MTR insists that "it would be wrong, nevertheless, to write off the potential application of the MTR to irregular operations," and that "many of the generic categories of MTR technologies, along with a few of its doctrines and organizations, could, with modifications, apply to lesser-intensity conflicts."[76] Most prominently, this report cites the promise of advanced sensors capable of searching for small groups of infantry, lightweight communications gear, and nonlethal weapons.

Diffusion of RMA Technologies: "Peer Powers" and "Niche Powers"

Perhaps the most vexing and important question here has to do with whether, to what extent, and with what rapidity the RMA technologies demonstrated by the United States in the Gulf War, and those projected for the future by such observers as Libicki and Bernstein, will be diffused to other nations along a spectrum from major to second-tier to lesser powers. In other terminology these matters are discussed according to the criteria of "peer powers" (potential rivals to the United States such as Japan, Germany, and Russia), or "niche" situations in which a smaller nation might be able to advance in a narrow sector amid the panoply of technologies and tactics developed by the United States across the board.[77]

The arguments on these subjects, specifically those having to do with the prospects for diffusion, seem to run in both directions, or all along a spectrum. Some analysts, noting that many of the key RMA technologies (computers, communications equipment, radars, infrared sensors) are freely available in civilian markets (actually, they are for the most part products of these markets), see the diffusion of these technologies as probably being rapid and largely unstoppable, even if the United States is able—in cooperation with some of its allies—to revive or broaden COCOM (coordinating committee for multinational export controls) so as to direct it to these issues. A nation such as India, with an extensive in-place capability in computers and armies of computer scientists and technicians, is seen as being in a good position to move rapidly into the RMA. In short, the growing trend toward military production being rooted in dual-use technologies is seen as virtually ensuring the rapid diffusion of these technologies, both to potential peer rivals of the United States and to midlevel powers. From a theoretical perspective, Keith Krause, in a recent book on the history of arms production and trade, has pointed out that historically arms technology has inevitably and rapidly diffused away from the originators of that technology.[78]

Other observers are more skeptical and, in the process, more sanguine. Some point out that with the cold war over and with outlays for military research and development down all over the world, the dangers for the United States of peer development of RMA technologies has declined (but in the long run, China and Japan remain wild cards in this respect). Other analysts have cautioned that the acquisition of key RMA technologies is only a first step and that it remains to be seen whether nations other than the United States will have the wherewithal to develop integrated systems of the technologies, amounting to a credible or serious reconnaissance-strike complex or sensor-shooter system.

In one recent publication that discussed the possibilities of peer force competitors for the United States it was noted that, "if Russia reconstructs itself politically, especially in the form of a military dictatorship, its participation in the RMA could bring it to peer status with the United States relatively quickly."[79] This seems to assume that Russia can indeed compete in high technology, despite some indications to the contrary. The same writer states that the only other nations that can participate at least partially in the RMA are France, Germany, Britain, Israel, and Japan. China is pointedly omitted as are Korea and India, rightly or wrongly. This author points to the importance of a combination of culture, wealth, and access to technologies, particularly the flexibility of military cultures. The main technological barriers are said to be advanced data-processing systems, space-based sensors, and access to usable stealth technologies.

No RMA without Space?

Robert L. Butterworth has pointed to what he sees as the primary bottleneck for new aspirants to RMA status (this is, of course, a relative matter). He says,

Table 6-1. Independent Space Capabilities, 1992

Satellites and launchers	Satellites
United States	Canada
Russia	Sweden
Western Europe	Pakistan[a]
China	*Starting*
Japan	Argentina[a]
India	Brazil[a]
Israel	South Africa[a]
	Taiwan

Source: Robert L. Butterworth, "Economic Constraints on the Revolution in Military Affairs: There Are No Competitors without Space," Arlington, Va., Aries Analytics, Inc., April 18, 1994, prepared for the National Defense University, Institute for National Strategic Studies.

a. Launcher development interrupted by MTCR sanctions.

in effect, that "there are no competitors without space." He thus brings up the earlier question of the military role of space, which in itself is a key spatial, that is, geographic issue:

Thus one approach to estimating the rate and extent by which the revolution in military affairs might diffuse is to examine the efforts of other countries to develop space systems, which are relatively accessible to observation and inter-pretation (at least in terms of gross capabilities). Space development cannot provide a total picture of a country's plans for and progress toward acquiring broader high-technology military capabilities, but the effort to develop space capabilities of certain kinds, like the failure to do so, can provide strong indications about the type of new forces being pursued and the capability they are likely to have.[80]

Butterworth points out that only five countries have dedicated military space programs (United States, Russia, China, France, Israel), but that much can be done with a civil space architecture, noting the use made by coalition forces in the Gulf War of Landsat and Satellite Probatoire d'Observation de la Terre (SPOT) imagery. He says that "most satellite applications are intrinsically dual-use," but that "commercial communications satellites do not support ad-vanced low-probability-of-intercept and antijam technologies, for example, and navigation and remote sensing is much more precise for military applications." Table 4-1 summarizes the state of independent space capabilities as of 1992, indicating that of the countries in the greater Middle East only India, Israel, and Pakistan have taken some steps in the direction of satellite capabilities. In the case of Israel that now involves the Ofeq 3 intelligence satellite launched in April 1995 with a Shavit launcher, whose orbit apparently passes over or near Damascus, Teheran, and Baghdad every ninety minutes.[81]

Butterworth concludes that "only the United States has fielded military capabilities approximating a theorist's vision of the 'reconnaissance strike complex,'" and that "only recently has it been able to make space systems tactically meaningful to ongoing battlefield operations."[82] The latter point pertains not only to the Gulf War, but also to the near real-time counterbattery strikes against Bosnian Serb artillery, which involved an evolutionary step beyond those capabilities. He sees smaller programs, including the peer situations of Europe, China, and Japan within the constraints of their current budgets, as being able only to "provide information useful for strategic planning and order-of-battle estimates" and to deal with stationary targets (but not with the details of a moving battlefield); hence one would be speaking here of something more like an evolving modernization rather than an RMA. Furthermore, the evolving capabilities—perhaps to be made available to middle-range states—of remotely piloted vehicles (RPVs) and of JSTARS-type aircraft are claimed to be not even remotely substitutable for satellites, in large measure because they cannot handle the required volume of communications, but also because of the spatial limitations with regard to coverage.[83] Of course where the conflict is within a limited area, as was the case for the Israeli operations in the Bekaa Valley in 1982, RPVs can provide reconnaissance coverage and real-time sensor-shooter capabilities roughly equivalent to those of satellites. In addition, some systems that the United States does not entirely control will probably become widely accessible, for example, an improved version of GPS, which will eventually be available to all users.

To acquire a serious real-time, sensor-shooter capability, these countries would need U.S. satellites and associated downlinks and other technology. Nothing else will suffice. Rather (see below) they will likely be forced to resort to asymmetric responses. On the other hand, widespread access to GPS may enhance various nations' capabilities in, for instance, missile accuracy. Israel, however, is potentially somewhat of an exception. Its new Ofeq satellite can apparently provide order-of-battle-type information in an area from Pakistan to Sudan, hence dealing at least with the issue of strategic surprise, which came to the fore in 1973, without American assistance. Ofeq can also presumably provide warning of missile launches, from land or sea, offering a capability similar to that of the U.S. Defense Support Program satellites, which were designed to detect the missile plume signatures of Soviet intercontinental and submarine-launched ballistic missiles. In a war that capability would perhaps allow the Israelis to get aircraft off the ground to prevent persistent nerve gas attacks on airfields and to disperse tanks and their transporters. But in the context of a real MTR-RMA capability, Israel too lacks (for the foreseeable future) satellites and related systems that could provide real-time sensor-shooter capabilities in a fast-moving battlefield situation over a broad area such as the Sinai Desert.

In light of the points made in this discussion, the predicted wholesale diffusion of dual-use technologies notwithstanding, pop-up and fire ant warfare appears to be a long way off for nations such as Iran, Iraq, Syria, Algeria, and Pakistan.

There is a consensus in the current literature that the evolving RMA will inherently favor the defensive side, despite the fact that this may somewhat contradict the parallel assumption that RMA developments will (or may) increase the temptation for preemption in escalating regional crises. Oddly this follows upon a period in which at first, after the lessons of the 1973 war were initially absorbed, it was thought that a new cycle of warfare had been inaugurated in which the newer PGM weaponry (antitank missiles and SAMs) would be giving an advantage to the defensive side. Subsequent revisionist interpretations of the 1973 war, and the later presumed lessons of the 1982 war in Lebanon, appear to have reversed these assumptions. But the RMA has revived them, presumably in large measure because of the expected vulnerability of massed forces and of rear echelons traditionally needed to conduct sustained offensive operations.

The possibilities for large-scale simultaneous operations, involving the concentration of fires using long-range precision strikes, constitute a major assumption of today's RMA literature. Many writers see in this regard merely an elaboration on what the United States was able to achieve at the outset of the Gulf War, abetted in the future by something much closer to real-time fusion between the reconnaissance and strike components of the RMA architecture. If so, but modified by the countervailing questions about the possibilities for dispersion, concealment, mobility, and other countermeasures (such as camouflage and the large-scale use of dummies), it is this assumption that gives rise to predictions of a future of hair-trigger preemptive situations, in which he who hesitates could be lost.

Battle Data and Measuring Comparative Military Strength

Order-of-battle data of the sort presented annually in *The Military Balance* (published by the International Institute for Strategic Studies) have long been used as an approximate gauge of the military strengths of contending countries and of regional balances, such as those between Israel and Syria, India and Pakistan, and Iran and Iraq. Sophisticated analysts have always known, of course, that such "objective" data—for example, numbers of tank divisions, combat aircraft in various categories, and ships, added to the broader set of macromeasures of national power, such as population, per capita GNP, and defense expenditures—must be supplemented by attention to "subjective" factors such as morale, leadership, and diplomatic capabilities in order to produce a real, bottom-line net assessment.

However, the advent of the RMA has raised some serious questions about even the approximate or indicative validity of such measurements. Indeed, the failure by the Iraqis to translate their imposing raw numbers, as measured by troop and weapons deployments, into any kind of combat capacity against the U.S. high-tech military machine illustrated as well as anything could that the old ways of measuring military power are no longer as suitable as they once were. Or are they?

Presumably the looming huge advantage that will accrue to those countries able to master, produce, and integrate the new RMA technologies will be rooted in the "facts" of scientific and educational levels (levels of socioeconomic development). We have long been accustomed to making an approximate estimate of a country's power by using its GNP, as a product of population and per capita GNP. Is it possible that the latter alone will now be a better measure, within limits, of national power, that is, that technological capacity will transcend the population factor? Or is it possible that population will retain its importance for military power even in the face of the development of new weapons that appear to nullify the importance of mass? After all even in a pop-up and fire-ant environment, attrition will presumably remain an important factor, so that it is possible that huge numbers of forces, albeit not concentrated or massed in traditional style, may continue to be important. Presumably the order-of-battle data that measure military power largely in terms of numbers of platforms will decline in importance with the passage of time.

Asymmetric Strategies

Recently some analysts of the RMA have turned their attention to various asymmetric strategies available to countries that are not among the small group having the capability for peer competition.[84] This scrutiny involves looking carefully at what an Iran or an Iraq could do to counter its overwhelming inferiority in MTR technology so as to avoid a massive and sudden defeat in a contest with the United States or an American-led coalition. Options can include a variety of low-tech approaches (use of concealment and camouflage), terrorism, use of weapons of mass destruction against nearby states (see our subsequent discussion of triangular second-strike capability), hostages, and various forms of information warfare, among others. These considerations, in turn, raise some interesting questions about the geographical dimensions of such offset or asymmetric strategies. How much easier is cover and concealment in wooded or mountainous terrain? How much more difficult are various forms of terrorism or information warfare at transoceanic distances? To what extent can all of these strategies be pursued indirectly in relation to client states or allies of the United States in nearer proximity to the nation facing an overwhelming MTR disadvantage?

Future Implications for Geographical Aspects of
Regional Warfare

The above review of current but inherently tentative thinking about the emerging MTR-RMA raises a number of intriguing questions about how some of the key problems of military geography—at the tactical and strategic levels—will be affected. Those questions, of course, must be viewed somewhat separately with regard to different types of warfare, that is, conventional and various forms of low-intensity conflict, and intraregional wars versus various possibilities pitting external powers against regional countries in possible repeats of the Gulf War. Cognizant of those variations, it is appropriate to point out that there is a growing tendency among military analysts dealing with the RMA to rely upon the overall generalization that the geographic factor in warfare is about to be minimized if not largely eliminated. Whether that generalization is based primarily on the Gulf War experience and may therefore be questioned on the basis of too extensive a reliance on inference from one case remains to be seen.

In line with the above, the following set of generalizations appears to emerge from the current literature.

—The relationship of MTR to some of the aforementioned geographic criteria, for example, sizes of countries and of battle areas, frontal width, strategic depth, and location of key barriers such as mountains and rivers in relation to battle lines, is vital but unclear.

—Concentration of forces becomes a liability, dispersion and mobility an asset and advantage; massing of fires, not formations, becomes a rule of warfare.

—In line with the nonlinear nature of the modern battlefield, traditional topographical barriers become less vital and meaningful, and so too the forward defense of border regions. Generally, traditional notions about the importance of strategic depth are called into question.

—Closing with the enemy and proximity to the enemy become more hazardous, as the advantage shifts to long-range, indirect fires (doctrine of less proximity).

—MTR-RMA applies more effectively on deserts and plains; less so in jungles, cities, forested mountains, that is, wherever it is relatively flat and uncovered and where clearer weather prevails.

—RMA provides the technological basis for round-the-clock warfare, increases the importance of night warfare; warfare is more easily fought in all weather and all seasons.

A number of points can be made with regard to these generalizations, specifically concerning the various possible warfare scenarios in the greater Middle East region. First, the future impact of MTR-RMA on the conduct of warfare—or, rather, the joint impact of geography and MTR-RMA on the conduct of warfare—will depend greatly on what mix of combatants we are talking about. Outside the region (but potentially involving fighting within the region) that could involve the partial symmetry of major nations fighting each other, for example, the United States versus China or Russia, it being assumed that the United States would retain a large lead over such rivals when it comes to critical RMA technologies. It could, in a repeat of the Gulf War, involve the United States (most likely) or another major power (or a multinational combination) fighting against a regional power such as Iraq or Iran, with those countries perhaps having acquired some limited MTR technologies. Or it could involve intraregional wars, for example, Arabs versus Israel, India versus Pakistan, Iran versus Iraq, or Algeria versus Morocco, with whatever degree of symmetry or asymmetry there is in the use of MTR systems. Whether one is talking about such possibilities in 1997, 2000, 2005, or 2010 is obviously relevant in that countries at all levels of development will be moving ahead, albeit at different rates, toward mastery of these technologies. What of a U.S. intervention against Iran in 1997, an Israel-Syria war in 2001, an India-Pakistan war in 2005, or a series of Russian counterinsurgency operations in the Caucasus and Central Asia in 2005, or a war between Islamic Algeria and France in 2010? The answers here are murky and indeterminate, and it is not clear whether current military analyses over- or underestimate the rate of impending change.

Particularly intriguing is the question of whether geographic barriers— mountains, rivers, deserts, marshlands—will retain their traditional importance or be transcended by the RMA technologies.[85] Such barriers have long been assumed to provide an advantage to the defensive side and to act as a counterbalance to numerical advantages. Some examples: the mountain passes and sand dunes of Sinai, the Suez Canal, the marshes of the lower Tigris-Euphrates Valley, the Zagros Mountains (primarily on behalf of Iraq), the Rajasthan Desert, and the mountains of Kashmir.

Now it is said that such barriers may be less important, as MTR massed fires can destroy an enemy's forces as well as infrastructure well behind barrier areas. But then, too, massed fires can cause havoc among troops concentrated to defend barriers, just as with assault troops attempting to reach them, as happened to Iraq's defensively positioned forces in Kuwait.

This often becomes an argument that states, in effect, that control of terrain is no longer important. Hence, according to Admiral David Jeremiah, "future weapons will be able to strike enemy forces at great distances," and "in mid- or high-intensity combat, it may not always be necessary to physically occupy key terrain on the ground, vital airspace, or critical choke points at sea in order to control them."[86]

Concentration of forces is now predicted to be a liability in the age of RMA. But that may not necessarily mean that numerical advantages will have become a thing of the past, the Desert Storm example notwithstanding. If there is relative symmetry in MTR capabilities, the more numerous side may be more likely to survive and win, even at a huge cost, the matter of preemption momentarily left aside. There may, indeed, be a parallel here to earlier arguments over the use of tactical weapons in cold war Central Europe, which compared the relative efficiency of linear and balanced defenses versus the use of staggered defenses intended to channel concentrated opposing forces into killing grounds.

In the past Israel has preempted (or when it did not as in 1973, it paid for it) in crises, as a way to counterbalance its numerical disadvantage. Pakistan, in a similar situation, contemplated a preemptive doctrine but was never really able to implement it. In a future that sees MTR technological development on both the Arab and Israeli sides (and the same for Pakistan and India and Iraq and Iran), it is interesting to speculate as to whether that will enhance the level of preemptive imperative on the Israeli side. Or would advantage over Syria (and maybe also Egypt if the latter should reappear as a confrontation state) regarding RMA capabilities allow Israel to wait longer in a crisis before preempting, being perhaps less concerned about control at the outset of a war over key terrain areas such as the Golan Heights, Sinai Desert, or (perhaps) the hills overlooking the Jordan Valley? On the other hand, a Syria or Egypt faced with a relative disadvantage in RMA technologies might be tempted to preempt in a crisis in the spirit of "use 'em or lose 'em."

Depending upon the circumstances, it may indeed be true that there will be less of a premium put on defending certain forward border areas and terrain features at all costs. In the event of resumed hostilities with Egypt, Israel might even welcome the opportunity to deal with Egyptian forces massed in the Sinai. But it must also be stressed that a United States fighting against Iraq or Iran in desert border areas might more easily contemplate a war fought with long-range indirect fires just because then it need not be concerned about protecting civilian populations. With or without MTR Israel must be concerned about its people in the Hula Valley and their fate in the event of a nonlinear, dispersed war along the Israel-Syria border. Similarly, that would obtain for Pakistan in relation to its populous Punjabi area and for both Iran and Iraq in the Shatt-al-Arab estuary region. This is a problem that has not been discussed very much in connection with intraregional warfare, maybe because the RMA seems still a distant mirage for nations well removed from the prospect of pop-up and fire ant warfare.

On balance it is not clear just how the new technologies will affect the issue of strategic depth. It is tempting to say that the importance of such factors may decline in line with the previous comments about the nonlinear battlefield and the expected more dispersed and mobile nature of the fighting. If indeed the RMA gives an advantage to the defensive side, that might help such nations as Pakistan and Iraq laboring under asymmetrically disadvantageous strategic depth problems, but only if they have at least an even capability with their rivals.

It is hard to say from the vantage point of the mid-1990s what all this adds up to for a "bottom line," projected somewhat into the future. It appears that, with the possible exception of Israel, diffusion of the RMA into the region will be slow, somewhat dependent on the willingness of major power suppliers to give or sell the critical components to regional nations or (given the importance of dual-use technologies) to allow their military-industrial firms to contribute to that end. Apart from Israel, some other countries, particularly India, will move toward partial RMAs, perhaps being able to specialize in specific areas such as electronic countermeasures, limited satellite reconnaissance of real-time tactical value, and a variety of PGMs. Vis-à-vis the United States, nations such as Iran or Iraq may be forced further toward consideration of so-called asymmetric strategies, ranging from more effective use of camouflage and decoys to terrorism, to possible use of weapons of mass destruction, and (see later) triangular deterrence strategies in imitation of the Iraqi missile bombardments of Israel and Saudi Arabia. For the present, the nature of the terrain and weather in the greater Middle East should continue to amplify the U.S. advantage in contests with countries such as Iran and Iraq; similarly, that should obtain for Israel vis-à-vis its several potential rivals. For Iran and Iraq and India and Pakistan, although less so, potential intratheater warfare should not be significantly altered by very limited partial RMAs, wherein in both of these pairings, the force multipliers effectively provided by desert terrain and normally clear weather will be of lesser moment in tougher climes and terrain.

Previously we had noted the expanding interest in nonlethal weapons, now a staple of the MTR literature as well as a subject of a recent high-level Pentagon study. For the most part such weapons would appear to be beyond the reach of regional powers, and one might suspect that they would be less likely to use them in place of more lethal weapons merely to circumvent the inhibitions wrought during CNN wars by collateral damage. Presumably, the main future possibilities for use of nonlethal weapons will be by the United States or an American-led coalition against regional powers where there are such inhibitions and where abjuring the use of more lethal weapons would not alter the outcome of a war.

There is, however, one type of nonlethal weapon that, if it were transferred to or produced by some regional states, might be particularly applicable in some specific geographic contexts—antitraction technology (or "slickum"). This technology could be highly effective used on roads (maybe also against railways) in cases where road networks are sparse or there is sole or predominant reliance on one or two roads. An Iraqi army moving toward Israel over the one main road from Baghdad to Amman could, hypothetically, be drastically impeded by use of such antitraction materials. A similarly effective use could be made of such materials on the few roads traversing Sinai, particularly where off-road movement for vehicles on the desert is not very feasible. The major portion of the Ethiopia-Somalia war in 1977 was fought back and forth along one road leading through the Kara Marda Pass in the mountains leading from Somalia toward the city of Harar, and would have been highly susceptible to such an agent if it had been available. In other areas of the world, such as in the jungles of Vietnam, such materials would be less useful.

In instances of low-intensity conflict, it would not appear that looming RMA developments will have much impact on the geographic aspects of warfare, or at least not yet. Neither the United States nor any other potential RMA aspirant (such as Europe, Japan, or China) is likely to be engaged in counterinsurgency operations in the region in the foreseeable future. Russia's performance in Chechnya provides sufficient evidence that that nation, once potentially such an aspirant, has not even approached the capability for using such technologies in what basically is a counterinsurgency operation. Nations such as Iraq and Iran are very far from the use of such technologies in counterinsurgency operations against their respective Kurdish minorities.

Some of the earlier speculations about the possible efficacy of American or European air strikes against Serbian forces in Bosnia had raised some interesting questions in connection with geography. Indeed, some spokesmen for the U.S. Air Force had opined that "we only do deserts," and had waxed pessimistic about its ability to deal effectively with Serbian irregular forces in what is mostly mountainous and somewhat wooded terrain often masked by inclement weather.[87] But others, begging to differ, had pointed out that Serbian artillery and light armor is largely roadbound in Bosnia, and that such military targets would thus have been particularly vulnerable to air strikes backed by advanced satellite-borne sensors and PGMs. It is interesting to speculate, for instance, how much better the Russian air force could now do in Afghanistan or elsewhere in Central Asia if it had access to a higher level of RMA technology in the context of low-intensity warfare.

One point stands out: Of all the regions of the world, and particularly those where future combat may be anticipated, the greater Middle East region features terrain that is, relatively speaking, very suitable for the application of MTR-RMA technologies. Much of it is desert in one form or another and even where there is mountainous terrain, as in Iraq, Iran, Afghanistan, Sinai, and the Kabilya in Algeria, there is very little cover and concealment. There are, of course, some forested or heavily treed mountains in Afghanistan, Turkey, the Caucasus, and Lebanon, and some of the rugged terrain in, for example, Afghanistan, western Sahara, and Ethiopia features defiles and canyons that do provide some cover. But for the most part there is little cover and the weather tends to be fairly clear. Throughout the entire region it is perhaps only in peripheral Sri Lanka that there are jungles where the effectiveness of various sensors in locating men and vehicles under jungle canopy might come into question. Of course, in some future regional conflicts, fighting might be concentrated in cities such as Beirut, Jerusalem, Lahore, and Abadan (or future low-intensity conflicts in places similar to Grozny), still another milieu where RMA technologies might remain relatively ineffective.

Summing up the relationship of MTR-RMA to military geography, Libicki projects the following:

> On a battlefield where machines command others, foot soldiers—whose relative ranks have been dwindling for a few hundred years—may be the only humans left. Platforms already dominate low-density environments such as air, sea, plains, and deserts with their ample running room; these platforms, in turn will be supplanted by the Mesh. High-density environments such as cities, jungles, and mountains remain the preserve of the foot soldier; the Mesh will take over much more slowly in such realms.[88]

The RMA and Nuclear Weapons

One other highly conjectural set of questions is raised in the context of the geographic impact of looming RMA technologies, which has to do with their impact on the possible use of nuclear weapons. In other words, there is the question of whether RMA technologies can, in some circumstances, be used in lieu of nuclear weapons. Indeed, just that is being advocated by a growing number of military analysts in the United States, including large numbers of senior military men, who are advocates of the delegitimization of nuclear weapons in the post–cold war period. Such questions become particularly pertinent in connection with a possible U.S. intervention in the Gulf area against a future Iran or Iraq armed with nuclear weapons, where the United States, in line with a no-first-use doctrine, might wish to try to eliminate the nuclear forces of a regional foe solely with the use of conventional weapons. That might particularly be the case if such weapons were emplaced amid urban populations where preempting them with nuclear weapons would cause a lot of collateral damage. Cities are cities, but generally speaking, such operations would presumably be easier in the types of terrain that prevail throughout most of the Middle East. Just because such weapons would be more difficult to hide in that region, their emplacement within cities might be all the more likely.

SEVEN

Getting into the Middle East: Power Projection and Forward Presence

MILITARY FORCES have historically faced major challenges getting into and fighting in the greater Middle East because of its formidable geographic barriers. Even the Soviet Union, at the height of the cold war and within close proximity to key Middle East theaters, faced enormous logistics problems contemplating an invasion of Iran or intervention in the Arab-Israeli conflict to assist Egypt or Syria. For the other major external powers, Britain, France, and especially the United States, the factor of distance has to be added to the topographical barriers. Yet during the course of the centuries these constraints have been overcome. In this chapter we examine the problems of power projection and forward presence and some of the impressive new technologies that provide rich countries such as the United States with the means to overcome geographical and political obstacles. We first review the concepts of power projection and forward presence and several types of intervention by outside powers. We then look at the issue of basing access for the United States and other powers and the impact that new technologies may have on forward presence. Finally, we survey the status of the various strategic mobility forces of the outside powers and several of the regional powers and conclude with a discussion of the impediments to power projection.

Power Projection and Forward Presence

The massive logistical effort that underpinned the successful operation by the U.S.-led coalition during the Gulf War illustrated the importance of power projection into the Middle East as a core element in the security politics of the region. The tremendous requirements for moving a large land army and accompanying airpower units half-way around the globe[1] can be measured in ton miles, but raw data alone ignore the importance for the United States and its allies of en route and in-theater basing assets, the availability of overflight corridors for air transports, the necessity for forward prepositioning of war materiel, and the fortuitous proximity between the earlier massive but now reduced U.S. military presence in Western Europe and the Middle East.

As we outlined in chapter 2, the projection of power into the Middle East–Indian Ocean arena and the establishment of points of access in that area have been mainstays of global power rivalries for a very long time. Since the conclusion of the Gulf War and continuing to the present, the U.S. military has had a central focus on how to sustain power projection into the greater Middle East region at a time of declining defense budgets and the withdrawal of many U.S. forces from Western Europe.

The phrase power projection is a relatively new one, dating back only a couple of decades or so. Earlier, however, theorists such as Albert Wohlstetter dwelled upon the concept of a "power-over-distance gradient"—the effectiveness of long-distance military power projection is diminished the further it needs to be extended from its source, if for no other reason than that much of the effort has to be dissipated on logistics or on "tail" rather than "tooth"[2]; hence Britain's long-term problem regarding the projection of land power into Europe (and the European powers' reverse problem when contemplating an assault on Britain) and, on a larger scale, the American problem of extending its power reach into Europe and Asia. The American-Soviet rivalry over the Persian Gulf region and its peripheries during the 1970s and 1980s focused attention on the concept of power projection,[3] which is really a modern shorthand for what earlier had been conveyed by a Confederate cavalry general when he spoke of the need to get there "fustest with the mostest."

Alternatively, in some recent publications, the concept of power projection has been replaced by, or subsumed by, the concept of forward presence. The latter had long been used to describe overseas bases or large-scale forward or overseas troop deployments, as in the cases of U.S. deployments in, for example, Germany, Japan, and Korea. Now forward presence is routinely being used to pertain not only to ongoing, existent bases or deployments, but also to the modalities for moving large-scale forces into an overseas region, heretofore discussed under the rubric of power projection, one important element of which has always been basing access.[4]

Even though these two terms, forward presence and power projection, are often used interchangeably even if not defined synonymously, the U.S. military and particularly its central command has for the most part become accustomed to a distinction. Forward presence essentially is kept in the theater routinely, what is put into the theater in peacetime, what is there today, including bases, other infrastructure, and forward-based personnel and equipment. Power projection, on the other hand, refers to what can get to the theater if there is a crisis (munitions, bombers); that is, it has to do with the time-distance problem. The U.S. Navy, meanwhile, is inclined to define power projection more in terms of long-range interdiction or strike capability.

The concepts of power projection and forward presence subsume several important categories of overseas or long-range operations, along a spectrum from all-out, major military efforts such as Desert Storm to more minor expeditions, such as the U.S. interventions in Somalia and Haiti in 1992–94. However, several major categories can be delineated: outright intervention, arms resupply during conflict, coercive diplomacy, showing the flag or symbolic acts of "presence," force interposition and peacekeeping, and extended deterrence.

Outright Intervention

Outright military interventions requiring long-range power projection can take any number of forms, primary among them large-scale military operations featuring some mix of air, land, and sea forces. They can be a response to aggression from within the region; invasion by a major power against a regional state, perhaps in response to unwanted internal developments; or some outright military activity on the part of an outside power on behalf of one side in a conflict where the designation of "aggressor" or "victim" may be ambiguous. Such interventions may be unilateral or conducted by multilateral coalitions. They could entail involvements in regional wars at the conventional level or with the use of weapons of mass destruction (WMDs). In some circumstances one might envision preemptive attacks against nascent or fully operational WMD programs or a preemptive takeover of oil fields in response to an embargo or unacceptable price hikes. The possibilities are legion.

Therefore it is not easy to lay out a comprehensive set of scenarios that might bring the power projection capabilities of outside powers into the greater Middle East into play. At one level it is (deceptively) easy to focus on some "base case" scenarios, such as those that might later be related to the U.S. policy of dual containment vis-à-vis Iran and Iraq. But wholly unexpected crises may arise, in some cases involving escalation from relatively minor episodes to confrontations of major proportions.

Much of the current planning in the Pentagon dwells heavily, not without good reason, on the possibility of a U.S. (or U.S.-led coalition) military intervention against either Iraq or Iran, in an approximate repeat of the Gulf War. At the moment Iran seems somewhat more likely a candidate, if only because its military buildup is proceeding without the restrictions imposed upon Iraq after its defeat in 1991, which would appear to lessen the threat from the latter so long as the sanctions are in place.

Such an intervention by the United States or a coalition could take any of several forms, depending upon the origins of the crisis and on the types of weapons involved, including WMDs and associated delivery systems. It could be something very close to a repeat of 1990–91, with either Iran or Iraq invading Kuwait or Saudi Arabia, or both, or in the case of Iran, involving an attack across the Persian Gulf against one of the other Gulf states. Or an American (or an Israeli) preemptive strike against either Iranian or Iraqi WMD sites could ignite a broader conflict resulting in large-scale outside intervention. Similarly, such a scenario could evolve from an Iraqi or Iranian preemptive strike on Dimona, triggering a complex crisis involving U.S. or other forces. Less likely would be a drawing in of outside forces in the case of renewed hostilities between Iraq and Iran, perhaps involving WMDs, with the United States intervening on one side or the other. Such a crisis could be initiated by an Iranian or Iraqi invasion of neighboring states, but also by conventional missile or bombing attacks on cities or oil fields, by naval activities or mining inside or outside the Strait of Hormuz, or by attacks on U.S. naval forces in the Arabian Sea. Not to be totally discounted would be a joint Iraqi-Iranian assault on Kuwait, Saudi Arabia, and the other Gulf Cooperation Council (GCC) states, or one or some among these. These scenarios involving the United States and Iran or Iraq or both are elaborated upon in chapter 10.

There are other possible—though less likely—scenarios that might bring U.S. power projection capabilities into play in the greater Middle East. Direct intervention on behalf of Israel if it were on the verge of a conventional battlefield defeat (and if it were on the verge of resort to nuclear weapons) is possible, albeit unlikely (though the probabilities rise if Egypt or Turkey or both were engaged as full-scale combatants). Also unlikely would be a U.S. intervention on behalf of Saudi Arabia (or Jordan or Syria) against Israel—that would require a radical shift in U.S. regional security policies. Even less likely would be a U.S. action in South Asia—intervention on behalf of Pakistan versus India, or India versus China—which scenarios would presumably require massive troop deployments and the specter of large-scale attrition. Perhaps only slightly more likely would be the reintroduction of U.S. troops, albeit on a larger scale, into Lebanon in order to expel Syria or to contest the latter's occupation of or activities in that fragmented state.

If we move further away from the base cases that occupy the attention of military planners, there are other possibilities. A recrudescent Russian imperialist thrust into the Caucasus or Central Asia (perhaps foreshadowed by current Russian military support to the incumbent Tajikistan regime) could conceivably trigger a U.S. intervention, though the continuing fear of escalation up to the nuclear level would appear to rule that out. A revived Russian imperialism could again threaten the Persian Gulf. A fundamentalist Algerian regime might have to be contested in Morocco or Tunisia or Western Sahara; a fundamentalist Egyptian regime might have to be countered in Sudan or in post-Qaddafi

Libya. The possibilities presented by an aggressive revolutionary regime in Turkey would, of course, constitute a dangerous wild card.

Not all the possibilities for—presumably reactive—power projection operations into the Middle East necessarily involve the United States or a U.S.-led coalition. One or more European nations—or some military manifestation of the European Union—could conceivably conduct military operations on their own in the region, for instance, against a fundamentalist Algerian regime prone to terrorist activities within Europe. For the long run there is speculation about China or Japan acquiring long-range power projection capabilities that could reach into the Middle East, based on their need to secure sea lanes for supply of oil and for purposes of intervention against regional states or to contest U.S. influence and access in the Persian Gulf region. Oil-hungry Japan would have to undergo extensive remilitarization to reach that point, and China, which denies the aspiration for such projection capabilities, would require much more modern air and naval capabilities. In order to intervene effectively in the Middle East, both China and Japan might require en route or within-region basing access, which is why China's earlier reported interest in access to facilities in Burma and its development of a blue water navy has been viewed with such concern.[5] Some long-range hypothetical scenarios have included speculation about Japanese access in India for such purposes.

Arms Resupply

Arms resupply during conflict represents an indirect form of power projection so long as the supplier's political aims are more or less congruent with those of the recipient, that is, if the latter is pursuing some political aims on behalf of the former. But even if that is not the case the two major elements of power projection—long-range logistics and en route basing access—are also the key elements of an arms resupply operation.[6]

In various conflicts major powers have had to deal with a range of motives concerning requests for arms resupply. The reasons for positive responses include a desire by the supplier to see its client prevail, to avoid the rupturing of an arms client relationship, to drag out a war so as to cause attrition on the other side, to forestall a client from moving toward nuclear acquisitions or even nuclear use in the case of desperation in facing defeat, to have its own weapons tested on the battlefield, to keep or expand basing access, and to earn money from sales. Negative responses have been impelled, for example, by a desire to stop or foreshorten the fighting, by fears of escalation of a conflict (the quagmire image), by desires to punish a client state for its "aggression" or to increase supplier leverage over a client, to hasten the prospects for peace settlements, or to adhere to embargoes mandated by the UN. To the extent that arms resupply is used as an instrument to tighten client relationships and to further a major power's influence in a region, it can be taken as an indirect form of power projection.

There are many scenarios for arms resupply operations, ranging from the obvious to the hypothetical, which tend to parallel those that might trigger direct intervention, namely war scenarios involving the three crucial regional conflicts, that is, the Arabs versus Israel, Iran versus Iraq, and India versus Pakistan. The necessity for, and likelihood of, major resupply operations will greatly depend on the duration of possible wars. For instance, the 1973 war and the long Iran-Iraq war involved resupply diplomacy, while the 1967 and 1982 Middle Eastern wars and the relatively short confrontations between India and Pakistan in 1965 and 1971 for the most part did not.[7]

The United States might have a serious dilemma regarding arms resupply in the case of a new Arab-Israel war; perhaps less of a problem if only Syria were involved on the Arab side (but that in turn might mean a quick Israeli victory and, hence, the absence of a requirement for resupply). But it would have a really serious dilemma if either or both Saudi Arabia and Egypt were fully engaged in combat with Israel, which would most likely be resolved by a complete embargo on all combatants (this has often and traditionally been a U.S. policy going back to the 1930's Neutrality Acts, as exemplified by the joint embargo on Pakistan and India in 1965).[8] The United States, if it did choose to resupply Israel, would—relative to the 1973 experience—suffer from the lack of basing access as well as the absence of forward stocks of equipment in NATO in Europe that could be drawn down for a resupply operation. On the other hand, the U.S. C-5 (and now C-17) fleet would be more capable than it was in 1973 of resupplying Israel without the use of en route bases.

There are other conflict scenarios that might involve such large-scale re-supply operations. The United States has eschewed such support for Pakistan in the past. It might or might not be more forthcoming if there were a nuclear threat against India by an endangered Pakistan. Another Iran-Iraq war might well see a repeat of the complex resupply patterns of the 1980–88 period. Indeed, the general shift in arms supply policies away from a geopolitical and toward a more commercial basis might foreshadow a commercial free-for-all, even as one or another major government "tilted" toward one side or the other for a variety of political and economic reasons entwined in oil diplomacy.

But with the cold war over, the issue of payment might become more important, and that might affect even oil-rich countries (note that Iraq went heavily into debt in the late 1980s, despite extensive financial aid from Saudi Arabia and Kuwait). No longer is it likely that Russia or China will empty their weapons stores in support of a "friend" without serious prospect of payment, the way the USSR did for its Arab clients during the 1973 war, and after the 1967 and 1982 wars. The U.S.-Israel relationship may or may not be altered in this respect. Overall the implications of these new emphases are far-reaching.

Generally speaking regarding resupply operations, it is clear that en route access (staging bases and overflight rights) would no longer be determined by cold war ideological alignments, as they had been in previous Middle Eastern crises or in the cases of the Horn war and Angola.[9] More likely either such access would be relaxed, or restrictions would be imposed on the basis of friendship with, or the leverage of, the combatants themselves. How this would work out in connection with largely commercial resupply operations is not clear.

Coercive Diplomacy

Coercive diplomacy has often been characterized with the colloquial term "gunboat diplomacy," which conjures up the image of British naval force demonstrations all around the globe during its imperial heyday, or the frequent use by the United States in the earlier part of this century of "station fleets" to intimidate wayward political forces in Central American and Caribbean "banana republics."[10] Terminologically speaking this is actually not an easy subject to deal with. James Cable, in his classic work on gunboat diplomacy, conjures up a range of activities under the general heading of "limited use of force," then further subsuming such categories as "definitive force," "purposeful force," "catalytic force," "expressive force," and "coercion."[11] These are ranged along a spectrum from limited uses of force short of outright war to various forms of signaling intent, demonstrations of "tilt," and symbolic acts of intimidation.

Studies of the recent history of coercive diplomacy make it clear that in various circumstances, on both sides of the cold war divide, naval, air, and ground forces were used for such purposes as coercion and signaling, or as expressions of political support. For the United States forward movements of aircraft carriers or attack aircraft or marine units were often the preferred instruments of coercion or support. In recent years Airborne Warning and Control Systems (AWACS) or Joint Surveillance Target Attack Radar System (JSTARS) aircraft, representative of the potentialities for force multiplication, have more often become the primary instruments for signaling.[12]

If the recent past can be used as a guide, it is easy to predict that there will be U.S. responses to further incidents involving Iraqi troop movements near the Kuwait border or Iranian moves in connection with the Persian Gulf islands or its threats to shipping, or any of a variety of possible provocations by Libya. If radical Islamic forces were to take power in Algeria, Egypt, or Turkey, there would be the expectation of sporadic but consistent conflict with the United States, involving the latter's use of coercive instruments of diplomacy. In the future any heightened frequency of terrorist incidents within the United States emanating from the Middle East can be expected to provoke some level of military response, perhaps in the form of threatening aircraft and ship movements, or the beefing up of marine and army deployments. Russia will no doubt continue to use its forces for purposes short of outright war all along its shaky southern periphery, though longer-range projections of coercive diplomacy in imitation of the period before 1990 appear increasingly unlikely. If the past is a guide France, Britain, and India, at least, will periodically be engaged in limited acts of gunboat diplomacy at various points of the greater Middle East. But coercive diplomacy and, relatedly, extended deterrence appear increasingly to be a predominantly U.S. game, without much of a challenge from another superpower actor.

Showing the Flag

Still further along the spectrum is the phenomenon of "presence" or symbolic acts of force-posturing earlier known as showing the flag. In the past this may have involved, even in periods of relative peace, routine port visits or the rotation of station fleets. At the peak of the U.S.-USSR naval race in the 1970s, as the Soviet blue water navy expanded, statistics indicated a constant rise in the Soviet naval presence, assumed to have major ramifications in connection with extended deterrence for client states and the perceptions by various nations of the changing global correlation of forces.[13]

With the cold war over and the Russian navy rusting away in its home ports, no longer do such data retain the same meaning, for example, competitive shows of strength, flexing of muscles, intimidation of "rival's" client states, and symbols of waxing power and shifts in the correlation of forces. Now perhaps the primary meaning of such presence, for the United States at least, has to do with a form of extended deterrence on behalf of regional client states threatened by regional neighbors as, for example, Saudi Arabia vis-à-vis Iraq and Iran. Presence then is largely defined in terms of what Paul Huth calls "extended general deterrence," referring to "political and military competition between a potential attacker and defender in which the possibility of an armed conflict over another state is present but the political attacker is neither actively considering the use of force nor engaging in a confrontation that threatens war."[14]

Force Interposition Peacekeeping

In the "new world order" of the post–cold war and post–Desert Storm era, there appears to be a greater emphasis, as an aspect of power projection, on peacekeeping (and peacemaking) and force interposition, though the permanence of the seeming trends of the early and mid-1990s remains to be seen. During this period the United States has been engaged in major peacekeeping operations in Somalia and in Bosnia (also in northern Iraq) and, as part of UN operations, has participated with interposition forces in the Sinai Desert along the Israel-Egypt frontier defined by the Camp David Accords. Something similar may yet eventuate on the Golan Heights. Elsewhere in and around the greater Middle East, multinational UN peacekeeping forces are stationed in Cyprus, Lebanon, Macedonia, Kashmir, and in the Caucasus. These operations, Bosnia somewhat excepted, do not depend upon massive logistical operations. But they do often depend on basing and overflight access: witness the U.S. use of forward bases in Italy, Hungary, and Albania in relation to the operation in Bosnia, and the use of various bases in Egypt and on the Arabian Peninsula in connection with the Somalia operations. Generally speaking there is usually permissive basing access in the cases of multilateral peacekeeping operations under the aegis of the UN or NATO or both.

The Problem of Basing Access

Over the course of the twentieth century patterns of basing access went through several successive phases. During the interwar period, actually stretching back before World War I, most basing access had been delineated by colonial possessions. Britain had its huge overseas empire and, concomitantly, a global network of coaling stations, dirigible mooring masts, underwater cable terminals, and such.[15] Hence for projecting power into the Far East, Britain could count on major naval bases at Gibraltar, Malta, Suez, Aden, Trincomalee, Singapore, and Hong Kong for a route via the Suez Canal, or bases at, for example, Freetown, Capetown, Durban, Mauritius, and Mombasa for a route around the African cape.[16] France had a smaller though still formidable network hinged largely on its possessions in North and West Africa, including Djibouti, Vietnam, Guadeloupe, and French Guiana. The Netherlands, Spain, and Portugal also were availed of access as a result of colonial holdings.

During the interwar period there was very little in the way of basing access afforded one sovereign nation-state by another; that had actually long been the case for multipolar global systems in which ideology played only a minor role in defining alignments and enmities, which, at any rate, tended to be shifting and ephemeral. The few exceptions in the 1930s (a period of growing ideological cleavages) pertained to Japanese naval access in Siam (Thailand) and some German access, particularly for submarines, to Spanish bases in the Balearic and Canary islands.[17] The U.S. acquisition of a string of British bases from Newfoundland to British Guyana was a harbinger of things to come, as was the U.S. acquisition during that period of access to facilities in Greenland, Iceland, and the Azores.

As the postwar period progressed most of the West's access to bases in former colonial possessions was lost; indeed, the USSR was granted access in many of them as, for instance, Angola, Vietnam, Yemen, and Somalia. But the United States in particular was able to construct a global system of basing access tied to its formal alliance structure and associated economic and security assistance. Then, too, the technological basis for power projection evolved over time, most notably with respect to longer-range aircraft, aerial refueling, and faster ships with roll-on–roll-off (ro-ro) capabilities.

During the cold war both superpowers made extensive use of staging bases and overflight rights in connection with a number of well-known interventions or arms resupply operations. There was the U.S. use of Portugal's Lajes air base in the Azores and others in Spain, where tankers were based, for the arms resupply to Israel in 1973.[18] Earlier the United States had used Turkish bases as jumping-off points for its intervention in Lebanon in 1958. The Soviets used numerous air bases in North and West Africa in supporting an arms airlift to Angola in 1975; later overflight corridors and bases in the Middle East were used to support the Soviet-Cuban-Ethiopian war effort against Somalia in 1977–78. In 1979 the Soviets overflew (and perhaps staged through) India in supporting Vietnam in its brief conflict with China.

Then came the end of the cold war and with it the withering of access networks for both superpowers. Effectively Russia withdrew from most if not all previous overseas bases in places such as Vietnam, Cuba, South Yemen, Ethiopia, and Angola, as well as from more contiguous areas where they had maintained a forward presence, for example, in Eastern Europe, Afghanistan, and Mongolia. Meanwhile the United States, though retaining a significant global basing network, also saw considerable loss of access in places long hospitable to U.S. forces, for instance, in the Philippines, Spain, and mainland Greece.

Several general factors may be noted in connection with the contraction of U.S. basing access after the cold war. Most important, numerous erstwhile U.S. allies and client states no longer saw a U.S. presence or routine U.S. access as necessary for deterrence against Soviet-backed regional foes. The U.S.-USSR competition in the third world had ended, and many former Soviet client states had turned toward the West, politically and economically. As a result and as illustrated in the case of the Philippines, national pride and sovereignty reemerged as important factors—a foreign military presence, no matter what the practical implications in security terms, still represented a degree of humiliation, even where there were long-term alliance relationships, as witness increased pressures in Japan for removal of the U.S. presence. Then, too, the end of the cold war brought increased pressures in Washington for reduction of security and economic assistance budgets long tied to basing access as a quid pro quo, explicitly or otherwise. The United States was less willing to meet such costs as the military rationales associated with the cold war evaporated.[19]

Still another obvious factor had to do with the tremendous reduction in U.S. forces overseas, particularly in Europe. As many of the land forces that had been stationed in Europe were redeployed at home or demobilized, the basing structure with which they had been associated declined. That translated into fewer air staging bases in Europe for operations into the Middle East. It also meant a much lower volume of in-place military equipment that could be drawn upon for use in the Middle East, as had been done on behalf of Israel in 1973 and, of course, in connection with the Gulf War in 1990–91.

Indeed, in some respects or at least as an apparent tendency, there was a return to patterns and practices of the interwar period when, as we have noted, there were few permanent basing arrangements involving pairs of sovereign nations. This accorded with a shift back to an international system not driven by ideological rivalries and by the disappearance of an overarching bipolar rivalry. Of course, the seeming facts of U.S. unipolar dominance over the world's security realm produced a significant difference, one reflected in the fact that most remnant basing access on a more permanent basis was held by the United States alone (France and to a lesser degree Britain also retained some bases in former colonial domains).

Increasingly basing access appeared to become ad hoc and situational, something to be granted if a given host nation agreed with—or at least did not disagree with—the specifics of a given military operation. During the Gulf War the widespread legitimization accorded Desert Storm because of its UN mandate and coalition-based effort resulted in virtually open-ended access for U.S. logistics operations, regarding, for example, air staging, port refueling, and overflights. The United States had access all over Western Europe and apparently was able to overfly former Soviet-allied nations in Eastern Europe, if not Russia itself. Similarly forthcoming were countries in or near the region of conflict such as Egypt, Turkey, Oman, and Kenya. Indeed, even India, long out-of-bounds for U.S. military forces, apparently allowed U.S. aircraft to refuel, setting off some internal political difficulties, as was also the case with long-time U.S. Asian ally Thailand.[20]

Later on, with the U.S.-led multinational peacekeeping effort in Somalia, the U.S. air and naval logistics forces were also availed of extensive, if ad hoc, access to several regional nations, each of which supported the basic principles of that operation.[21] Hence U.S. transport aircraft were provided access in, among other places, Egypt, Saudi Arabia, Oman, and even the long-time Soviet client states of South Yemen and Ethiopia. Meanwhile U.S. air operations in Bosnia, conducted within a NATO framework, were carried out from bases in Italy and Germany, mostly from the Aviano base in the former. Here, not unexpectedly, access was not a problem.

But gradually and inexorably the United States was experiencing a contraction of its permanent basing access, as well as some difficulty in acquiring new forms of access.[22] In Europe the United States had come to rely on six main air bases as its basing structure was consolidated—at RAF Lakenheath and RAF Mildenhall in Britain, at Ramstein and Spangdahlem in Germany, at Aviano in

Italy, and Incirlik in Turkey.[23] Other former U.S. Air Force hubs at RAF Bentwaters and RAF Upper Heyford in Britain, at Hahn AB in Germany, and at Soesterberg in the Netherlands had been eliminated. The U.S. military will also soon be returning Rhein-Main air base to the German government.

The United States does retain some bases in the greater Middle East area or nearby in Europe and around the Mediterranean. There are naval support facilities at Iskenderun and Yumurtalik in Turkey in addition to the air base at Incirlik. The naval bases at Greece's Souda Bay and Nea Makri (communications) have been retained, as have the P-3C antisubmarine warfare aircraft bases at Sigonella on Sicily and at Rota in Spain (nine aircraft each). Also in Italy the U.S. Navy's headquarters at Gaeta and an attack submarine base on Sardinia at La Maddalena have for the time being been retained. Newer points of access have been acquired in Albania. Further around the Indian Ocean with relevance to possible greater Middle East operations are the naval air station and support facilities at Diego Garcia and the communications facility at Australia's Northwest Cape.[24]

Overall, however, the U.S. Navy has had an increased presence in the Middle East since 1980 at the time of the buildup associated with the rapid deployment force in the wake of the preceding events in Iran and Afghanistan. It has actually had a continuous presence in the Persian Gulf since 1948 (earlier centered on Bahrain), and since 1980 has had at least four combatants in the Persian Gulf at all times. What is new (see later) is the presence of carriers and amphibious groups in the Gulf, a presence that was enlarged during the latter stages of the Iran-Iraq war with the Earnest Will reflagging operation, but then expanded during and after the Gulf War. The greater Middle East has actually been the one region of the world in which the U.S. military presence has expanded since the end of the cold war, even without the traditional rationales (as sold to host states) of cold war politics and the need for protection against the Soviet threat, though the events of September 1996 appeared to foreshadow a marked reversal.

Still, after the Gulf War, there were some signs that the United States would be experiencing increasing difficulty in acquiring or maintaining permanent access in the post–cold war period in relation to security problems in the greater Middle Eastern area, both within and outside the immediate region. While both Kuwait and Qatar agreed to allow the United States to preposition materiel for one combat brigade, Saudi Arabia refused to become a third party to such a program.[25] Meanwhile Thailand, which had long provided extensive access for U.S. air and naval facilities, refused to allow the United States to station offshore prepositioning ships near its coast.[26] That refusal was echoed by Malaysia and Indonesia, Muslim states sensitive to Middle Eastern political winds. The Thai position was less clear, but perhaps explicable in the light of a reportedly tightening political relationship between Thailand and China in the context of a potential long-term U.S.-China rivalry. Meanwhile restored U.S.

ties with Vietnam came with the hint of future provision of naval access for the U.S. Navy at Cam Ranh Bay.[27]

What then, for the future of U.S. (or U.S.-led coalition) basing access in connection with possible power projection operations into the greater Middle East region? What stands out, clearly, is the contingent or situational nature of basing access with respect to future possible U.S. interventions in the greater Middle East region. In other words, one can envision various scenarios in which en route or intratheater countries will decide on the basis of the comparative leverage of the United States and whatever regional foe is involved whether or not to provide access.

In situations involving Israel access for the United States in connection with arms resupply operations is likely to be more restricted than in the past, given the loosening of NATO ties, correspondingly lessened U.S. leverage over its NATO allies as the latter are no longer so dependent on American support for their security, and an unchanging propensity by the European powers to back the Arab side for pragmatic reasons related primarily to oil diplomacy and monetary policy. In a future Arab-Israeli war by and large regardless of the war's origins or the nature of the combatant coalition on the Arab side, it is unlikely that the United States will be able to move military supplies to Israel from Europe, or to use bases in Spain, the Azores, or Sicily for en route staging. Despite the drop in oil prices and the seemingly reduced menace of a 1973-type oil embargo, Arab leverage over basing diplomacy would appear to be higher than ever.

With regard to possible U.S. interventions in the Persian Gulf area, the prospective nature of its access is far from clear. The political conditions for such access during Desert Storm were optimal and are not likely to be repeated. One might speculate, however, that in a rerun of Desert Storm the United States would still be availed of access in the same places as in 1990–91, that is, Egypt, Oman, Saudi Arabia, Bahrain, and Turkey, as well as with any number of nations where air staging bases were required, that is, those in Western Europe and in Asia. This will be discussed in greater detail in chapter 9, where we shall note that from the vantage point of 1996, levels of U.S. access had never been higher.

New Technologies

By the mid-1990s there were some additional technological and politicoeconomic factors that had begun to alter, at least potentially, the prospects (and the implications thereof) of U.S. basing in the greater Middle East region.

In the technological dimension, there was the somewhat startling revelation that the U.S. Navy was developing, and planning to deploy by the early years of the twenty-first century, "a new class of navy ship that some believe could revolutionize maritime warfare as dramatically as the ironclad warships of the Civil War, the aircraft carriers of World War II, and the cold war's ballistic missile submarines."[28] Called "robo-ship" by admiring wags, and an arsenal ship by others, the new vessel will actually be a floating missile pad run by remote control, with 500 vertical missile tubes able to launch at targets on land, sea, and in the air, with a very small or even no crew. It could be commanded from another ship, from a plane, from far away on the ground or, presumably, from a satellite. It is expected that the first three ships of this sort will be deployed around 2001: in the Pacific near Korea, in the Mediterranean, and in the Persian Gulf, that is, two of the first three are designated for the greater Middle East. Their objective will be to slow down an invasion such as the one in 1990 of Kuwait long enough for reinforcements to arrive.

The robo-ship is, apparently, designed to be well armored (with a double hull) so as to be able to survive some serious hits, and therefore is also a bit sluggish. It is expected to be a white water system that hugs the shores. But overall, to the extent that it can play its expected role, its deployment will translate into reduced requirements for land-based aircraft and missiles as well as fewer earlier generation naval vessels. The projected capabilities of the robo-ship are tantamount to bringing to fruition all of the older ideas about floating islands in lieu of bases. Such a system would be useful not only for repelling full-scale invasions, but would also be expected to take on the primary role for coercive diplomacy played first by gunboats and battleships, later by aircraft carriers, and more recently by AWACs and JSTARS force multiplier or battle management aircraft. Whether such a system can be made relatively invulnerable to attack, particularly by submarines while in waters immediately offshore, is, of course, an additional question.[29] Generally speaking it is expected that the arsenal ship, particularly if it can be deployed in numbers, will greatly reduce the U.S. logistical requirements in the event that it has to conduct a war in the greater Middle East region.

On more of a politicoeconomic plane there has been considerable talk, particularly as it pertains to the U.S. Navy, of replacing "bases with places." That means worrying less about the kind of sovereignty issues that used to be associated with status of forces arrangements and permanent access, and shifting the focus not only to a priori agreed-upon situational access but also the increasing exploitation of commercial facilities. The U.S. Navy learned to get along without Subic Bay by relying on commercial repair facilities in Singapore and elsewhere. In the Persian Gulf area the Jebel Ali artificial port at Dubai can now serve many of the same purposes. But access during a crisis may still remain problematic.

Technical Bases

For the most part traditional air and naval bases are at the center of the relationship between power projection and basing access. But there are also now important technical facilities, encompassing a broad spectrum of activities subsumed under command, control, communications, and intelligence (C^3I) and space-related military endeavors.

Even before World War II there were technical functions that involved or required basing access, that is, communications nodes, receivers, intercepts, ground terminals for underwater communications cables, and even some early radars.[30] Since then there has been a profusion of such facilities, and the importance of access to them has come to rival that of air and naval bases.[31]

During the cold war, the United States built elaborate global networks in all of these areas, and most of their activities were directed at the Soviet Union. Some, for instance, provided monitoring of the telemetry of Soviet long-range ballistic missile tests or of the testing of Soviet naval missiles in the Black Sea, or provided the wherewithal to track Soviet missiles and satellites during a war, or to gauge the accuracy of the targeting of U.S. nuclear-armed missiles so as to assist retargeting, or to track the movements of Soviet nuclear submarines. As the cold war ended many of these facilities were closed down. The once massive U.S. global sound surveillance system network, central to the global effort at tracking Soviet submarines, has apparently been shut down.[32] In Turkey the once crucial and sensitive technical facilities at Sinop, Samsun, and Karamursel have all been shut down [but some signals intelligence (SIGINT) capability remains] and the seismic monitoring station at Belbasi is apparently still operational but about to close (the big U.S. space tracking radar station at Pirinclik is still apparently in operation). Several facilities on mainland Greece have also been shut down. But by way of compensation, the United States is now apparently acquiring extensive access to facilities in Albania, once host to Soviet submarines in the early cold war period.[33]

The United States still requires a variety of technical facilities to deal with newer security threats in the Middle East (Iran and Iraq), if not also the possibility of a resurgent Russia. There is the question of which U.S. technical facilities outside the greater Middle East region, or around its periphery, may be important in the future for the conduct of Middle East military operations. There is also the question of the importance of such facilities *within* the Mideast in the context of the possibility of conflict within or outside of the region itself, the latter pertaining to renewed rivalry between the United States and Russia, if not on a global scale then in regard to Russian attempts to regain control and power over its "Near Abroad" in Central Asia and the Caucasus.

Just outside the greater Middle East, there are a range of facilities that would be useful to the United States in the event of a Middle Eastern conflict: a satellite control station at Mahé in the Seychelles, a major communications and intelligence facility at Nea Makri on Crete, and SIGINT stations on Cyprus (jointly operated with Britain) and Diego Garcia and in Italy, Turkey, and Morocco. During and after the Gulf War, the United States was able to acquire enhanced access for SIGINT in one or more of the GCC states, supplementing two such facilities long hosted in Oman. Communications and other facilities on Australia's Northwest Cape (very low frequency) and in Diego Garcia are used to cover the Indian Ocean. It is clear that in many C³I technical areas satellites have taken over much of the role formerly played by ground stations and aircraft, hence lessening the need for ground access, all the more so if satellite-to-satellite communications can reduce the requirements for overseas satellite ground links.[34]

The United States is alone among the major powers in having some considerable basing access in and around the greater Middle East region. Just how far back the U.S. Navy will be cut and how much naval power can be directed at this critical region with what remains is unclear. Depending upon which of the various naval force options hold sway as the 1990s progress, observers expect that it will retain somewhere in the range of 300 to 400 ships, which would include 10 or 11 aircraft carriers, some 18 strategic submarines (SSBNs), from 40 to 60 attack submarines, maybe another 150 surface combatants, some 50 amphibious assault ships, and an increasing number of mine warfare vessels.[35] Maintenance of the capability to deploy several carrier battle groups to the greater Middle East in case of crisis or war will be crucial, particularly given the uncertainty, in some contingencies, for in-theater basing of land-based aircraft.

Table 7-1 portrays the changes in the U.S. Navy force structure from 1988 to 1993, spanning the end of the cold war, and one semiofficial projection of what that structure will look like in FY 2001.

The U.S. Navy does maintain a major and constant naval presence in the Persian Gulf-Arabian Sea area, using forces provided by the Atlantic and Pacific fleets, which normally includes a carrier battle group consisting of a carrier, two cruisers armed with surface-to-air missiles (SAMs) for fleet defense, two frigates also armed for fleet air defense, a tanker, and a couple of nuclear attack submarines. In the Mediterranean a typical ongoing deployment consists of a carrier battle group with six surface combatants and a couple of fast support ships, four attack submarines, and an underway replenishment group consisting of four support ships and two escorts.[36]

Some skeptics might prefer to rely less on carrier aircraft. Donald Rice, secretary of the air force under President Bush, emphasized that his service, with one refueling and large conventional payloads, could cover the globe from three bases (Barksdale in Louisiana, Diego Garcia, and Guam) and that six B-2s operating from the United States could carry out an operation similar to the 1986 Libya raid, which involved two aircraft battle groups, an air force F-111 squadron, and extensive supporting assets.[37] So there are arguments for emphasizing land-based airpower over sea-based airpower, with obvious applicability to the greater Middle East just because of the contingent nature of basing access. Whether overflight access would be available—or could be

Table 7-1. Navy Force Structure Costs

Type of ship	FY 1988	FY 1993	FY 2001 (projected)
Carrier	14	13	11 (+1 NRF)
Battleship	3	0	0
Cruiser	38	52	29
Destroyer	68	37	69
Frigate	85	35	32
Patrol hydrofoil	6	0	0
Submarine (non-nuclear)/ nuclear-fueled submarine	100	88	52
Nuclear-fueled ballistic missile submarine	37	22	18
Large deck	12	12	11
Other amphibious	49	42	32
Mine countermeasures/ mine hunter (coastal)	4	13	11
Logistics	58	50	41
Fleet support/tenders	66	53	21
Naval Reserve Force	25	18	15
Total	565	435	343
FY 1988–2001		−222 ships (−39%)	
FY 1993–2001		−92 ships (−21%)	

Aircraft category (navy and marine corps)	FY 1993	FY 2001 (projected)
Power projection	1,213	942
Tactical support	737	582
Antisubmarine warfare	566	428
Electronic warfare	154	136
Logistics support	491	408
Training	1,236	1,025
Research, development, testing, and evaluation	220	174
Other	87	11
Total	4,704	3,706
FY 1993–2001 reduction	−998 aircraft (−21%)	

Source: *Force 2001: A Program Guide to the U.S. Navy* (Washington, D.C.: Deputy Chief of Naval Operations, 1995), p. 118.

circumvented or even violated—in some cases is another question applicable, let us say, for a B-2 flying from Louisiana to the Persian Gulf.[38]

Other Nations' Basing Access in the Region

For the foreseeable future no other aspiring maritime power, or for that matter, a combination of them, can match U.S. naval power in the Indian Ocean–Mediterranean Sea area. None indeed might even be envisaged as capable of deterring U.S. or U.S.-led coalition activities in this area, either with naval power per se or a ground-based deterrent.

France continues to have some capacity. It deploys some troops and an air support unit in Mayotte and La Réunion, and its Indian Ocean squadron (frigates, patrol boats, amphibious vessels, and support ships) makes extensive use of a facility at Djibouti, where troops and combat and transport aircraft are also deployed. Other such deployments are maintained in the Central African Republic. In Saudi Arabia France bases nine Mirage 2000Cs and a C-135; in Turkey, as part of Operation Provide Comfort, it has four Mirage F1-CRs, eight Jaguars, and a C-135.[39] Britain also has some modest presence and access in the greater Middle East, including air bases, some troops, technical facilities on Cyprus, and a presence at Gibraltar that involves periodic rotations of Jaguar aircraft as well as a naval base. It also has a small military presence on Diego Garcia, six Tornadoes and a VC-10 in Saudi Arabia (Southern Watch), and eight Harriers and two VC-10s in Turkey (Provide Comfort).

Russia's formerly formidable overseas basing network has collapsed as its fleets have rusted. It no longer maintains access to, nor presence at, such former Soviet naval strong points as those in Cuba, Vietnam, Cambodia, India, Iraq, Syria, South Yemen, Algeria, Yugoslavia, Guinea, Angola, and Mozambique. Thus the United States no longer faces a potential naval rival in the Atlantic, nor for the moment in the Pacific except for areas along the East Asian littoral. It thus can and will focus on regional naval dominance in the Indian Ocean–Red Sea–Mediterranean area, where much of the world's oil is at stake along with the core problem of potential regional warfare involving WMDs.

For the future both China and Japan are viewed as potential threats to U.S. hegemony, naval and otherwise. But neither seems likely any time soon to achieve the capability for projecting significant naval power into the greater Middle East, specifically, the northwest Indian Ocean. China does now have one SSBN, forty-eight tactical submarines, and fifty-five principal surface combatants, in addition to a force of some 870 patrol and coastal combatants.[40] As mentioned in chapter 4 China is in transition from a largely coastal defensive orientation to one that might allow for more extended power projection, even beyond the South China Sea.[41] There have been rumors about possible Chinese access to Burmese naval facilities; there is also a possible purchase of an aircraft carrier from Ukraine or Russia. Pakistan and Iran both have had close security ties to China and could, under some circumstances, be envisaged as basing hosts for the latter. But it will be a long time before China can challenge the U.S. Navy in the greater Middle East. It is a matter of technological development as well as of quantitative measures of naval force levels. Japan, meanwhile, has developed a strong navy, with seventeen submarines and sixty-two principal surface combatants, comprising seven destroyers and fifty-five frigates.[42] But there are no current indications that Japan plans to develop the capability for projecting naval power into the greater Middle East region.

Regional Naval Forces

India has the second strongest naval presence in the greater Middle East, and much has been written about its aspirations to become a predominant naval power in the Indian Ocean, with a reach extending, at least hypothetically, to southern Africa or Australasia.[43] Its naval forces comprise fifteen submarines, twenty-five principal surface combatants including two carriers (carrying Sea Harriers and helicopters), fifteen missile-armed corvettes and six other missile patrol boats, twenty mine warfare ships, and nine amphibious vessels.[44] They are based close to the northwest Indian Ocean and Persian Gulf, with major bases on its west coast at Bombay, Goa, and Karwar. India has conducted some minor offshore operations in the Maldives and Sri Lanka but at present its defense planning is referenced to possible conflicts with either Pakistan or China or both. Regarding the latter, India's forward deployments at Fort Blair reflect concerns about future Sino-Indian naval rivalry in Southeast Asia, specifically in the area around the Indonesian straits and in connection with oil in the South China Sea. Meanwhile, Indo-American relations have warmed considerably since the end of the cold war, which saw neutral India as, effectively, a Soviet client state and quasi-ally. Basically, however, India is concerned about Chinese deployment in the West Indian Ocean as both countries increase their demand for Middle East energy.

In the greater Middle East itself there are only minimal naval forces that could not easily challenge U.S. naval power. Iraq has limited facilities at Basra, Az Zubayr, and Umm Qasr (the last currently closed), and deploys one frigate plus eleven patrol and coastal combatants and five mine warfare ships (the patrol craft are armed with older-version Soviet missiles). Iran's navy has much better access to the open sea from bases at Bandar Abbas, Büshehr, and Chah Bahar. It also has three newly acquired Soviet Kilo-class submarines as well as two destroyers, three frigates, and some forty-five patrol and coastal combatants, some armed with Harpoon missiles. The submarines are a hypothetical threat to the U.S. Navy, as are Iran's minelaying capabilities. Iran's Revolutionary Guard naval force operates out of five island bases in the Persian Gulf with forty Swedish Boghammar Marin boats armed with antitank guided weapons and other weapons, plus its Chinese-supplied Silkworm missile battery. But, for the foreseeable future, there are few signs that Iran could mount a really serious challenge to the U.S. Navy in the Persian Gulf area, that is, something more than a limited capability for mine warfare or for the use of shore-based missiles.[45]

Power Projection within and from the Greater Middle East

We have until now focused on the problems of forward presence and power projection in and to the greater Middle East, which, as we have indicated, is mostly pertinent to U.S. military operations, less relevant for France, Britain, Russia, or China. But power can also be projected and forward presence established within as well as outside of the region by its various contending nations. This involves some interesting geographical relationships and questions.

There has been little if any actual granting of forward bases (air, naval, technical) within the region. Iran has been rumored seeking some access in Sudan. Iraq was earlier rumored (whether accurately or not) to have tested missiles in Mauritania. More in the way of presence than anything else, one could point to Pakistan's earlier deployment of a sort of Praetorian Guard in Saudi Arabia, once 20,000 strong. On a more covert basis Israel had some presence in the form of advisors in the Shah's Iran. In the 1973 war the movement of Moroccan troops to aid Syria on the Golan could be viewed as a form of power projection. Currently Israel's reported use of Turkish airspace for training its air force constitutes another example of presence. The movement of a variety of Islamic "fighters" from Egypt, Saudi Arabia, and Sudan, among others, to fight the Soviets in Afghanistan in the 1980s is also germane. Different kinds of access are also provided some nations in the region by others for such purposes as clandestine and covert operations and terrorist activities, a subject beyond the scope of this work. Arab use of Cyprus in this connection might be noted as an example, or Iran's penetration of Lebanon. Forward presence and the application of force beyond contiguous borders has been in evidence in a number of instances.

Regarding the serious long-range or out-of-area projection of military force, however, perhaps only Israel and to a lesser degree India have displayed such capabilities. Israel's navy and air force have projected power as far afield as Entebbe, Uganda, Ethiopia's Dahlak Islands, and Tunisia. In the Entebbe raid that required some access for landing rights and use of airspace in Kenya. India's counterinsurgency operation in Sri Lanka and its overturning of a coup in the Maldives by a naval task force are worth noting. Further, over the past decade there has been a lot of speculation about India's ambition to field a blue water navy that, representing a regional hegemon, might project power and influence throughout the Indian Ocean from southern Africa to southeast Asia.

Thus far forward presence or power projection outside the region for the region's nations has been virtually a nonsubject. Sometimes in Washington think-tank scenario-writing in connection with the possibilities for an Iran or Iraq to target the United States with ballistic missiles, the (fanciful) possibility for access for those missiles in the Western Hemisphere is bruited. But for the most part long-range power projection by regional states, directed against the

United States or Europe or both, involves the subject of terrorism, one of several components of what in the revolution in military affairs (RMA) literature have come to be dubbed "asymmetric strategies."

Though terrorism per se is beyond the scope of this work, several points, obvious to the point of banality, can be made about its geographic aspects in the context of power projection. The large distances from the greater Middle East to the United States, relative to Europe, seem to have shielded the former from large-scale terrorist violence, the World Trade Center episode notwithstanding.[46] Whether that is entirely because of distance and an intervening ocean or, rather, because of more stringent barriers to entry in the United States or because of tacit compliance with terrorist groups by some European governments in connection with the political leverage of oil-rich states is not easy to say. But a major looming question is the ability of regional states such as Iran or Iraq to conduct WMD terrorist operations in the United States, whether as part of an intrawar (asymmetric) deterrence strategy, or as acts of revenge in connection with previous wars, as has been said of the World Trade Center bombing. Obviously the distance factor and the lack of overland routes for movement of materiel and personnel will greatly complicate this kind of long-range power projection as will the advent of better sensors for WMD materials. The distance and contiguity aspects of information warfare may constitute a more difficult subject with which to deal.

The Future of Strategic Mobility Forces

Throughout the Gulf War the United States conducted a truly massive logistics operation, combining a sealift and an airlift, which was a primary reason for the successful outcome of that war. That seemed to prove again that U.S. power projected over intercontinental distances can prevail over regional powers despite the vaunted power-over-distance gradient. As shown in table 7-2, the U.S. logistics effort from August 1990 to March 1991 was staggering, and it is clear that U.S. military capabilities are the American answer to greater Middle East geographic realities.

Since the 1990–91 Desert Shield–Desert Storm operation, the United States has continued to plan for the contingency of an approximate rerun, presumed to come in the form of another war with Iraq or Iran. This planning, however, has had to be done with the assumption that such a war might have to be fought under altered conditions relative to 1990–91 in some fundamental respects. First, the drawdown of the U.S. defense establishment and cuts in its defense budget might result in reduced logistical resources in the key areas of air- and sealifts. Second, the United States would no longer be availed of the number of men and the quantity of materiel that were forward in NATO Europe at that time, which were used as a springboard and forward depot. Another problem

Table 7-2. Desert Shield and Desert Storm Strategic Airlift and Sealift
Summary (August 1990–March 1991)

Aircraft type	Number of missions	Number of passengers	Tons of cargo
C-141	8,537	93,126	155,955
C-5	3,770	84,385	222,024
C-9	209	. . .	. . .
KC-10	379	1,111	12,129
Commercial	3,309	321,005	145,225
Other	. . .	1,093	10,219
Total	16,204	500,720	543,552

Ship type	Number of voyages	Tons of cargo
Fast sealift	33	321,941
Prepositioning	46	464,289
Ready reserve	131	691,048
U.S. flag	65	317,193
Allied	206	646,315
Total	481	2,440,786

Source: "Strategic Mobility, Forward Presence, and the Defense of American Interests," Report, School of
International Affairs, Georgia Tech University, Atlanta, September 1991.

would be the willingness of key European nations such as Britain and Germany
to allow the United States full freedom for such operations, particularly in
circumstances where the United States was trying to operate on a relatively
more unilateral basis. Finally, in-theater bases for U.S. transports and ports for
its transport ships might not necessarily be available in another conflict, nor
might en route staging bases and overflight corridors. In short, there has been a
concern about the possible uniqueness of the logistical conditions under which
Desert Storm was fought, even aside from the advantage given the UN coalition by Iraq with the some five-month-long period for an in-theater buildup, a
form of "phony war."

There are, of course, some trends in the opposite, more favorable direction.
For instance, the more lethal qualities of the emerging MTR-RMA technologies will presumably ramify into reduced logistical requirements for platforms
and munitions, that is, there will be "more bang for the buck," or, more
relevantly, bang for a given ton-mile. It is generally claimed by analysts of the
RMA that the emerging technologies will make it much easier for a small,
highly capable hi-tech military force to defeat a much larger but more traditional foe.

Ever since the Gulf War the U.S. military has concentrated on maintaining
its strategic mobility capabilities even to the extent of worrying about two

simultaneous regional conflicts. Realistically those aspirations have been scaled back to the idea of fighting one such war while deterring another. Even more realistically most planning is focused on one such war, most likely in the greater Middle East and in the face of the changes we have described since 1991, most notably with respect to the loss of much of the former forward presence in Europe plus the aging of some mobility assets, in a climate of budgetary stringency.

As is commonly discussed in military journal articles devoted to logistics, there is the crucial triad of airlift, sealift, and prepositioning.[47] Table 7-2 has shown the relative importance of each during the Gulf War, indicating the predominant role of sealift for cargo (but airlift is important for urgent cargo), and of airlift for moving personnel. Depending upon the nature of a future conflict—the amount of lift required and the speed criteria for delivery—one has little reason to think that the relative roles of the legs of the triad will change greatly, though it is clear that enhanced prepositioning is being stressed.

Airlift: The United States and Other Powers

The U.S. Air Force maintains a formidable and unchallenged airlift capacity. According to *The Military Balance* that involves 76 C-5As and 50 C-5Bs, 239 C-141Bs plus 18 in store plus 529 C-130 tactical transports for in-theater uses, with 78 more in store, Additionally, some of the 59 available KC-10A tankers can alternatively be used as transports, and this transport armada is supported by 553 KC-135 tankers with an additional 73 in store.[48]

But the C-141 fleet is aging and procurement of C-17s has come to be viewed as critical to maintenance of a viable airlift capacity. The C-17 is seen as particularly necessary because of the emerging strategic requirements for transport aircraft that can carry heavy loads (even if not equal to the C-5) and land on the shorter and rougher runways of airfields in third world conflict regions such as the greater Middle East, as competently as the lower-payload C-130. That means a payload of about 160,000 pounds that can be carried 3,400 nautical miles without refueling, and the ability to land on rough airfields with roll-out lengths of a little over 3,000 feet.[49] It also requires an aircraft that can be operated by a two-man crew if someone is available to help handle cargo on the ground; by contrast the C-141, C-5, and C-130 use crews of five to eight with significantly higher operating costs. The C-17 has more or less met these specifications, with a payload mission of 157,000 pounds, but with an unrefueled range of over 3,200 nautical miles and a cargo box as large as the C-5, which can carry outsized cargoes such as M-1 tanks, Patriot missiles, and some helicopters from the continental United States (CONUS) to remote third world airstrips.

Originally the U.S. Air Force planned to purchase 210 C-17s, but that was whittled down to 120 in 1990. To date 17 C-17s have been delivered, 13 of which are part of an airlift wing in South Carolina, the other 4 being used for developmental and operational testing. The aircraft is, however, part of the air force's operational inventory and is ready for worldwide missions. But the C-17 program has been plagued with cost overruns, developmental problems, and close political scrutiny in a post–cold war environment. The Pentagon's Defense Acquisition Board has directed the air force to acquire up to 40 operational aircraft, expected to be available in 1998 or 1999. After that, as we have noted, decisions appear to have been made to move ahead to a 120-aircraft fleet. The alternative would be for the procurement of nondevelopmental airlift aircraft, an apparent euphemism for either additional C-5Bs or a modified commercial transport to replace part of the aging C-141 fleet.[50]

With all the problems of the C-17 program and the aging C-141 fleet, the United States does still maintain a formidable long-range airlift capability, one still at or near the levels in evidence at the time of the Gulf War. The United States is unmatched by any other power for projection into the greater Middle East, even though Russia and the European Union nations would have much shorter distances to contend with.

The Soviet Union was once capable of substantial airlift operations, as it demonstrated in the airlifts to Egypt and Syria (1973), Angola (1975–76), Ethiopia (1977–78), Vietnam (1979), and throughout the Afghanistan conflict. No doubt its airlift assets have aged and decayed along with the rest of the once formidable Soviet military machine. but the most recent International Institute of Strategic Studies (IISS) data reveal a remaining (paper) inventory for Russia's Military Transport Aviation Command (VTA) of some 350 aircraft, including An-12s, Il-76M/MD Candid Bs (replacing the An-12s), An-22s, and An-124s. These are backed up by some 300 additional transports outside of VTA: Tu-134, Tu-154, An-12, An-72, Il-18, and Il-62 planes, plus a huge civilian Aeroflot fleet of 1,700 medium- and long-range passenger aircraft, including some 220 An-12s and Il-76s. Russia has only some forty tankers (Mya-4 and Il-78), but that has to be taken in the context of the shorter distances from Russia to the greater Middle East.[51] Russian capabilities remain formidable in regard to possible intervention or arms resupply contingencies in the region, but that eventually would, presumably, require a priori a reorientation of Russian security policy in general.

The European Union countries have some limited but significant long-range airlift capacity in relation to the greater Middle East, again within the context of (relative to the U.S. case) the much shorter distances involved. France has for many years had its Force Action Rapide rapid reaction force, comprising some 42,500 personnel—a paratrooper division, an air-portable marine division, a light armored division, and an air-mobile division, plus various support units.[52] Britain and Italy have similar forces. Proposals for new Western European Union military structures or a revamping of NATO involve a groping by the Europeans for a new post–cold war security strategy and identity that would entail some separation from the U.S.-dominated NATO structure. Still it would appear that the European Union nations, singly or in various combinations, lack the capability for large-scale interventions in the greater Middle East —that is, anything involving more than 50,000 troops, hardly sufficient to contest Iraq or Iran if perhaps sufficient to deal with a minor crisis elsewhere.

China is not now capable of mounting large-scale airlifts into the greater Middle East (the factor of air staging access in a variety of possible contingencies, for instance the possible use of staging points in Pakistan or Iran, left aside).

Generally speaking none of the countries outside the greater Middle East, with the exception of the United States, seems capable, much less inclined, to project power into the region for the foreseeable future. European operations on any meaningful scale or in any conceivable scenario seem unlikely except as part of a U.S.-led coalition operation. Similarly, China has limited capabilities for such intervention. So too India, except perhaps for emergency removal of its nationals working in the Persian Gulf area. A revived Russia could become a factor, as it was in the 1973 resupply operation or as it was with the insertion of operational forces (personnel manning SAM installations and combat pilots) in the 1969–70 Suez War of Attrition. But there has been a marked deterioration of the Russian fleet since 1990, and its airlift capabilities have also been drawn down, with regard to equipment, training, and caliber of personnel, so that 1973 could probably not be repeated. In addition Russia's capabilities are now also constrained by new geographical facts, namely, the loss of former springboards toward the greater Middle East in the Ukraine, Caucasus, and Central Asia, which could extend the range requirements for airlift operations and perhaps raise some problems with overflights. A fully revived Russia, a.k.a. the Soviet Union, would be another story, despite all the questions about what that entity's newer political relations in the greater Middle East might look like.

Sealift

While merely sustaining America's airlift capacity since 1991 has appeared a standoff, there has been a focused effort to improve its sealift capacity, which, as we have already noted, accounted for the preponderance of the Desert Shield-Desert Storm logistical effort. At present, according to IISS, the U.S. Military Sealift Command operates and administers some 290 strategic sealift ships in addition to the 76 ships of the Fleet Auxiliary Force. That comprises an active force of some 39 ships (17 dry cargo including two roll-on–roll-off (ro-ro) vehicles and 4 ro-ro containers) and 22 tankers and a standby force of some 259 ships.[53] A central component of this force are the eight fast sealift ships (FSS). Also central is the expansion of the fleet of ro-ro ships, of which there were 17 before the 1992 Congressionally mandated Mobility Requirements Study.

Currently the navy has plans to build 19 large medium-speed ro-ros (LMSR) up through about 2001, of which eight would become part of the prepositioned afloat fleet involving a heavy brigade's equipment and combat support, and the remaining 11 would be berthed in various locations for surge strategic deployment of the CONUS-based contingency forces.[54] Of these 19 ships, 14 will be newly constructed, the other 5 will be converted container ships. Each of these vessels will carry 300,000–400,000 square feet of military equipment at a speed of 24 knots. In addition there is an ongoing upgrade of the Ready Reserve Force (RRF) Fleet, expected to result in some 31 reserve force ro-ro vessels. Thus there are plans for some 58 ships (fast sealift, LMSR, and RRF ro-ros).[55] These would be the nucleus of an enhanced U.S. sealift capacity, which is backed up by a large force of other RRF ships (some 97 regularly maintained vessels) and a naval inactive fleet that includes some 60 mothballed World War II Victory ships plus some additional tankers and other cargo ships. There is also the category of auxiliary strategic sealift, comprising, according to IISS, "about a further 247 U.S.-flag and effectively U.S.-controlled ships potentially available to augment those holdings."[56]

PREPOSITIONING The third leg of the U.S. effort at forward presence and power projection—following those of airlift and sealift—is forward prepositioning of materiel close to anticipated theaters of combat. It is, of course, not an entirely new concept or usage. For many years the United States had maintained large-scale Prepositioned Overseas Military Materiel Configured to Unit Sets (POMCUS) stocks in Western Europe intended to be married up to troops flown from the CONUS in the case of a large-scale engagement with the Soviets in Central Europe. Similarly, the United States had long maintained prepositioned ships at Diego Garcia in the Indian Ocean for the eventuality of a Middle Eastern conflict, at which point those military stores could be married up with marines moved from the CONUS to regional combat zones. That proved to be a viable strategy in 1990–91 during the initial phases of Desert Shield–Desert Storm. Afterward on the basis of that success U.S. prepositioning was further to be institutionalized in the context of possible regional contingencies in the greater Middle East or Far East, a strategy made all the more necessary because of the extensive drawdown of the U.S. cold war military deployments in Europe and (to a lesser degree) in the Far East.

In the period since the Gulf War there has been a major emphasis by the United States on prepositioning materiel, afloat and onshore, in proximity to possible conflict areas in the greater Middle East. Not to be ignored is the significant extension of POMCUS stocks in Western Europe, primarily geared to the resumption of problems with Russia or whomever in Europe, but nonetheless available for Middle Eastern contingencies. In Germany, Belgium, and the Netherlands (nearly 60 percent of it in Germany), equipment is stored for four armored and mechanized brigades, which comprises nearly 2,000 main battle tanks, over 1,200 armored infantry fighting vehicles, over 1,800 armored personnel carriers, almost 1,400 artillery pieces, multiple rocket launchers, and heavy mortars, and some 150 attack helicopters. Additionally in Italy there is the equipment for the Theater Reserve Unit–Army Readiness Package South that includes 133 main battle tanks, 165 armored infantry fighting vehicles, 116 armored personnel carriers, and 98 pieces of artillery, multiple rocket launchers, and heavy mortars.[57]

But the U.S. prepositioning program now has a more dispersed global focus, reflecting the basic shift in U.S. priorities away from a European war scenario (with the USSR) to various "regional" conflict contingencies, as outlined in the Bottom-Up Review.[58] At the end of the projected war reserves reorganization, the army plans to have five reserve stockpiles: AR-1 in the United States, AR-2 in Europe, AR-3, the prepositioning afloat stocks that are being enhanced with an afloat brigade, AR-4 in Korea, and AR-5 in Southwest Asia. Except for the one in Korea all of these stockpiles would presumably be available for a greater Middle East scenario.

Within the greater Middle East the army has prepositioned significant stores of materiel, as indicated on map 29. In Kuwait that includes equipment for a tank and mechanized battalion and an artillery battery, comprising fifty-eight tanks, seventy-two Bradley armored fighting vehicles, eight M109A6 Paladin artillery pieces, and a Patriot SAM battalion.[59] Additionally twenty-four U.S. Air Force A-10 attack aircraft are now continually based at Kuwait's Ahmed Al-Jaber air base. A similar deployment is being carried out in Qatar, involving the prepositioning of equipment for one U.S. Army mechanized brigade, including up to 110 M1A2 tanks. The United States is still negotiating for a similar arrangement with the United Arab Emirates (UAE) on the basis of a 1992 security arrangement; meanwhile, some U.S. Navy equipment is prepositioned at Jebel Ali. In addition despite Saudi reluctance to allow large-scale U.S. prepositioning, it does allow it to store spares and electronics for support of fifteen tactical fighter equivalents. Some F-15s and F-16s have been based near Dhahran to enforce the no-fly zones in Iraq. The United States also constantly deploys U-2 and F-117 stealth aircraft in southern Saudi Arabia and rotated some of the latter into Kuwait in September 1996. Since August 1995 ships loaded with army equipment have been anchored offshore near Dhahran. The United States also has prepositioned naval stores in Bahrain, where it has had access to ports and airports, has the headquarters for its new Fifth Fleet, and had rotated in eighteen combat aircraft in 1995 to make up for a "gap" constantly an temporary absence of a carrier in the Gulf.[60]

A large quantity of materiel has also been prepositioned for the marines at Diego Garcia and Guam. At the former from where material was moved to Saudi Arabia in 1990, there is a Marine Prepositioning Squadron (MPS),

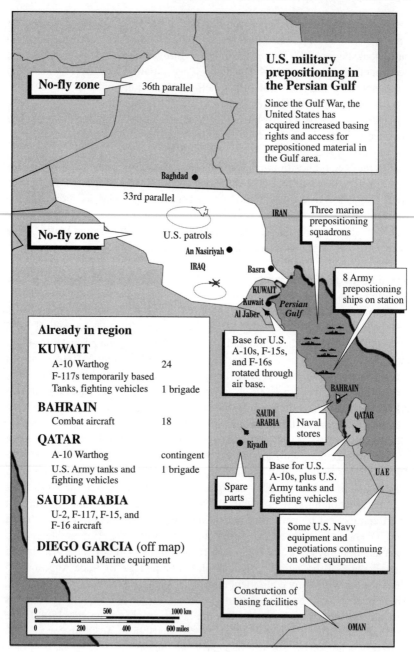

No-fly zone

36th parallel

U.S. military prepositioning in the Persian Gulf

Since the Gulf War, the United States has acquired increased basing rights and access for prepositioned material in the Gulf area.

Baghdad ●

33rd parallel

IRAN

Three marine prepositioning squadrons

No-fly zone

U.S. patrols

An Nasiriyah ●

IRAQ   Basra ●

8 Army prepositioning ships on station

KUWAIT

Kuwait ●   *Persian*
Al Jaber   *Gulf*

Base for U.S. A-10s, F-15s, and F-16s rotated through air base.

BAHRAIN

SAUDI
ARABIA   Naval
             stores   QATAR

● Riyadh

Base for U.S. A-10s, plus U.S. Army tanks and fighting vehicles   UAE

Spare parts

Some U.S. Navy equipment and negotiations continuing on other equipment

Construction of basing facilities

OMAN

Already in region

KUWAIT

A-10 Warthog	24
F-117s temporarily based	
Tanks, fighting vehicles	1 brigade

BAHRAIN

| Combat aircraft | 18 |

QATAR

| A-10 Warthog | contingent |
| U.S. Army tanks and fighting vehicles | 1 brigade |

SAUDI ARABIA

U-2, F-117, F-15, and F-16 aircraft

DIEGO GARCIA (off map)

Additional Marine equipment

| 0 | 500 | 1000 km |
| 0 | 200 | 400 | 600 miles |

Map 29

comprising five ships with equipment for one Marine Expeditionary Brigade (MEB).[61] These have recently been moved to the Persian Gulf. There is a similar deployment at Guam, involving only four ships. There is also prepositioned materiel in Norway for an MEB, which also could be moved into the greater Middle East region. Hence altogether there are thirteen such ships organized into three MPS squadrons, one each in the Atlantic Ocean, Indian Ocean, and western Pacific, each loaded with a thirty-day supply of combat equipment and sustainment for a force of some 17,300 marines.

In connection with prepositioning the United States has expanded and improved the basing infrastructure of some of the Gulf nations so as to allow for rapid deployments of air and naval forces in the event of a future crisis. In Oman, for instance, that has involved construction of barracks, shelters, and warehouses at four air bases at Masirah, Seeb, Khasab, and Thumrait. Overall the United States is working on arrangements with the southern Gulf states for facilities to base and sustain some 800 combat aircraft in the event of a crisis in the region.[62]

Actually, the marines are organized, potentially, in varied and larger units for overseas expeditions. There are three Marine Expeditionary Forces (MEFs) located, respectively, in the eastern United States in the Carolinas, in California, and in the western Pacific in Okinawa and Japan. These MEFs, made up of some 40,000 troops, are capable of sustained operations for up to sixty days without resupply of ammunition, food, water, or other supplies. A smaller and more basic building block of the U.S. Marine Corps is the Marine Air-Ground Task Force, described as an "integrated, combined-arms force comprising command, ground combat, aviation combat, and service support elements."[63] The prepositioning stocks are an aid and supplement to this process, and can be married up to these expeditionary forces in the event of overseas combat situations.

Not only weapons are prepositioned in and around the Gulf area, but also all kinds of support equipment and basic supplies needed to sustain forces ashore. For instance, even water supplies and desalination equipment are now part of this operation. A big effort has been made to ensure that everything—ports and airfields, access arrangements, weapons, and support equipment—fits together and can be made operational quickly.

The value of such prepositioned stocks and naval capabilities in ro-ro ships has actually been underscored in some recent crises after the Gulf War. In Operation Vigilant Warrior (October 1994), when the United States again had to deploy forces in the Persian Gulf area to deter Saddam Hussein from further mischief, eighteen afloat prepositioning ships based at Diego Garcia and in the Guam-Saipan area arrived within seven days. They delivered two million square feet of ammunition and equipment, including 9,000 military vehicles. During Operations Sea Signal (June 1994 to January 1996) in Guantanamo Bay and Uphold Democracy (September 1994) in Haiti, twelve ro-ro ships activated from the RRF were critical to timely logistical operations. So it would appear that in the time since the Gulf War the United States has had additional valuable experience in bringing to bear maritime support of land forces, particularly with regard to prepositioning and the use of modern ro-ro cargo ships.

In the crisis of October 1994, based on a combination of air and naval access and a broad prepositioning program, the United States had come to deploy a significant military presence in the Gulf region on a more or less stable basis. It had almost 20,000 personnel stationed in the area, of which some two-thirds were navy and marine personnel. It had two carrier battle groups in the area (one in the Red Sea) with 15 ships and 200 combat aircraft. There was prepositioned armor in Kuwait, five U.S. Marine Corps prepositioning ships moved from Diego Garcia, eight U.S. Army brigade afloat ships, and another six U.S. Air Force and Army prepositioning ships. In August 1995 in another minicrisis, it deployed twelve prepositioning ships from Diego Garcia and elsewhere, with enough materiel for an MEF of 16,500 and a 15,000–17,000 U.S. Army Corps. By the year 2000 the apparent goal is to have enough prepositioning and forward-basing capability to quickly deploy eight to ten tactical air wings and two army divisions, followed within eight weeks by four to five army divisions, an MEF, and additional air wings. In addition to the prepositioning of armored equipment in Kuwait and Qatar, there are plans for prepositioning several armored and mechanized battalions on five new ro-ro ships in the Gulf. That is in addition to the MPS (able to support an MEF) normally deployed at Diego Garcia, capable of supplying some 15,000 troops for thirty days. The U.S. Air Force and Army have additional prepositioning ships.[64]

Generally speaking (and ironically) by the time of the bombings that killed U.S. servicemen in Saudi Arabia in 1995–96, the level of U.S. military access to the Persian Gulf area had actually reached an all-time high. By mid-1996, there were 5,000 U.S. troops in Saudi Arabia and 4,000 combined in Kuwait, Bahrain, and the UAE, and perhaps that many again afloat in the region, in its periphery, or transiting through. Dubai has now become the U.S. Navy's biggest (measured by sailors ashore) port of call. Looking back, historically speaking, the turning point was the Earnest Will reflagging operation in 1987–88, at which juncture the United States was given significantly higher levels of access, particularly in the UAE, Bahrain, and Oman, perhaps largely because of the demonstration by the United States that it could make a commitment and stick to it. That confidence was presumably enhanced by the Gulf War experience and also by the military dominance achieved by the United States after the collapse of the Soviet Union. Whether the access acquired in the region by the United States will survive further terrorist incidents such as that in 1996—i.e., whether terrorism will cause the U.S. Congress to demand withdrawal of U.S. forces, as occurred earlier in Lebanon—remains to be seen.

For the long run, however, as it pertains to a possible future greater Middle East conflict involving the United States, prepositioning may actually come to play a less vital role. Some writings have begun to refer to prepositioning as a "sunset" system, that is, an incipient anachronism, to be rendered as such by arsenal ships that would reduce the need for the kinds of military weapons and stores used in Desert Storm.[65] In the light of such predictions, it is possible that long-range airlifts would also become a "sunset" phenomenon.

Impediments to Power Projection: Choke Points, Straits, and Air Overflight Corridors

Staging bases en route to the Middle East and ports and airfields within it are not the only important factors in power projection into the region; access through critical maritime choke points and overhead access for aircraft are also at least potentially significant. Access in these categories may vary considerably depending upon political conditions, as we shall see when we outline future conflict scenarios for the region, running the gamut from a repeat of the almost entirely unrestricted access seen during the Gulf War, to a far more limiting situation if the United States were attempting to conduct a more unilateral intervention, without support from key erstwhile regional friends. It is necessary in this context to underscore the importance of three strategic maritime choke points: the Suez Canal, the Bab el Mandeb Strait, and the Strait of Hormuz.[66]

Uninterrupted access through these three choke points is a vital component of U.S. Southwest Asian strategy, and the threats to access through these choke points are predominantly political and military, rather than legal or environmental. The key littoral states are either close allies of the United States or overt adversaries. The United States is unlikely to have six months of preparation time as it did in 1990–91. Denial of access through these points, or interrupted access, would seriously degrade the U.S. ability to reinforce and deploy in the Gulf from the CONUS and Europe, including the Mediterranean.

If any of these three choke pionts were closed, there would be other ways (besides prepositioning) of getting limited amounts of equipment to the northern Gulf, but each alternative poses enormous problems. If Hormuz is closed supplies could move overland from Al Fujayrah in the UAE to its Gulf ports such as Dubai. Though there is a road that runs into Saudi Arabia, this would be very time-consuming and is an inadequate route primarily because of limits on the capacity of the port. Furthermore, closing Hormuz would keep both warships and logistics vessels out of the Gulf, which would rule out a Desert Storm type of operation unless there had been massive prepositioning. Alternatively, if the Suez Canal were closed, the equipment could be moved by land across Egypt or, if Israel's relations with the Arabs were good, from Haifa to Eilat. The U.S. Navy currently uses Haifa for repairs and routine port calls and has agreed with Israeli shipyards to expand the available facilities. Moving goods across Egypt is limited by the inadequate ports on the Red Sea. With the closure of Bab el Mandeb one could go overland from Jidda to Riyadh, as the road network is good.

Alternative routes to the Gulf via the Cape of Good Hope or Indonesian straits face very serious inhibitors, especially time. The legal, political, and environmental constraints are likely to bedevil access through the Indonesian straits. The Islamic presence in the area could also create problems for U.S. warships. The Cape route poses serious weather problems, as sea conditions during the Cape winter are extremely poor. While this is not a problem for ordinary warships or large logistics vessels, it could have a serious effect on amphibious craft with troops on board.

The Suez Canal extends 121 miles from Port Said on the Mediterranean Sea to Suez on the Gulf of Suez (and the Red Sea). In the northern section the breadth of the canal is 1,197 feet at water level with 738 feet between buoys defining the navigable channel. The breadth of the southern section is 1,000 feet and 525 feet, respectively. The canal's permissible draught is 53 feet and can be increased to 56 feet by the Suez Canal Authority. This miniextension would enable a fully loaded vessel of 180,000 dead weight tons to traverse the canal, an increase of 30,000 dead weight tons.

As far as legal constraints are concerned, the canal is totally under the jurisdiction of the Egyptian government. U.S. access, whether commercial or military, is determined by the Egyptian government not the Law of the Sea (LOS). The Egyptian record on this is mixed. In the past there have been major limitations including bans on nuclear vessels, especially U.S. aircraft carriers. There was open access during Desert Storm and Desert Shield, but it is believed that very high prices were charged, especially for warships.

In 1993 Egypt approved a 3 percent fare increase. As in the past there is a 25 percent surcharge on warships, but nuclear-powered warships do not face any additional costs beyond the surcharge. In return for the higher fees, warships move to the front of the line, receive better and more experienced pilots, and are not required to pay transit fees in advance. The tariff for a SAM cruiser would be slightly more than $71,000, that for an aircraft carrier and five escorts would be $901,250. Egypt can use and has used all sorts of justifications for interference and fee increases, for example, environmental costs. This has occurred under an Egyptian regime closely tied to the United States. A less friendly regime resulting from an Islamic revolution could stop access overnight.

As far as physical inhibitors are concerned, there is a possibility of direct military action to prevent U.S. transit even if Egypt were on the U.S. side in a conflict. The canal is vulnerable to blockage, and one or two ships sunk, as was the case during the 1967 and 1973 wars, would be sufficient to close the canal. In addition ships going through the Gulf might be vulnerable to terrorist attacks if there were chaos or confusion within Egypt itself at the time.

The Bab el Mandeb, which links the Red Sea and the Gulf of Aden, is fourteen and one-half miles wide at its narrowest point. The coastal area is dry and barren plains with a few hills in the interior, and the surrounding water has coral reefs. Yemen is on the eastern side, and its closest land area, which is four to six miles wide, is covered by rocky, volcanic plains and several hills that are 600–900 feet high. The "Ras" is also surrounded by coral reefs.

At the narrowest point the strait is divided by Perim Island, which is roughly five square miles and dry and rocky. Shaped like a horseshoe with the ends facing southward and enclosing a small bay, the island, which once served as a British coaling station and is now controlled by Yemen, at times has accommodated small airfields.

Perim divides the strait into two distinct passages. The Small Strait, between the Asian coast and the island, is one and one-half to three miles wide and thirty-three to seventy-two feet deep. Strong and irregular tidal streams make navigation dangerous. The Large Strait, between the island and the African coast, is nine and one-quarter miles wide at its narrowest point. It is some 1,020 feet deep in the middle and is the preferred route for ocean-going ships. The entire strait, in the narrowest areas, is within the territorial waters of Djibouti and Yemen. Passage by ships of all nations is guaranteed under the LOS provision governing rights of transit passage.

Newly independent Eritrea has yet to delineate any maritime policy. Djibouti has several excessive claims, including an excessive straight baseline around the Seba Islands (1981), which are south of Bab el Mandeb. Djibouti also wants nuclear-powered warships (NPWs) and ships carrying nuclear materials to give prior notification (1979). To date the U.S. Navy, relying on its neither-confirm-nor-deny policy, has ignored this request.

Excessive Yemenite claims were initiated by the two governments prior to unification. For North Yemen this included prior notification for warship passage through the Bab (1967); prior notification for warships to pass through its twelve-nautical-mile territorial sea including the Bab (1978); prior authorization for NPWs (1967, 1982); and an eighteen-nautical-mile maritime security zone (1967). Claims by South Yemen include prior permission for warships to enter its twelve-nautical-mile territorial sea (1967, 1982, 1987); prior notification for NPWs and ships carrying nuclear materials (1978, 1987); and a twenty-four-nautical-mile maritime security zone (1978).

The United States rejects these restrictions as unlawful and operates its ships in accordance with international law. This applies to Yemeni claims before and after unification. In the past the United States also resisted similar Ethiopian claims. If anti-American groups gain control in Yemen, the government could attempt to enforce excessive claims or impose other restrictions to passage or both. In addition a conflict with the Saudis could destabilize the area. Yemen's relatively friendly relationship with Iraq is also troubling from an American perspective.

The Strait of Hormuz, which links the Persian Gulf and the Gulf of Oman, is twenty and three-quarter miles wide at its narrowest point. The port of Bandar Abbas and Iran lie to the north. Cape Musandam is part of the isolated Omani enclave that sits on the southern side, which is forty-two miles north of the rest of Oman. A strategic cluster of islands that includes Little Tunb, Abu Musa, and Great Tunb lies west of the strait; Iran and the UAE share control of these three islands. Two other islands in this area, Jazireh-ye Bani Forur and Sirri, are owned by Iran. The strait is 600 miles southeast of Abadan, an Iranian city at the northern end of the Persian Gulf.

The water is generally deeper on the Arabian side than the Iranian side of the strait. The inflow of water passes along the northern shore and the outflow along the southern shore. There are strong tidal currents near the Gulf entrance; storms, fog, salt, haze, and dust can all result in poor visibility; and relatively shallow water restricts submarine operations. The mean water temperature ranges from 70–90°F (21–32°C). Owing to the narrowness of the strait, oil tanker traffic is congested and easily interdicted by direct attack or mining.

From the legal perspective the strait at its narrowest point is within the overlapping territorial waters of Iran and Oman. The United States generally ignores excessive maritime claims. As of early 1992 Iran's excessive claims include declarations that seek to restrict transit in the Strait of Hormuz to LOS signatories (1982) and ask for prior notification for warships to enter its twelve-nautical-mile territorial sea (1982). The former declaration theoretically excludes American vessels, but the United States has rejected this position based on its compliance with customary international law. Excessive Omani claims include an excessive territorial sea baseline (1982); prior permission for warships to engage in innocent passage (1989); and prior permission for NPWs and ships carrying nuclear materials to engage in innocent passage (1989).

In 1993 a traffic separation scheme that transits traffic through Iranian waters was introduced. The Strait of Hormuz alternative routing system, which uses a deep channel unobstructed by oil platforms, has been a success, and five additional routes have been added. The U.S. Navy, which had made 110 transits as of May 25, 1993, has traditionally ignored or disputed excessive claims through its freedom-of-navigation program.

A number of political issues are relevant. While Oman may have tacitly shared the U.S. goals during the Iran-Iraq war, its public face was neutral, unlike Bahrain and Saudi Arabia. Oman fully supported Desert Shield–Desert Storm. Naval relations were very good between the two countries, and this made both operations easier. While access does not depend on good relations with Iran or Oman, good relations with Oman are helpful. U.S. relations with Iran are very poor and there is little room for naval compromise or cooperation.

There is a fourth, albeit highly hypothetical, possibility regarding choke point transit problems, and that involves the Straits of Gibraltar. British retention of a major base at Gibraltar presumably is a major plus in terms of future U.S. or Western access. But the possibility of a radical Islamic revolution in Algeria and the associated danger of conflict between the latter and France and other European states could raise some questions about access through Gibraltar in the case of an intended U.S. intervention in the Middle East, at least as it pertains to the possibility of terrorist actions if not quasi-legal governmental restrictions on transit. Of course, Algeria does not sit astride the strait, but it is in near proximity and does straddle a relatively narrow part of the Mediterranean. Far more hypothetical is the possibility of a revolution in Morocco that would lead to attempts to block access through Gibraltar, or political pressures on Spain from the Middle East that would also militate in this direction, perhaps in the context of worsening problems between Britain and Spain over Gibraltar.

Elsewhere, closure of one or more of the Indonesian straits could also affect possible U.S. power projection toward the Middle East. A Turkish revolution could lead to a change in the politicolegal situation governing access through the Turkish straits that could affect Russian (and Ukrainian) egress as well as U.S. ingress if the latter, hypothetically, might ever have reason to want to move its navy into the Black Sea.

During the Gulf War the United States was availed of overhead access for aircraft all over Europe and the Middle East, apparently including over Eastern Europe and perhaps over Russia. But such unrestricted access has not always been so assured during military interventions, for either the United States or other major powers. For instance, in 1986 the United States was not allowed overhead access by France or Spain for its raid on Libya mounted from Britain and was forced to fly around through the Gibraltar Strait, a route that wearied aircraft crews and stressed fuel capacities, probably to the extent of costing the United States some lost aircraft and crews. The United States had similar problems in its 1973 airlift on behalf of Israel. The Soviets in resupplying Ethiopia in 1977–78 were reputed to have ignored some nations' restrictions on overflights and to have gone ahead anyway, with an implicit warning of retaliation if access were interfered with.

In the future the United States could in some circumstances have major problems in projecting power into the Middle East if overhead access were denied. It can, presumably, again fly from the CONUS through the Gibraltar Strait and on to the eastern Mediterranean without hindrance. If Egypt were a problem regarding overflights, Israel could provide a corridor through to Eilat, leaving the problem of Saudi overflight permissions en route to the Gulf, lest the United States have to fly all around the Arabian Peninsula. Corridors across various parts of Africa might be available in some circumstances, particularly with the wholesale reorientations of political positions that have followed the end of the cold war. Overflight restrictions in Southeast Asia could lengthen some U.S. logistics routes and could affect future movement of prepositioned U.S. materiel out of Europe toward the Middle East. At the extremes, if U.S. access were highly restricted in a unilateral action, the combination of tanker capability and access to British-controlled Ascension Island and Diego Garcia, in addition to Guam, would allow the United States to conduct logistics operations into the greater Middle East, albeit with a big penalty in terms of cost, time, and cargo load reduced by extra fuel requirements.

EIGHT

Geography and Weapons of Mass Destruction

AS THE GULF WAR demonstrated, the further proliferation of weapons of mass destruction (WMDs) in the greater Middle East region is a reality (see map 30). In a region in which primordial hatreds abound and military budgets and arms acquisitions continue on at high levels, this has profound dangers. Indeed, the generic problem of proliferation is now centered almost entirely on this region, with the sole exception of North Korea. It has become a central focus of U.S. national security policy, all the more so as it is so closely linked to the continuing problem of containing Iraq and Iran.

WMDs, long present in the Middle East, are assuming an increasing level of importance owing to the growing strategic reach of several regional states and evidence of the use of these weapons in recent wars. This geographic factor could change the relationships and even the types of confrontations within the region.

The Iran-Iraq war, the Gulf War, and the collapse of the USSR have had major but differing impacts upon the prospects for proliferation. The massive and overt use of chemical weapons in the Iran-Iraq war loosened long-held international norms restricting their use, making it more likely they will be used again in the future.

In a related vein the extensive use by Iraq of conventionally armed Scud missiles for bombardment of Israel and Saudi Arabia, itself a follow-on to the missile wars between Iraq and Iran, and the use of Scuds in the Afghanistan conflict set a grisly precedent. The missiles fired at Israeli and Saudi cities were very primitive, old systems. Future conflicts might see the use of far more sophisticated, accurate, and longer-range missiles, perhaps armed with "un-

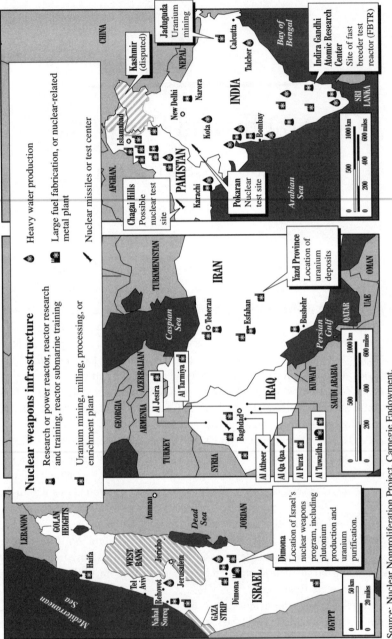

Map 30

Source: Nuclear Nonproliferation Project, Carnegie Endowment.

conventional" warheads.[1] In particular the growing range capabilities of missiles entering the inventories of several of the region's states raise all manner of possibilities for threats to countries not only within the Middle East, but around its peripheries. Defense planners are now beginning to draw increasingly broader concentric circles centered on India, Pakistan, Iran, Iraq, Israel, and Libya, trying to gauge the ever-expanding limits of each of these nations' growing strategic reach (see map 31). Some of these countries, such as Israel and India, have already deployed and used space launch vehicles that could be reconfigured into surface-to-surface missiles with intercontinental reaches.

The Gulf War taught another geography lesson. Countries not directly involved in conflicts could be drawn in as targets for a form of indirect aggression. Iraq, unable to strike back at the United States in response to the latter's all-out strategic campaign against its infrastructure and command and control centers, instead launched a missile attack against Israel. That may have seemed anomalous in the context of the Arab-Israeli conflict and, moreover, Iraq was no doubt deterred from the next step of use of chemical weapons against Israel by the prospect of having to deal with the latter's nuclear weapons, but this episode may be the harbinger of a new form of strategic warfare. Missiles become the weapon of the weak, wherein regional countries threatened by overwhelming force may be availed of a form of deterrence by the ability and the will to strike nearby states friendly to the great power adversary (or perhaps even neutral countries). As the strategic reach of some of the regional powers expands, this concept may come to apply to threats not just against contiguous (Saudi Arabia) or nearby (Israel) countries, but indeed to vulnerable countries further away.

The collapse of the USSR has also had a major impact on regional strategic calculations and the growing specter of proliferation. For one thing the hitherto once massive and impenetrable frontier barrier between the southern USSR and the Middle East has not only become freer and more permeable in a general sense, uniting heretofore separate parts of the Islamic world, but has become a potential sieve for the transfer of military technology, perhaps to include nuclear, chemical, and biological weapons and missile technology.[2] Even in the depths of the cold war, the United States and the Soviet Union had had convergent concerns and policies when it came to nuclear proliferation; indeed, the Soviets were obviously more threatened by the prospect of new nuclear states near their borders than were the Americans. Now, however, not only are the Russian borders far more permeable, but commercial considerations compete with geopolitical ones so that both Russia and China have become potential sources for the transfer of sensitive technology and materials related to WMDs.

Then, too, the old Soviet Union had earlier provided some form of extended deterrence—nuclear or conventional or both—on behalf of its regional friends and clients such as Syria, Iraq, and Libya. Now that umbrella has been re-

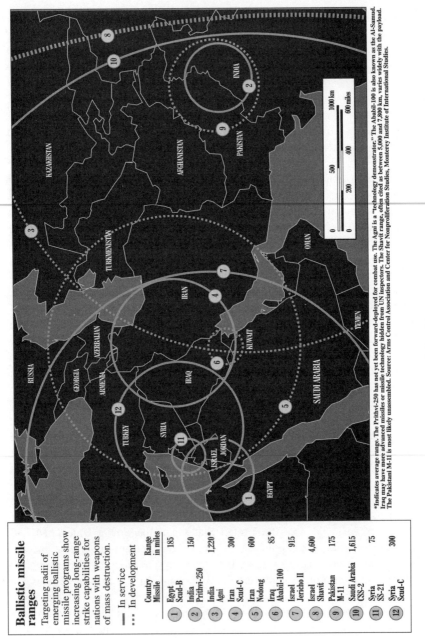

Ballistic missile ranges

Targeting radii of emerging ballistic missile programs show increasing long-range strike capabilities for nations with weapons of mass destruction.

— In service
··· In development

	Country Missile	Range in miles
1	Egypt Scud-B	185
2	India Prithvi-250	150
3	India Agni	1,220*
4	Iran Scud-C	300
5	Iran Nodong	600
6	Iraq Ababil-100	85*
7	Israel Jericho II	915
8	Israel Shavit	4,600
9	Pakistan M-11	175
10	Saudi Arabia CSS-2	1,615
11	Syria SS-21	75
12	Syria Scud-C	300

*Indicates average range. The Prithvi-250 has not yet been forward-deployed for combat use. The Agni is a "technology demonstrator." The Ababil-100 is also known as the Al-Samud. Iraq may have more advanced missiles or missile technology hidden from UN inspectors. The Shavit range, often cited as between 5,000 and 7,300 km, varies widely with the payload. The Pakistani M-11 is most likely unassembled. Source: Arms Control Association and Center for Nonproliferation Studies, Monterey Institute of International Studies.

Map 31

moved. The effect has been that the United States has had a freer hand for military operations in the region—as it exhibited in the 1991 Gulf War—and that Israel's and maybe Pakistan's threshold for use or threat of use of nuclear weapons might have been lowered. Regarding the former assumption one Indian military expert, in being asked his opinion about the lessons of the Gulf War, made the widely reported remark that its primary lesson was that one should not go to war with the United States unless one has nuclear weapons.[3]

There is, however, no consensus about the long-range implications of these developments, which are centered on the region's three major conflict pairings: Arabs versus Israel, Iraq versus Iran, and Pakistan versus India. Some writers, for instance, the noted Israeli military historian Martin van Creveld, are convinced that large-scale conventional warfare (and with it, strategy itself) has become a virtual anachronism and that the possession of nuclear weapons by a growing number of middle-range or developing countries is more likely to produce stability than the specter of regional nuclear war.[4] That view further perceives the future in the third world as dominated by low-intensity conflict in all of its manifestations—terrorism, ethnic conflict, and the like.

Creveld's thesis that nuclear weapons are likely to produce stability in third world contexts—nuclear stability plus the lowering of the probabilities for conventional warfare—is similar to that advanced by Kenneth Waltz, who long has expounded the virtues of nuclear proliferation as providing stability, both in the former superpower context and with respect to regional rivalries in the third world.

Others, however, strongly disagree, perceiving the newer, emerging nuclear rivalries as inherently *more* unstable than the previous superpower standoff.[5] This view is based on assumptions about the hatreds involved and also on the assumption that regional states such as Iraq and Iran are more prone to irrational decisionmaking, whether or not culturally based, even where there is the specter of nuclear doom. The factor of geographic propinquity is assumed to be crucial here. Neither the United States nor the Soviet Union ever had reasonable grounds to fear that their territories and populations would be overrun, literally, by the other side's armies. However, in the Middle East the national security policies of Israel, Pakistan, and perhaps Iraq are driven by fears of national dismemberment. Finally, the loosening of norms with regard to the use of chemical weapons may have implications for the lowering of the nuclear threshold, raising the probability that conventional wars might quickly get out of hand, moving up the ladder of escalation through the use of chemical weapons and then a reactive use of nuclear arms.

In this chapter we briefly review the current and projected statuses, to the extent known, of the region's nuclear, chemical, and biological warfare programs. Then we examine in more detail their associated delivery systems with a specific focus on geographical factors, that is, location and distance, and their relationship to the possibilities for preemption of nascent nuclear programs.

Finally, we take a brief look at the geographical aspects of the clandestine transfer of nuclear technology across borders.

Regional Nuclear Weapons Programs

Until recently most of the literature on nuclear proliferation focused on whether certain countries were capable of building nuclear weapons, or whether they wanted to achieve nuclear status (capabilities and intentions). Now there is growing concern with the strategic implications of a world in which there is more proliferation. In the Middle East there is presently a two-tiered breakdown: the new nuclear powers (Israel, India, Pakistan) and those who would like to achieve that status, that is, Iran and Iraq in the first instance, trailed by Libya and Algeria, and then perhaps in the long run by Syria and Saudi Arabia.

What is the status of these programs and what can we anticipate in the coming years? Until recently it was common in academic analyses and in statements issued by the U.S. government to treat the Israeli, Pakistani, and Indian nuclear programs as "threshold" situations. There were frequent referrals to "bombs in the basement" or to the idea that operational weapons could be produced by a "turn of the screw" with relatively short warning time or in some more protracted interval (say, a few months), given appropriate strategic warning. Now it appears very probable that all three of these powers have on occasion deployed fully operational nuclear units, primed to function at short notice.

The details of what is publicly known about the region's various ongoing nuclear, chemical, and biological warfare programs are reviewed in appendix 4, but they can be summarized briefly here.

By the mid-1990s there was the widespread assumption that Israel, India, and Pakistan all had acquired operational nuclear weapons. Depending upon the source Israel was reported to possess anywhere from 50 to 300 nuclear weapons, perhaps comprising a mix of Pu-239 and U-235 fission weapons, fusion weapons, and related neutron weapons for tactical purposes. India and Pakistan were widely believed to possess much smaller but growing nuclear arsenals, perhaps in the range of 10 to 20 weapons, though some experts suspected a much larger arsenal in India, which is now more than twenty years beyond its test of a nuclear device and is known to possess several unsafe-guarded plutonium production facilities. India has refused to go along with a nuclear test ban; Israel may or may not have tested in the ocean below South Africa (but no doubt has conducted computer simulation tests minus the bomb material); and Pakistan may have had its weapons tested by China.[6]

Iraq and Iran are both assumed to be moving toward nuclear acquisitions. The former may have been only a year or two away at the time of the Gulf War, after which its nuclear infrastructure was largely—but presumably not to-tally—dismantled by the UN. But the knowledge and much of the wherewithal remain. Iran appears to be striving mightily to acquire nuclear weapons, and the amount of time that may require is estimated variously from five to ten to fifteen years. It will if it can, or cannot be stopped, and it may be getting crucial help from sources in Russia, China, and Pakistan. Algeria and Libya have given indications of wanting atomic bombs, but in neither case is such an eventuality feared for the short run. Egypt, Syria, and maybe Saudi Arabia and Turkey loom as other long-run regional candidates for nuclear status.

Virtually all of the major actors in the region have been reported developing chemical or biological weapons or both: Syria, Egypt, Iraq, Iran, Israel, and Libya primarily, but also India and Pakistan. Chemical weapons have already been used in Yemen, Sudan, Afghanistan, and in the Iran-Iraq war, and in the hands of Iraq, Syria, and Egypt vis-à-vis Israel are commonly discussed in the context of a "poor man's nuclear weapon," that is, an asymmetrically weak but still potentially potent counterdeterrent. Biological weapons may have the potential of causing widespread casualties roughly on a par with nuclear weapons, and it is obvious that large-scale inventories of anthrax and other biological warfare materials have been acquired by Iran, Iraq, and perhaps other regional nations, weapons that can be delivered by aircraft or missile.

Delivery Systems

Each of the countries in the greater Middle East that has developed or is developing nuclear, chemical, and biological weapons has also been moving ahead with corresponding long-range delivery systems; variously and usually in combination these systems involve surface-to-surface ballistic missiles, cruise missiles that can be launched from the ground and from aircraft and ships, and, of course, combat aircraft ranging from tactical strike aircraft to larger bombers. Beyond the simple question of the possession of such systems, there are also the problems, indeed difficult ones for developing nations, of producing adequate fusing systems for the weapons themselves and, in the case of missiles, such technology as guidance systems and ablative shields to prevent warheads from burning up upon reentry into the earth's atmosphere. But each of the countries we have discussed has had assistance from major powers including Russia and China across the board, from the purchase of military and related civilian dual-use technology to the loan of military technicians and consultants.

Use of Missiles during the Iran-Iraq and Gulf Wars

During the long Iran-Iraq war Iraq greatly escalated its ability to use conventionally armed missiles as terror weapons against Iranian urban centers. At the outset of the war it was only capable of using some shorter-range FROG-7 missiles, acquired from the USSR, against nearby Iranian staging areas close to Dizful and Ahwaz, without much noticeable effect.[7] But as early as 1980 Iraq began using unmodified Scud missiles against Iranian cities near the border. These weapons did not have the range to reach Teheran, which is 315 miles from the border, some 140 miles beyond the range of the Scud B, which then had a range of 175 miles, a circular error probability (CEP) of about 3,000 feet, and a 2,200-pound warhead. Iraq was later able to modify the Scuds, with assistance from various sources (China, Egypt, France, Germany, and the Soviet Union are all possible suspects), and by 1987 it was able to test a new missile, the Al-Hussein, with a range of about 400 miles and a CEP in the range of 6,600–9,800 feet.[8] Now Iraq had the range capability to reach Teheran and Qom from firing positions south of Baghdad, and it began to use that capability. Some estimates are, according to Cordesman, that Iraq fired more than 160 missiles at Teheran between late February and mid-April of 1988, even if with smaller warheads than the unmodified Scud Bs, so as to allow for the extension of range.[9] These terror weapons, along with the use of chemical weapons on the battlefield, were considered among the major reasons for the Iranian collapse in the summer of 1988.

After the Iran-Iraq war and up to the beginning of the Gulf War, Iraq worked furiously on new and improved missile programs and on its "super gun." The scale of these programs, when they were unearthed after the Gulf War, stunned even close observers of Middle Eastern security affairs. After 1991 UN officials were able to determine that Iraq had had some fifty-two missile storage, assembly, and maintenance facilities, in addition to numerous others related to its chemical, biological, and nuclear weapons programs. It began to produce and test a number of new missile systems, with the assistance of scientists and companies from several nations, including Egypt and Argentina (the Condor program), China, Germany, Switzerland, Brazil, and Britain. By late 1989 Iraq had announced that it had tested the new Tamuz 1 missile with a range of 930 miles, which claims were not verified; there were also reports of a solid-fueled missile with shorter range perhaps intended for mounting nuclear warheads. In addition Iraq improved its missile deployment capabilities, using its own manufactured launchers as well as TELs (transporter-erector launchers) acquired from the USSR. By February 1990, according to Cordesman, just before Iraq's attack on Kuwait, U.S. intelligence was aware of five fixed missile launcher complexes in western Iraq, with twenty-eight operational launchers armed with missiles that could easily reach Israeli cities as well as targets in Syria and Turkey. By the beginning of the Gulf War Iraq apparently had at least 1,000 Scud missiles of all types, plus several hundred extended range Al-Hussein and Al-Abbas missiles. To fire these it appears to have had something like 225 of its indigenously produced and cheaper launchers, as well as some 36 Soviet-style TELs, numbers well above those estimated by Western intelligence sources before the war.[10]

During the Gulf War Iraq used its conventionally armed missiles as terror weapons against Israel and Saudi Arabia, firing a total of ninety-one missiles, forty, forty-eight, and three, respectively, against those two countries and Bahrain, causing relatively little damage and few casualties, with the exception of the one shot that killed a number of U.S. military personnel in Saudi Arabia.[11] The missile fusillade against Israel almost succeeded in drawing the latter into the war, but Iraq did eschew the use of chemical weapons, no doubt in large measure because of the fear of an Israeli nuclear counterstrike.

After the war UN inspection teams uncovered at least seventeen sites where Iraq had conducted research, production, testing, and repair of its missiles, and destroyed most of the missile stocks that Iraq had declared in addition to much related ancillary equipment (Iraq had declared only fifty-two ballistic missiles and thirty-eight launchers). Actually the UN destroyed sixty-two missiles, eleven missile decoys, numerous fixed and mobile launchers, and a small number of transporters. It also found some thirty chemical warheads for Scuds, most of them nerve gas warheads. But it was clear that Iraq probably had succeeded in concealing a large number of missiles and launchers, as well as much of the manufacturing and testing equipment and parts, that it had purchased before the war. It may have retained many underground storage facilities built before the war, and have built more of them after the war. In January 1992 the director of the CIA, testifying before Congress, averred that Iraq may have retained "hundreds" of its missiles after the war, even in the face of the UN inspections.[12] It is reported that U.S. satellite photos have revealed that Iraq has rebuilt its missile research facility at Al-Kindi and that it has continued to produce and test short-range missiles. That, of course, could be the basis of a missile inventory that could be made usable in a new round of an Iran-Iraq war, not to mention in an Arab war against Israel. Similarly, the CIA director testified that Iraq had probably succeeded in hiding much of the production equipment for its chemical weapons program before the coalition bombing attacks. It appeared that much of this capability had been retained, and that Iraq could, after the lifting of sanctions, easily resume its chemical weapons production and be back to its earlier production capability within a year. It may also have retained much of its VX gas production capability and stockpile as well as its biological weapons capability, UN efforts notwithstanding. That capability had earlier involved laboratory and industrial capacity to produce various biological warfare agents, including anthrax, botulism, some mycotoxins, camel pox virus, and others. Further, Iraq had conducted an extensive weaponization program in this area, including live firings of 122-mm rockets with anthrax and botulin toxin. At the time of the Gulf War it is reported to have possessed bombs and missile warheads containing botulism, anthrax, and aflatoxin, a natural carcinogen. Some of these warheads could have been fitted to advanced Scud missiles.[13]

Iran also used Scud missiles as terror weapons against Iraq in the earlier war, using the Scud B, with a range of around 190 miles, with 1,760-pound warheads. Iran used its first such weapon in March 1985, fired about forty of them between 1985 and 1987, and then seventy-seven more in 1988 during the "war of the cities."[14] Initially Iran actually had an advantage in this respect because of the strategic depth factor, wherein it could target Baghdad from western Iran, but Iraq could not target Teheran up to the latter stages of the war because of the limited range of the Scud B. But Iran's Scud attacks were not all that effective, largely because it lacked the quantity of missiles required for sustained attacks, and their inaccuracy was such that they could not be targeted effectively at Baghdad. Unlike Iraq, Iran was not able substantially to improve its Scud Bs during the war, with regard either to range or accuracy.

After the war Iran continued to purchase Scud Bs from North Korea, some 200–300 of them between 1987 and 1992, and, more important, has also been attempting to acquire M-9 and M-11 missiles from North Korea as well as a new Chinese missile with a range of some 600 miles. With North Korean aid and Syrian cooperation, Iran has also been working on an improved version of the Scud, the so-called Scud C, and Syria and Iran may be working on acquiring and producing a longer-range North Korean missile, the Nodong I, with a range of 600 miles. All of this is, of course, in conjunction with ongoing chemical and nuclear weapons programs, in this case unhindered by UN inspections, albeit also perhaps without the kind of massive technological input from Western firms equivalent to what Iraq had in the period before the Gulf War.[15]

Missiles under Development

As map 31 shows, the ranges (and payloads) of some of the in-service missiles in the greater Middle Eastern region, not to mention those of some missiles under development, are impressive and ominous, particularly when one begins to draw omnidirectional radial arcs from around the borders of these nations or from known or suspected missile or weapons storage sites. India, for instance, is moving toward the deployment of the 1,500-mile-range Agni missile, but already has the PSLV and ASLV space launch missiles in service, with ranges of 5,000 and 2,500 miles, respectively, missiles assumed easily convertible into an intercontinental ballistic missile (ICBM). Those ranges allow for coverage of all of China and most of the Middle East and Central Asia. Pakistan has already deployed short-range missiles, and has others with up to 370-mile ranges under development, which will put much of India under threat. Israel's Jericho II (900-mile range) has been tested downrange to the

Mediterranean waters north of Libya; its Shavit space launcher, assumed convertible into an ICBM, is rated at a range of 4,660 miles.[16] Iran is acquiring 370-mile-range Scud Cs, but has a 660-mile missile under development (in effect, North Korea's Nodong I), which can target Israel, the Gulf, and Turkey.[17] Indeed, in one of the hypothetical scenarios in the RAND Corporation's recent "The Day After . . . Study," Iran was assumed by 1996 to have completed a successful series of test flights of a new 1,500-mile-range intermediate-range ballistic missile (IRBM) "that is clearly derivative of the Chinese CSS-X and may be identical to the new Pakistani Hatf III IRBM."[18] Those ranges would, of course, easily put Israel at risk, as well as large parts of Russia and India. Iraq, now hamstrung by the destruction of its known missile arsenal, is assumed to aspire again to similar capabilities, and is claimed to have new missiles under development. Saudi Arabia has in service 1,700-mile-range CSS-2s acquired from China. Syria, meanwhile, with shorter range requirements seemingly for the most part restricted to targeting Israel, has concentrated on shorter range but more accurate missiles, such as the Soviet SS-21, but has acquired the 370-mile-range Scud C from North Korea and may buy the 370-mile-range Chinese M-9, which would allow for deployment some distance from the Israeli borders. There is no denying the ominous, cumulative impact of these developments.

As indicated in some of the accompanying maps, the ranges associated with some of these missiles are of critical strategic importance, particularly if they are to be armed with WMDs. Indeed, such calculations have been important for some time. When Israel first deployed the Jericho I in the late 1960s, presumably in parallel with its having deployed operational nuclear weapons, it was important to note that those missiles—whose apparent accuracy did not allow for hard-target strikes—could reach to Damascus, Amman, Cairo, Alexandria, and the Aswan Dam, but *not* to Baghdad, Riyadh, Teheran, or the Persian Gulf oil fields, nor to any of the cities in the old USSR.[19] The Jericho II, however, with approximately three times the range of its predecessor, can reach to all of the latter targets, and an ICBM based on the Shavit space launcher can reach still further than that, including within the region such potential deterrent-value targets as Tripoli (1,300 miles), Karachi and Islamabad (1,990–2,175 miles), Teheran (930 miles), and the Persian Gulf oil fields (930–990 miles), as well as Moscow at 1,550 miles. Iraq's Al-Hussein upgrade of the Soviet-supplied Scud was able to target all of Israel's major cities from firing positions in western Iraq, and its mobility made it hard to locate and destroy. Its possible follow-on missiles, such as the Al-Abbas, which was under development before the UN dismantling operation, had a predicted range of some 560 miles, which would have allowed for targeting Israel from the eastern side of Iraq and knocking them out would have been a much tougher job for Israel's air force. Iran, on the other hand, would not be able to target Israel (though it could easily target all of Iraq) with its Scud Cs, though it apparently could do so with its reportedly under-development Tondar-68 system, which has a predicted range of about 620 miles. If it should ever develop or acquire missiles with ranges up around 2,500 miles, it would be able to strike targets in Central Europe. Syria, meanwhile, if it should succeed in deploying 370-mile-range Scud Cs or Chinese-origin M-9s of similar range would be able to target Israel from far back in northeastern Syria, which could provide for more effective hiding and defensive measures against Israeli air attacks.

The same kinds of calculations are germane to an understanding of the evolving strategic equation in South Asia and the relationship between India and China. Pakistan is reportedly working on a Hatf-3 missile with a range of some 370 miles. That would allow it to target New Delhi, but not Bombay (500 miles), which would still apparently require airstrikes. But it is now on the verge of deploying, perhaps with nuclear warheads, the M-11 missiles of Chinese origin and design, built with blueprints and equipment supplied by China. Those missiles, claimed to be in storage at Sargodha airbase near Lahore, are said to have a range of more than 185 miles, almost able to reach New Delhi.[20] India can easily cover Pakistan with the about-to-be-deployed Agni system, but to achieve counter-cities deterrence against Beijing and the remainder of the Chinese heartland, it will need to convert its ASLV space launch vehicle with a range of around 2,500 miles or the PSLV with a range of 5,000 miles to an ICBM mode. But by 1996 India had test-fired its most advanced Prithvi missile with a range of 150 miles, which, given Pakistan's lack of strategic depth, would allow India to strike most major Pakistani cities in less than five minutes after launch.[21]

Libya, meanwhile, is reported to have under development an Al-Fatah missile with a 590-mile range which, if it should ever become operational, would, for instance, be able to target Italy up almost to Rome. Israel would still be out of reach, requiring presumably aircraft with refueling capability. Algeria is not now among the nations reported to be working on long-range ballistic missiles, but if such systems were added to its seemingly nascent nuclear ambitions, it would require systems of 500- and 800-mile ranges to reach Marseille and Paris, respectively. These are not intended as predictions of future strategic requirements or ambitions, but are seen merely as the outer limits of some scary future possibilities, that is, worst-case scenarios from the perspectives of those nations that could possibly be targeted.

It should be stressed that the acquisition of long-range ballistic missiles by the countries of the region is by no means totally dependent upon indigenous production or assisted regional programs. There are now reports of numerous possible acquisitions of such missiles from Russia, China, North Korea, and Ukraine (possessor of a large remnant arsenal of Soviet missiles) as well as reports of India's attempts to sell missiles in the Middle East. Iran has been shopping in North Korea and Russia, which may yet sell its SS-25 and SS-19 ICBMs. Ukraine has been reported as establishing a secret strategic cooperation relationship with Libya that may entail transfers of missile technology. China is selling CSS-2 medium-range missiles to Pakistan. Iraq, Algeria, and Libya are in the market for Indian as well as Russian and North Korean missiles. If such deals are consummated, they would result in Libya and Algeria being able to target southern Europe with chemical or biological warheads.[22]

As we have indicated above the ranges that are expected to be achieved from the developing ballistic missile programs in the greater Middle East are one critical aspect of the new strategic geography of the region. But there may be others. As was underscored by the massive but somewhat futile efforts of the U.S. Air Force to locate and destroy the Scud missiles that were bombarding Israel and Saudi Arabia during the Gulf War, the topography of the region as it relates to cover and concealment of both fixed and mobile missiles is important. The terrain of western Iraq is not too rugged and the weather is normally clear, but Iraq was able to make use of wadis, gullies, culverts, under-road passes, and perhaps some dug-in bunkers to render the American targeting operation from overhead very ineffective (perhaps somewhat compensated for by the clandestine operations of Special Forces moving about in dune buggies).[23] With several of the region's states now beginning to deploy long-range missiles, some mobile and some not, these developments will become increasingly important in connection with the deployment in some cases, such as Israel, of indigenous satellites. Whether Israel, for instance, can successfully hide its Jericho missile firing batteries is an important question—they are now apparently clustered in hardened silos dug into mountainsides, particularly in the area between Jerusalem and Tel Aviv near the village of Al Zakharia, among other places.[24] Another important and interesting question is whether Iran will be able to hide its missiles in open terrain from the Israeli air force, if it should come to that. Generally the cover and weather of the region appear to favor the side that has good surveillance capability, but the lessons of the Gulf War indicate that topography is also important.

The Cruise Missile Threat

Although most of the attention—as it pertains to delivery systems for WMDs and long-range strategic reach—has been devoted to ballistic missiles, analysts such as Seth Carus have reminded us of the potential additional or alternative importance of cruise missiles—"air breathers."[25] These can actually be used for long-range strategic purposes, as well as for tactical land attack or as harassment drones to fool air defense radars.

Cruise missiles have been developed and deployed by many nations, apart from the United States and the Soviet Union the list includes Brazil, Argentina, China, France, Germany, Israel, Italy, Japan, North Korea, Norway, South Africa, Sweden, and Taiwan. Numerous former Soviet client states, including those in the Middle East, are in possession of long-range air-to-surface cruise missiles, some of which could be armed with unconventional weapons.

Although, generally speaking, cruise missiles are less efficacious vehicles for long-range delivery of nuclear weapons than ballistic missiles or combat aircraft (because they are relatively slow and vulnerable to air defenses), they are in some respects advantageous for delivery of biological or chemical weapons, as their payloads are not subject to the stresses of exoatmospheric trajectories and descents.[26] Cost is another plus for cruise missiles. In addition the advent of effective antitactical ballistic missiles (ATBMs) plus the impending possible advent of stealth capabilities for cruise missiles may shift the advantage back to the latter relative to ballistic missiles. On a more tactical level—but with strategic purposes—the worth of the cruise missiles was underscored by the incredible accuracy and reliability displayed by U.S. Tomahawks (fired from surface ships and submarines) during the Gulf War.

With regard to the Middle East several points are worth noting. While most of the nations in the region probably look to cruise missiles for tactical rather than strategic purposes, some are also looking at strategic applications, for example, Iraq's reported interest in a ramjet system with a range of 400 to 500 miles, which could be used (as its earlier ballistic systems) against noncontiguous rival nations. Then, too, the onrush of technology has apparently solved one major technical problem—there are now terrain-guidance systems that can operate over the flat and featureless Middle Eastern deserts. (At the outset of the Gulf War the United States appeared to lack adequate terrain-matching maps for its Tomahawks, resulting in extensive use of an attack route along the Kuwait-Basra highway.) It was also demonstrated that global positioning systems can now be used in place of TERCOM, the terrain-matching technology, which removes a major cost factor for third world development of cruise missiles and also makes their use much easier in flat, uninhabited desert areas. Newer satellite systems such as the Global Navigation Satellite System seem likely to enhance these capabilities, and this is one area where third world countries may easily be able to rely on commercial or dual-use technology transfers. Finally, it is claimed that there are impending developments that may result in widespread possession of terminal guidance technologies that will increase the accuracy of cruise missiles.[27]

One further thought that might be added is that some nations might wish to deploy both ballistic and cruise missiles in connection with WMDs just to provide a more diverse array of offensive options to complicate planning for the defensive side.

Strike Aircraft

The extensive concentration on missile delivery systems for WMDs should not cause us to forget that the alternative use of attack aircraft remains a very viable option,[28] in some cases perhaps a preferred one. Aircraft are, of course, vulnerable to interceptor aircraft and surface-to-air missiles (SAMs). But they may also be more reliable as well as recallable. Whereas nonmobile ballistic missiles may provide hard targets for a preemptive attack, aircraft can be placed on round-the-clock aerial alert or on quick-response runway alert. Aircraft can also be provided with standoff capabilities, using cruise or other missiles, sometimes with a reach of 1,000 miles or more. But, also, the newer precision strike technologies involving laser designators and television-guided bombs have greatly increased the accuracy of aircraft-delivered weapons, and these technologies are expected to be widely diffused throughout the third world in the coming years. Finally, although defensive systems may render attack aircraft carrying WMDs quite vulnerable, it is obvious that in most or many cases involving aerial assaults by numerous aircraft, some of the attackers will get through, that is, a leakproof defense against such an assault is an unlikely proposition.

Table 8-1 shows the attack aircraft inventories of the countries in the region under discussion that, as we have noted previously, either now have, or are expected to have, WMDs.[29] It is obvious that some of these nations have attack aircraft, which—depending not only upon flight profile but also on the availability of aerial refueling capability—have reaches equivalent to some of the region's more impressive recently developed ballistic missiles. Egypt's lumbering old Tu-16s, which can presumably be armed with standoff missiles, have a combat radius of nearly 2,000 miles. Israel's F-15s have a radius of over 1,600 miles, which can be further extended by their refueling capability. Syria's Soviet-origin Su-20, Su-24, and MiG-23 aircraft have much shorter legs, but they are long enough to cover all of Israel. Libya has a few Tu-22 bombers with combat radii of nearly 1,500 miles, but it is now acquiring refueling capability for other aircraft that may allow for a reach all the way to Israel. Pakistan's Mirages and F-16s can reach well into India, and India's various indigenous and British and Soviet aircraft can do the same in the opposite direction.

These are not merely hypothetical matters. Iraq actually loaded some chemical weapons on aircraft during the Gulf War. Pakistan may have readied some of its aircraft for nuclear use in recent crises with India.[30] According to Burrows and Windrem, Israel has long operationally deployed a quick reaction aircraft strike force, based in underground hangers at its Tel Nof air base, which aircraft earlier might have been vectored in the direction of the Soviet Union as well as of nearby Arab states.[31]

Torpedoes

In a recent book Eric Arnett reintroduced another possibility for delivery of nuclear weapons—submarine-delivered torpedoes—that had been ignored in

Table 8-1. Attack Aircraft in States Capable of Using Weapons of Mass Destruction

Designation	Maximum speed[a]	Combat radius (miles)[b]	Ferry range (miles)	Bomb load (pounds)	Quantity
Egypt					
Tu-16	Mach 0.85	1,930	3,980	19,840	9
Mirage 5E2	Mach 1.14	750	2,490	9,260	16
F-4E Phantom II	Mach 1.19	400 l-l-l	2,600	16,000	33
J-6 (G-19)	Mach 1.04	430 w/tanks	860	1,100	76
Alpha Jet	Mach 0.76	560	1,830	5,500	15
MiG-17	710 mph at 2 miles		1,230	2,700	30
Iran[c]					
F-4DE Phantom II	Mach 1.19	520	2,600	16,000	35
F-5E/F Tiger II	Mach 1.6 at alt.	140	1,780	7,000	45
J-6 (MiG-19)	Mach 1.04	430 w/tanks	860	1,100	24
Iraq[d]					
Tu-22	Mach 1.4 at 7 miles	1,490	4,040	22,050	8
Tu-16	Mach 0.85	1,930	3,980	19,840	8
H-6D (Tu-16)	Mach 0.85	1,930	3,980	19,840	4
MiG-23BN	Mach 1.1	370 h-l-h	1,740	9,700	70
Mirage F-1EQ5	Mach 1.2	620	2,340	13,900	64
Su-7	Mach 0.7	200–300	900	3,300	30
Su-20	Mach 1.05	420 h-l-l	8,800	50	
Su-25	Mach 0.72	340	. . .	8,800	30
J-6 (MiG-19)	Mach 1.04	430 w/tanks	860	1,100	40
Israel					
F-15 (A,B,C,D)	Mach 2.5 at alt.	1,680	2,900	16,000	53
F-4E Phantom II	Mach 1.19	520	2,600	16,000	112
Kfir C2/C7	Mach 1.14	480	. . .	. . .	95
F-16 (A,B,C,D)	Mach 1.2	340	2,420	20,450	145
A-4H/N Skyhawk	Mach 0.88	920	2,000	9,150	121
Jordan					
A-4KU	Mach 0.88	920	2,000	9,150	24
Libya					
Tu-22	Mach 1.4 at 7 miles	1,491	4,040	22,050	4
Mirage 5D	Mach 1.14	400 l-l-l	. . .	8,820	55
Mirage F-1AD	Mach 1.1	265 h-l-h	. . .	. . .	12

Table 8-1. Attack Aircraft in States Capable of Using Weapons of Mass Destruction (*continued*)

Designation	Maximum speed[a]	Combat radius (miles)[b]	Ferry range (miles)	Bomb load (pounds)	Quantity
Libya (*continued*)					
MiG-23BN/U	Mach 1.1	370 h-l-h	1,740	9,700	36
Su-20	Mach 1.05	420 h-l-h	...	8,800	90
Su-24	Mach 1.1	200	4,000	17,640	6?
Saudi Arabia					
F-5E Tiger II	Mach 1.6 at alt.	140 l-l-l	1,780	7,000	63
Tornado IDS	Mach 0.9	640	1,790	19,800	20
F-15 (C,D)	Mach 2.5 at alt.	1,680	2,900	16,000	84
Syria					
MiG-17	710 mph at 2 miles		1,230	2,700	38
Su-7	Mach 0.7	200–300	900	3,300	15
Su-20	Mach 1.05	420 h-l-h		8,800	35
MiG-23BN	Mach 1.1	370 h-l-h	1,740	9,700	60
Su-24	Mach 1.1	200 l-l-l	4,000	17,640	
India					
Canberra	Mach 0.68		800	8,000	10
MiG-23BN/UM	Mach 1.1	370 h-l-h	1,740	9,700	60
MiG-21PFMA	Mach 2.1 at alt.	300	3,300		15
Jaguar IS	Mach 1.1	530	2,610	10,500	70
MiG-27	Mach 1.1	240 l-l-l	1,550	9,900	72
Ajeet	Mach 0.88	560	...	1,870	80
Marut	Mach 0.91	150	900	4,000	20
Pakistan					
Mirage III (DPand EP)	Mach 1.14	750	...	11,350	18–36?
Mirage 5	Mach 1.14	400 l-l-l	...	8,820	58
Q-5	Mach 1.09	400	...	2,200	135
F-16C	Mach 1.2	580 l-l-l	2,420	20,450	40?
Afghanistan					
MiG-17	710 mph at 2 miles	...	1,230	2,700	40
MiG-23	Mach 1.1	370 h-l-h	1,740	9,700	15

Table 8-1. Attack Aircraft in States Capable of Using Weapons of Mass Destruction (*continued*)

Designation	Maximum speed[a]	Combat radius (miles)[b]	Ferry range (miles)	Bomb load (pounds)	Quantity
Afghanistan (*continued*)					
Su-7B	Mach 0.7	200–300	900	3,300	60
Su-22	Mach 1.05	270 l-l-l		8,800	30

Source: Gordon M. Burck and Charles C. Flowerree, *International Handbook on Chemical Weapons Proliferation* (Westport, Conn.: Greenwood Press, 1991), p. 508.

a. Figure given is for flight at sea level, unless otherwise noted. The Mach number is a unit of speed equal to the speed of sound in the medium in which an object is moving. At sea level, Mach 1 equals 760 miles an hour.

b. Figure given for round-trip with maximum weapon load, unless otherwise noted. The "l-l-l" and "h-l-h" refer to the flight profile to which the given figure corresponds. The former means the bomber approaches the target at low altitude (in order to evade radar), makes the bombing run at low altitude, and departs at low altitude. Since the atmosphere is denser at lower altitudes, the aircraft's mileage would be increased if it chose a flight pattern that brought it in to the target at a high altitude, dropped down only for the bombing run, and then departed at high altitude ("h-l-h").

c. These estimates are likely overreporting Iran's present capability. *The Military Balance: 1989–90* says that, of the F-4s, only twenty are serviceable. Twenty of the F-5s are also thought to be in a usable condition.

d. This listing does not incorporate aircraft losses which Iraq suffered in the 1991 war.

the literature since earlier work done at the Hudson Institute. According to his speculations, "diesel submarines with nuclear-armed heavy torpedoes could provide a secure retaliatory force for countries whose principal adversaries had important coastal cities or installations."[32] Arnett discusses time fuses that would allow submarines to escape as well as contact fuses, and he notes that because of slow progress in shallow-water anti-submarine warfare (ASW) and torpedo defense, such weapons could easily penetrate to their target. He sees torpedo netting as an ineffective countermeasure against a warhead of even nominal yield.

This specter of nuclear torpedo warfare raises a host of questions about geographical features that are significant in terms of the undersea snorkeling ranges of some submarines now in the inventories of regional powers or those that might yet be acquired, the depths of waters near coastal cities in relation to ASW, and routes of transit and the relevant maritime geography. Submarines might or might not be put on station at the outset of a conflict. There is also an unanswerable set of questions in regard to such issues as communications and the need for surfacing to conduct them and fail-safe procedures. A look at the map will reveal various possibilities. A beleaguered Israel shorn of its air force and missiles (or one nearly overrun on the ground) could launch torpedoes against Tripoli, Algiers, Port Said, Alexandria, Latakia, Tartus, Izmir, and maybe even Karachi. But such a deterrent would not be useful against Iraq or against Iran, except perhaps at Bandar Abbas. Egypt, Libya, Algeria, and possibly Syria might conceivably (if they acquired nuclear weapons) use such a weapon against Tel Aviv and Haifa, particularly if Israel's air defenses and antimissile defenses blocked the obvious alternatives. And India and Pakistan, specifically regarding Karachi and Bombay, would appear to be mutually vulnerable in this area, though for the foreseeable future there do not appear to be impediments to alternative uses of aircraft and missiles.[33]

Preventive War and Preemption in a Regional Nuclear Context: Geographical Considerations

Although the concepts of preventive war and preemption are often used interchangeably in the strategic literature, there is a clear distinction between them. Preventive war refers to a situation in which one country, usually one that considers itself to be ahead in a regional arms race or balance of power situation, fears that the balance is shifting against it and decides to initiate war. The idea, in other words, is that if war is more or less inevitable, better now than later when the rival will be stronger. In this sense, too, preventive wars are usually planned for and premeditated, sometimes over a fairly long period of time. Preemption, on the other hand, refers to an escalating crisis situation in which both antagonists come to feel that war has become inevitable and there appears to be a clear advantage to getting in the first blow.[34] It could well be in a situation in which neither side has premeditated a crisis or war, perhaps an instance in which a major crisis has erupted from a relatively minor incident that has escalated out of control. In the traditional nuclear literature, the term "crisis stability" refers to the extent to which a nuclear balance is structured so that neither side has an incentive to preempt in such a crisis.

There are numerous historical examples of both preventive war and preemption, though the characterizations may be arguable and, in some cases, there may actually be a gray area between them. The opening move of World War I is often characterized as a preventive war by Germany with the Sarajevo incident as a pretext, but initiated because of the long-run implications for Germany of the gradual but ineluctable growth of Russian power. Japan's attack on Pearl Harbor has been similarly characterized, with Japan seen as having recognized the inherently stronger nature of American power once the United States had begun to mobilize. Israel's attack on Egypt in 1956, aided by Britain and France, was clearly a preventive war in the context of escalating Egyptian and Syrian weapons purchases from the Soviet bloc, which threatened to shift the power balance against Israel. Israel's attack on June 5, 1967, however, is more aptly characterized as an example of preemption, flowing from the escalating crisis over the Egyptian closure of the Straits of Tiran. On the other hand, in retrospect, it has become more common for analysts to characterize that crisis as one of an impending preventive war by Egypt to forestall the fruition of the Israeli nuclear program, it being thought by Nasser that once the Israelis had acquired atomic weapons, it would become much more difficult and certainly costly to destroy the Jewish state.[35] Iraq's attack on Iran in 1980 had both the character of a preemption (in connection with escalating border violence between the two states) and of a preventive war, in the latter context having to do with Iran's then seemingly short-run disadvantage because it had been cut off from its traditional source of weapons in the United States after the Shah was overthrown.

There had long been speculation about India's obvious temptation to destroy Pakistan's nuclear facilities, to the extent that it might ever have been feasible, given the dispersion of that program and the location of key facilities underground. Some of the (vague) speculations were about the prospects for a joint India-Israel operation, even before the now partial rapprochement between those two long-hostile nations, with Israel concerned about the "Islamic bomb" implications of the Pakistani program. More recently New Delhi and Islamabad have signed an agreement promising not to preempt each other's nuclear facilities, which by now is probably a moot point in view of the seeming impossibility of either to do so at this late date when both are widely suspected of fielding operational nuclear arsenals. Finally, one might point to periodic speculation that Israel might attempt to knock out Syrian chemical warfare infrastructure and arsenals in a preventive strike, prior to the latter's acquisition of more sophisticated missile delivery systems that might pose a serious threat to Israel. Further afield in our extended region of interest, there had been the interesting question of whether, in some circumstances, Russia might have attempted a preemptive strike on the nuclear weapons stockpile left over in Ukraine from the old USSR, and what the impact of that might have been on the Middle East and South Asia. That situation now appears to have been defused, and Ukraine appears willing to give up the nuclear weapons on its soil.

What then are the prospects for nuclear-related preventive wars in the currently evolving greater Middle East milieu? The nuclear programs of Israel, India, and Pakistan are all probably, and in the eyes of their rivals, beyond the point where they could be wiped out by a preventive strike. Their nuclear infrastructures and those of Iraq and Iran are shown in map 30. Given Pakistan's lack of strategic depth and its inferiority in conventional military power, India might contemplate instigating a land war in order to capture and then destroy the Pakistani nuclear infrastructure, banking on its presumed nuclear superiority, that is, escalation dominance, over Pakistan (although Pakistan has substantial second-strike deterrence) to keep the war at a conventional level. Such an assault might have to be coupled with assurances to Pakistan regarding the latter's political survival and a promise to prevent civilian massacres. A similar attempt by the Arabs to take over the Israeli nuclear infrastructure, if the Arabs including Egypt were ever to become confident about conventional superiority over Israel, is far less likely. The Israelis would assume massacres of their civilians; indeed, that is largely what their nuclear program is all about.

That leaves Iraq and Iran, both obviously aspiring to nuclear status, and indeed, perhaps now racing each other toward the goal of deploying nuclear weapons. If one or the other acquired operational nuclear weapons first, would it then be tempted to launch a preventive war with the aim of stopping the other from getting that far? Would such a preventive effort be conducted only with conventional weapons (missile or bombing attacks on nuclear facilities, or a march on Isfahan or on the facilities surrounding Baghdad), or would nuclear weapons come into play? Or could chemical warfare weapons be used to deter attacks on nuclear facilities? Might Israel be tempted to try another preemptive attack on rebuilt Iraqi nuclear facilities, or even on Iranian facilities, despite the daunting distances involved? In the latter case would that be carried out with conventional weapons only? If nuclear weapons were used, what might be the reaction of nuclear-armed Pakistan, Iran's neighboring ally of sorts?

There are, indeed, still other possibilities for preventive wars and preemptive strikes in this broad regional context, some of them hypothetical and long term. What would Israel do if Egypt should at some point make a move toward nuclear status? What would Egypt do if Libya moved seriously in that direction? What would be the reaction of France if a postrevolutionary Islamic Algeria moved further in the direction of nuclear acquisitions, particularly in the context of a worsening political situation in France with respect to immigrants and refugees?

Of course, preventive action could be taken by powers outside the region. The United States has, apparently, at least considered a preemptive strike against North Korea's nuclear program, and it might later consider such action against either or both Iran or Iraq, either on a unilateral basis or something with multilateral trappings. The possibility of such a strike against Libya's new Rabta II chemical plant is now being widely discussed.

The problem for those contemplating preventive strikes is, of course, that after there has been one Osirak-type raid, the necessary element of surprise is no longer there. Osirak could be done only once, and all of the countries now conducting clandestine nuclear programs will be sure to have dispersed and redundant facilities, well dug in and disguised.[36]

Geography and the Strategic Role of Weapons of Mass Destruction

The previous review of extant and projected WMD programs and arsenals, and of the various related delivery systems—ballistic missiles, cruise missiles, aircraft, and torpedoes—leads to a variety of questions rooted in geographic factors, concerning, for example, delivery system ranges, concentrations of populations and economic targets, the vulnerability (as a function of distance and terrain) of delivery systems to preemptive or first strikes, the geographical aspects of possible tactical use of nuclear weapons, the vulnerability of coastlines, and the impact of seasons, weather, and prevailing winds on WMD effects. It should be stressed, however, that answers to (or analyses of) all of these questions depend greatly on the nature of the WMD arsenals fielded by the various regional powers, that is, how many nuclear weapons and of what megatonnage, what types of biological and chemical warheads and bombs and what kinds of delivery systems. These arsenals, as we have noted, are now being developed, and in most cases we do not have answers. Publicly available sources (but probably also the CIA) do not have figures on how many nuclear weapons Israel, Pakistan, or India possess, or the types of warheads, or their explosive capabilities; this is also true regarding the biological warfare arsenals of several regional nations. But some estimates can be laid out and some issues bruited. Another unknown factor is the future of U.S. defensive systems in the region that might block the use of WMDs, even by one regional nation against another. Potentially that could involve boost-phase intercepts from aircraft or satellites, defensive systems such as the Theater High-Altitude Area Defense placed within the region, or perhaps one or two robo-ships (the arsenal ship) armed with weapons that could knock down missiles.

The previous section's discussion indicated that in the three major conflict pairings in the region, Arab–Israel, Iran–Iraq, India–Pakistan, each side can now effectively target the other with WMDs, by ballistic missiles or aircraft or both. With Jericho II and later Jericho III, and with aerial refueling, Israel can reach all of its adversaries' cities, the matter of defensive systems aside. The contiguous Arab states can also easily reach Israel now; Iran's capability looms nearer; Pakistan's and Libya's capability is much more questionable. India and Pakistan can easily reach each other, though Pakistan would require long-range air strikes to get into India's interior. India is on the verge of a reach to China's coastal heartland. Iran and Iraq are mutually vulnerable. The United States can strike anywhere in the region; Iran and Iraq can only strike back at nearby U.S. allies such as Saudi Arabia and Israel via triangular second strikes. As we have noted Libya and Algeria may eventually be able to reach Italy and France with missiles.

In short, in all of these cases the facts of contiguity and limited strategic depth for capital cities and other major urban centers and industrial areas produce a hair-trigger situation of mutual vulnerability, which is a function of the short distances involved. We have also noted a number of situations in which coastal cities are vulnerable to attack by nuclear torpedoes.

But distance factors are only part of the story. The accuracy of missiles over those distances, and their reliability, may matter, particularly if the numbers of warheads and missiles are limited, and the ability of aircraft to penetrate rivals' airspaces to core cities is in doubt. Here there are potentially great asymmetries. Israel's F-16s, even if the country were caught by surprise or abjured preemption of its rivals' air bases, could presumably do a good job of protecting its small air spaces, though that would become more questionable if the Egyptian and Saudi air forces, fielding U.S. aircraft, were to enter the fray at the outset in a coordinated attack. Israel's ability to fly over Arab cities, or to approach them with "toss bombing" tactics used with nuclear weapons, is presumably not in doubt, though long-range strikes on Baghdad, Teheran, and Riyadh, for example, using tanker refueling might not be a certainty, particularly given the Saudi airborne warning and control systems and similar Egyptian capabilities. India's air force is by now much stronger than Pakistan's, but whether it could maintain a leakproof combat air patrol over New Delhi or Bombay against Pakistani Mirages, even after a preemptive strike on Pakistan's airfields, is questionable. Iran and Iraq would appear wide open to each other's air strikes as well as missile attacks.

In all these cases the warning times are short. This is then very different from the early cold war period, when the United States anticipated hours of warning against bomber attacks over the Arctic, and even later with the twenty minutes or so of warning expected through Defense Support Program (DSP) satellites with infrared sensors that could "see" the takeoffs of Soviet ICBMs (the Soviets had more limited capacity in reverse, perhaps dependent also on satellite downlinks in Cuba). But in the several greater Middle East contexts, the warning times are reduced to minutes (only Israel will anytime soon have a DSP-type capability, begging questions of the provision of such satellite information to other regional powers by the United States or Russia).

What the implications are of such short warning times with regard to the probabilities of war or deterrence is not easy to say. There is a lot of talk about hair triggers, implying a relatively ominous danger of preemptive instability (still begging overall questions about reactive second strikes). Contiguity plus primordial hatreds is a dangerous combination. There is a larger danger of accidental war, perhaps to be triggered by sudden large-scale aircraft takeoffs or missile tests, raising all manner of questions about command and control (paradoxically, the hair triggers and small sizes of nuclear forces may drive decentralization of control in some cases to limit vulnerabilities to disarming first strikes). The short distances and times involved also somewhat remove a deterrent factor on the side of the "weak" from the cold war era, namely, launch-on-warning or launch-under-attack.

All of the above raise questions about the possible geographical bases for providing second-strike invulnerability in the several regional cases, begging questions about what happens in the event of all-out counter-city or counter-value strikes at the outset. Israel has little space for hiding missiles (or moving them around), but it has rugged mountains in which to hide and harden them. It also has limited air space, but has the refueling capability to mount a continuous nuclear airborne alert in a crisis. Iran, by contrast, has much space and rugged terrain in which to hide missiles and (vis-à-vis Israel) considerable spatial depth versus air attack, raising questions about Israel's ability in the future to locate and target missile sites, perhaps mobile missiles. Iran and Iraq both have space and favorable terrain in their dyadic context, and mutual invulnerability is assured in the short run by the inability of both sides to target either fixed or mobile missiles. India appears to have a huge advantage over Pakistan in these spheres, but its ability to locate and destroy a missile force plus airfields plus aircraft on airborne alert to the point of vitiating Pakistan's second-strike capability is in question.

The seemingly important geographic-demographic factor in this region—or at least some part of it—is the concentration of populations in a relatively few urban areas. But that brings us to some very vague and unanswerable questions that were often addressed in the cold war deterrence literature under the heading of "acceptable losses." Indeed, the very core strategic concepts of "first-strike deterrence" and "second-strike deterrence" were largely defined by the idea of acceptable losses, that is, one had second-strike deterrence vis-à-vis the other cold war power if one could absorb a first strike and still have enough remaining retaliatory power to inflict unacceptable losses on the opponent. What was or was not acceptable in the eyes of the American or Soviet leadership or publics was never made clear: Would it have been worth 10 million lives to win the cold war by a first nuclear strike? Would the Soviets have deemed a decapitating first strike that took out its capital city and leadership unacceptable? Would the U.S. ballistic nuclear strategic submarine (SSBN) force, even if whittled way down, still have been a viable deterrent as a revenge force?

In the greater Middle East, of course, with the exception of India, national populations are much smaller than those of the United States or the Soviet Union. There is an enormous range from India's 1 billion or so people to Israel's 5 million. It is hard to say what kinds of (Strangelovian) calculations can be made (or are made by governing elites) about acceptable losses. Is 5 million unacceptable to India, even if it is a small percent of its population? In what conceivable circumstances? What about Pakistan with 120 million? How about 1 million of Iraq's some 20 million people, or 2 million of Egypt's some 60 million? Then there is the nature of regimes. Some of the nations in the region have authoritarian governments that may deem the loss of themselves and their capitals as unacceptable, even if nothing else was lost.

With regard to most of the countries in the region, there are only a small number of major cities whose destruction (and with them a concentration of elites) would be tantamount to national destruction. For Israel, one good-sized nuclear weapon on each of Tel Aviv, Jerusalem, and Haifa (in the first two with substantial Arab populations in Jaffa and East Jerusalem) would effectively spell the end of national existence. Similarly, for Jordan regarding Amman, Irbid, and Az Zarqa; for Syria with Damascus, Aleppo, Homs, and Hama; and Saudi Arabia's Riyadh, Jidda, Mecca, Medina, and Dhahran. Egypt's Cairo, Alexandria, and Port Said in the Nile Delta are the core of the country, yet with a population of 60 million and rising, it would have a lot of people left over from a three-bomb attack. India, of course, might see the destruction of Delhi, Bombay, Madras, and Calcutta as devastating, but it would not spell the end of national existence. Toward the end of the cold war, it appeared that such "calculations" had come to be viewed as merely crazy, if not irrelevant, as it came to be understood that the idea of several nuclear blasts in urban areas was, literally, unthinkable. Whether that kind of rationality will be maintained in the greater Middle East, particularly because genocidal fears related to conventional battlefield defeat abound, remains to be seen.

Some analyses have recently been conducted by the Office of Technology Assessment and the Center for Strategic and International Studies (CSIS) on the possible "effects," that is, casualties, that might result from attacks with various kinds of nuclear, biological, and chemical weapons.[37] The effects are seen to vary widely depending upon the type of WMD (nuclear versus biological versus chemical), the size of the warhead, the weather and the time of day, and population densities within cities. At a gross level of generality these data show that, assuming use of one Scud-sized warhead with a maximum payload of 1,000 kilograms, nuclear weapons are the most deadly (with larger warheads), biological weapons next, and then chemical weapons. Assuming, for instance, a city with 3,000–10,000 people per square kilometer, a Hiroshima-sized nuclear weapon (12.5 kilotons) would be expected to kill 23,000–80,000 persons, the approximate magnitude of fatalities at Hiroshima and Nagasaki. A 1-megaton weapon (the Vanunu revelations hint that Israel has developed a hydrogen bomb, whether or not of that explosive power and untested) would, according to these sources, kill in a range of 570,000–900,000 persons, that is, could obliterate Cairo, Tel Aviv, or Damascus. The figures for biological warfare (using anthrax spores, perhaps the most likely weapon of choice) are in the 30,000–100,000 range, that is, slightly higher than a Hiroshima-sized nuclear weapon. Those for chemical weapons such as Sarin are much, much lower, in the 60- to 200-person range. CSIS also extrapolates to larger quantities of anthrax and Sarin that might be dropped by an aircraft dispensing the agent as an aerosol. For anthrax that results in much higher kill rates than with a smaller missile warhead. Those results also vary greatly on a range from "clear sunny day, light breeze" (130,000–460,000) to "overcast day or night, moderate wind" (420,000–1,400,000) and "clear calm night" (1,000,000–3,000,000).[38] The comparable figures for Sarin are much smaller. Hence it appears that a few aircraft that succeed in penetrating air defenses with nominal size bombs carrying 100 kilograms of anthrax spores can cause casualties higher even than a hydrogen bomb. Chemical weapons may be less overwhelmingly devastating, yet CSIS reports that a successful attack by eight MiG-23s, or four Su-24s, each delivering 0.9 tons of VX gas, could under optimum conditions kill 125,000 civilians, a result that no doubt would engender a murderous response, perhaps at the nuclear level, if the attacked nation was so equipped.[39] Implied in these figures is the utter devastation that would be wrought by a handful of nuclear or biological weapons delivered either by aircraft or missiles in a region where urban populations are concentrated and where devastation of just a few cities would, effectively, put an end to national existence.

The geography of tactical nuclear use in the greater Middle East (rarely mentioned in the region's proliferation literature) presents some daunting hypothetical problems for analysis. By analogy one might dwell on the discussions of hypothesized U.S. or USSR use of such weapons in cold war scenarios. In the 1980s there was considerable talk about U.S. use of tactical nuclear weapons (TNWs) to halt a Soviet thrust toward the Persian Gulf if the latter were to break through the Zagros Mountains and threaten to overwhelm a small force of defending U.S. Marines and airborne troops supplied from Diego Garcia and Germany. In the late 1950s the United States planned to "channel" invading Soviet forces in Central Europe into killing grounds such as the Fulda gap or Hof corridor, though it was recognized that that would have engendered a Soviet TNW response. The purpose was deterrence by uncertainty. Late–cold war scenarios often involved a nuclear "lay-down," gruesomely referred to as "barraging the FLOT (forward line of troops)," from Lübeck to Passau, variously involving nuclear mines, artillery, aircraft-dropped weapons, and even the use of U.S. SSBNs in Atlantic or North Sea firing positions. With prevailing winds from west to east, it was assumed that the Soviets, along with German civilians, would get the worst of it. Both the United States and the Soviet Union had thousands of nuclear warheads for such purposes, and there was no trade-off, numerically speaking, among weapons use at the tactical, theater (deep military targets), and strategic (homeland strikes) levels.

In the Middle East, at the moment and for the foreseeable future, such trade-offs do exist. Or do they in all cases? If Israel possesses a number of nuclear weapons at the higher end of the range of estimates running from 50–60 (IISS, among others) to 200–300 (Seymour Hersh), and if it is assumed that only 20 to 30 would be needed as a counter-cities reserve (taking into account the possible loss of delivery systems at home or en route to targets), that would appear to imply a large number available for tactical use. The requirements for the narrow Golan would not appear to be great. Those vis-à-vis an attack by Iraq-Iran-Jordan and allies into the current West Bank would be larger; and a "lay-down" in Sinai (potentially over 100 miles frontal width) would require a considerable amount of weapons, modified by the fact that there are only a few routes suitable for armor. The winds from Sinai could, presumably, carry radiation over Israel, which may account for the (unsubstantiated) rumors about Israeli development of neutron weapons that result in little downwind radiation. To the extent that this was not a factor, however, Israel might contemplate a postattack radiation curtain down Sinai or across Golan or both that might render these areas impenetrable to armies for a long time.

Iran and Iraq would, presumably, require similarly large arsenals to pose a tactical nuclear threat backed by a counter-cities reserve to deal with the problem of escalation and escalation dominance. But then use of only a few TNWs on the battlefield, even if not providing a lay-down defensive curtain, could still be very useful in creating widespread panic and terror, in impelling the big powers to step in and halt a conflict (presumably desired by a losing side), or in opening up a gap for a counteroffensive through a devastated area. The same is true for Pakistan and India, hypothetically, with their own long borders comparable at least to the old Central European FLOT. Iran and India have the downwind disadvantage in these respective dyads.

Some of the same geographic criteria apply to the massive use of chemical weapons in the same places, or along the same defensive frontiers. But as the CSIS report shows, that would require a truly massive volume of chemical warfare ordnance. Hence it is claimed that to maintain a 300 meter deep strip of VX contamination in front of a position defending a 60-kilometer-wide area for three days would require 65 metric tons of agents delivered by 13,000 155-mm artillery rounds![40] For a wide swathe of the Iran-Iraq or Pakistan-India FLOTs in a war, that adds up to a mountainous volume of chemical warfare munitions. But in World War I and to a degree in the Iran-Iraq war, sufficient quantities were used to produce meaningful tactical effects.

The New Geography of Illicit Nuclear Transfers

The collapse of the Soviet Union and the melting of the iron curtain, not just in Europe but all around the former Soviet Union, has, as we have noted, had enormous consequences, for instance, regarding new patterns of oil pipeline routes and wholly new patterns of trade relations. So, too, has there been a major impact with regard to the possibilities for illicit transfers of nuclear technology and materials; perhaps also of chemical and biological weapons materials and precursors. For a long time the old Soviet border with the northern tier countries of the Middle East—Turkey, Iran, and Afghanistan— and its proximity to others such as Iraq, Syria, and Pakistan was characterized by tight, iron-clad controls, perhaps a bit weakened during the Afghanistan war. Those old borders, which today are borders with Georgia, Armenia, Azerbaijan, Turkmenistan, Uzbekistan, and Tajikistan, plus the borders be- tween some of the latter and Kazakstan, and with the now truncated Russia, are far more porous, freed of the controls of the old secret police and border guards. One result is a vastly increased potential for illicit transport of all sorts of merchandise, including nuclear technology and materials.

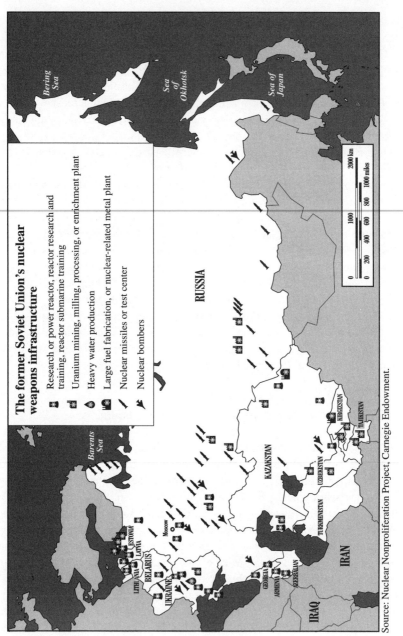

The former Soviet Union's nuclear weapons infrastructure

Research or power reactor, reactor research and training, reactor submarine training

Uranium mining, milling, processing, or enrichment plant

Heavy water production

Large fuel fabrication, or nuclear-related metal plant

Nuclear missiles or test center

Nuclear bombers

Map 32

Source: Nuclear Nonproliferation Project, Carnegie Endowment.

As shown in map 32, a substantial part of the former Soviet Union's nuclear infrastructure, including its nuclear weapons manufacturing facilities, were located around Russia's southern and western rim.[41] These included nuclear submarine training reactors, research reactors, power reactors, nuclear fuel fabrication plants, and uranium mining and processing plants. Many of them were in the Caucasus and Central Asia, close to the nuclear aspirant states of Iraq, Iran, and Pakistan. Although Russia has made promises to adopt nuclear safeguards and to control the movement of nuclear materials, such controls are widely rumored to be ineffective. Hence according to one recent press report:

> In the meantime, aspiring nuclear states such as Iran are shopping for opportunities, scattered smugglers are moving radioactive and sensitive nuclear-related materials to the West, and the original vision of a region free of nuclear weapons on Russia's borders is becoming more and more blurred, they said.
>
> Western-style nuclear export control systems in the new, poor and sometimes chaotic republics are embryonic or nonexistent. Little has yet been done to tackle the huge challenge of adopting formal nuclear accounting procedures to ensure that civilian nuclear materials are fully secure, according to interviews with U.S. officials, international nuclear regulators, outside specialists, and local officials.[42]

That article goes on to refer to "shopping runs by aspiring nuclear states and get-rich-quick schemes by loose networks of criminals with access to nuclear materials."[43] Kazakstan has been asked by the United States and Germany to investigate smuggling involving low-grade nuclear fuel and dual-use metals such as beryllium and zirconium, which can be useful to a nuclear program but also have industrial applications. There were reports of nuclear warheads being transferred from Kazakstan to Iran, but it has never been proved. Not only Middle Eastern routes for illicit transfers were involved, as there was some evidence of shipments from Ukraine through Estonia to Europe of the controlled dual-use metals hafnium and zirconium.

The fact that during the Soviet period, ethnic Russians, particularly in Kazakstan, generally excluded the ethnic minority groups on their southern border from high-level participation in nuclear programs has compounded the problem. This legacy is said to have made it difficult for independent Kazakstan and the other Central Asian states to maintain effective nuclear regulatory control systems, particularly in an area where smuggling had long been a way of life.

This issue was amplified in 1992 in a report by a Task Force on Terrorism and Unconventional Warfare, prepared under the aegis of the House Republican Research Committee in Congress, which noted.

> Indeed, since late 1992, the flow of militarily useful nuclear materials, military technologies, and various weapons systems have become a security concern of some significance. As military goods from the former Soviet Union and Warsaw

Pact are shipped to the Middle East via Western and Central Europe, where the financial transactions are worked out, a potential strategic threat to the West has evolved. Taken together with the drugs, counterfeit money, and terrorist equipment that are used as payment for black market military goods, and the prospect of a serious threat of terrifying proportions begins to take shape.[44]

This report detailed the roles of numerous former KGB and Stasi officials in enabling the trade in nuclear materials, as well as the involvement of a banking network, much of it in Western Europe. Not only nuclear materials but also chemical and biological weapons materials were involved, as well as conventional weapons. There were reports of demands in the black market for super-rare materials and components, including gaseous lithium used to boost the power of nuclear weapons. Hence the following in a report on materials moved to Iraq from Serbia, Germany, Italy, and Croatia:

> The trucks' route ran from Serbia to Bulgaria to Turkey and into Iraq. On their way back they carried mainly tanks of fuel, but also fruits. The trucking operation between Serbia and Iraq was run by a Vienna-based Bulgarian named Bozhilov. As a result of the great success of shipping via Bulgaria, its role in the transportation network has expanded, and Varna, long a major port for illegal weapons and terrorist equipment, has now become a primary port for illegal nuclear materials and weapons.[45]

This report concluded that "there is no doubt that there is a major and increasing flow of militarily useful materials and equipment, forbidden military technologies, and numerous weapons systems throughout the black market."[46] Later, in 1996, there were additional reports related to a powerful private company called Nordex, reputedly the largest Russian-owned firm in the West, and one described as having originally been set up as a spy asset to earn hard currency for the KGB. Nordex was alleged to have been involved in smuggling nuclear components and fissile materials from Russian national laboratories to Iran and North Korea. Reportedly in 1993 and 1994 the U.S. National Security Agency found indications that Nordex was involved in nuclear smuggling.[47]

In short, the breakup of the old USSR has had some unexpected and possibly baneful side effects rooted in the new fluidity and porousness of long-closed borders. That too is an element of a new strategic map.

Part Four

CONCLUSIONS

This last section provides both a review and a summary of the main conclusions derived from the earlier chapters. It reflects the view of the authors that the greater Middle East remains a highly dangerous region yet holds enormous promise for growth and change. Chapter 9 outlines the new vulnerability of the region to renewed conflict and two particularly dangerous scenarios—a new Arab-Israeli war and a major American intervention in the Persian Gulf with and without allies. Chapter 10 reviews the pressures on regional leaders to end their bitter quarrels and work together to solve the enormous economic problems of the region by cooperation rather than conquest. Chapter 11 is a summary of the study.

NINE

Geography and New Military Conflict

HOW WILL the new geostrategic features discussed in the previous chapters influence the nature and course of future Middle East conflicts? In this chapter we first review the impact of new technologies on conflict escalation and economic warfare. We then examine two scenarios for war: a renewed Arab-Israeli conflict and a second major U.S. intervention in the Persian Gulf.

In most Middle East wars there has been evidence of limitations on the full use of force. Those limitations have usually been tacit ones backed by the threat of retaliation, and often involving some form of mutual deterrence. The Israelis and Arabs have heretofore eschewed extensive bombing of civilian populations, the Iraqi Scud attacks against Israel in 1991 notwithstanding as a contrary example. The Arab-Israeli wars have also seen geographical limitations as, for instance, with the Israelis having denied themselves the option of attacking through Jordan to get around the Syrian defenses on the Golan in 1973. In 1967 as well as in 1973 the Israelis did not follow up on the possibilities for moving into Arab capital cities, something that was militarily within reach, but perhaps politically infeasible, particularly because of pressure from the external big powers. The Iran-Iraq war was fought almost all-out in all respects, but one might have imagined far greater use of chemical weapons against both military forces and civilians and perhaps also more extensive killing of prisoners of war. In the major India-Pakistan wars, air strikes have for the most part been limited to military targets. In the Gulf War, the U.S.-led bombing raids were carefully designed to minimize collateral damage (civilian casualties) and, of course, the war ended up being geographically limited to the reestablishment of control over Kuwait.

But now in the greater Middle East region, with the proliferation of weapons of mass destruction (WMDs) and their associated delivery systems, there are growing possibilities for the escalation of conflicts, a matter particularly precarious in situations in which the impending loser of a conventional conflict fears dire consequences and thus may be willing to venture escalation. The combination of the expansion of vulnerable industrial complexes and other forms of infrastructure and the diffusion of military technologies that allow targeting of such infrastructure has broadened the possibilities for unlimited warfare in the region. The spread of such military technologies has also changed the requirements, mostly for the United States, of extended deterrence on behalf of client states in the region, especially those that possess significant reserves of oil and gas.

The Impact of New Technology

The combination of long-range air systems with much greater accuracy and sophisticated communications has revolutionized the capacity of certain countries, particularly the United States, to project power and to override traditional constraints of topography and distance. If surface-to-surface and cruise missiles and long-range combat aircraft continue to proliferate throughout the region, the strategic reach of all major players will be similarly enhanced—assuming that they are able to penetrate the increasingly sophisticated air defense systems. Thus if Iran were to modernize its armed forces with missiles and aircraft equivalent to those now found in Western inventories, it would have strategic implications way beyond the Gulf and would threaten both Israel and Egypt. Similarly, Israel's strategic reach today is noted throughout the Gulf, particularly by the Iranians and the Iraqis. India, in turn, has given the existence of long-range strategic missiles in Saudi Arabia—purchased from China in the 1980s—as one reason why it must develop its own intermediate range system, the Agni.

In other words, the nominal strategic reach of long-range systems now transcends much of the area we call the greater Middle East region. But how effective are these systems likely to be in overriding the traditional constraints of geography? In terms of terror attacks or mutual deterrence of each other's cities and urban areas and major industrial plants, the new technology overrides traditional geographic constraints. However, in the context of more formal warfare involving the movement of ground forces and the occupation and holding of terrain, many of the traditional constraints are still important. New equipment may be better suited to deal with terrain but as the Gulf War experience showed, and as we argued in chapter 6, there are still tremendous uncertainties about its functioning in the Middle East geographical environment, particularly in inclement weather.

One of the most dramatic impacts of the new technology on the strategic geography of the greater Middle East concerns the growing vulnerability of high-value economic targets to attack by adversaries equipped with advanced conventional munitions. This reality reflects the paradox of modern civilization: as countries develop economically they become more dependent on the infrastructure of the modern state. Villages in the oil-rich Arab states of the Gulf have evolved from primitive, undeveloped backwaters into bustling modern cities owing to the huge revenues from oil income. The impact of such phenomenal wealth on the region has had predictable results, most notably a dramatic increase in per capita and total requirements for electricity and fresh water. On a per capita basis the use of these two items approximates that found in industrial countries and reflects the very high standards of living that most residents in these countries enjoy.

The rate of modernization over the past thirty years has been staggering. For instance, as late as 1960 Abu Dhabi was a tiny underdeveloped oasis on the totally underdeveloped Gulf coast. Today Abu Dhabi and Dubai are two of the most modern cities in the world with huge thoroughfares, high-rise offices and apartments, international airports and ports, and as many cars per capita as there are in Western Europe. Israel, which began its life as a state putting a high premium on agriculture and socialism, now looks like a Western European country, and Tel Aviv compares well with its European Mediterranean equivalents.

The downside of this development is new vulnerability in event of attack by an adversary equipped with the sorts of technology used by the United States during Desert Storm. For the first time since the industrial revolution, the infrastructure of an advanced society is now at risk of major destruction while the civilian population may be spared high numbers of casualties. For example, it is difficult to think of a region as vulnerable as the Arabian Peninsula to this type of warfare, with the exception of Taiwan and Singapore. There is much more redundancy built into industrial plants and utilities in the advanced industrial societies of the north.

Seen in strategic terms what we have here is a vastly changed set of economic targets across the region, which, in turn, can influence the nature and goals of modern warfare. The first rounds in this new type of warfare occurred in the closing days of the 1973 Arab-Israeli war, when Israel launched a limited strategic air offensive against Syrian infrastructure targets. The damage was extensive and demonstrated the kind of damage airpower, operating in a relatively permissive environment, could inflict on an enemy. However, the next major war to involve countervalue targeting, the Iran-Iraq war, suggested a reverse hypothesis; namely that even the most vulnerable of installations and vehicles (oil tankers) could not be fully destroyed unless airpower was used aggressively and with accurate delivery systems.

But Desert Storm changed the equation again. It is obvious that certain operations carried out by the allies during Desert Storm will be difficult for any

Middle East country to undertake by itself for the foreseeable future. However, we have argued that if there is no significant change in the overall balance of military power between the regional adversaries, military operations against fixed-point, soft economic targets are likely to become more lethal and effective, in the absence of theater defenses against missile and air attack. Without good missile defenses, most of the greater Middle East countries will become increasingly vulnerable to precision air attacks.

Iraq's vulnerability was demonstrated during the Gulf War. It is useful to review the basic philosophy and operational plans adopted by the United States and its allies during Desert Storm in the conduct of operations against key Iraqi targets. The alliance settled upon five air campaign objectives when the war started. Each of these five had a list of targets to be attacked. The primary directives were to go after military objectives, and to the extent that civilian targets were programmed for attack it was because they were directly related to military operations.

The first objective was to isolate and incapacitate the Iraqi regime. This required attacking leadership command facilities, crucial electricity production facilities that provided power to military and industrial systems, and telecommunications and C^3I systems. The second objective was to gain and maintain air supremacy, which entailed attacking Iraq's strategic integrated air defense systems including radar sites, surface-to-air missile (SAM) and control centers, and air forces and air fields. The third objective was to destroy the nuclear, biological, and chemical warfare capability, and targets included research, production, and storage facilities. The fourth objective was to eliminate Iraq's offensive military capability by destroying major parts of its military production infrastructure and power projection capabilities. This required targeting military production and storage sites, Scud missiles and launchers and production and storage facilities, short-term oil refining and distribution facilities (as opposed to long-term production capabilities), naval forces and port facilities. The fifth objective was to render the Iraqi army and its mechanized equipment in Kuwait ineffective, which required targeting the Iraqi army based in Kuwait as well as the railroads and bridges used to supply it.[1]

While the logic of the targets made sense in the context of the Gulf War, for the purposes of this study it is worth noting the importance the allies placed on targeting electricity production facilities, reasoning that this would have a serious impact upon Iraq's capacity to manage a modern war as switching to backup generators not only causes computers to drop off-line but leads to other disruptions in the military command and control system, and even minor delays in processing information can have serious impact in wartime. Clearly, however, the targeting of electrical production facilities had collateral impact on the Iraqi civilian sector and eventually knocked out the country's capacity to function in a modern industrial mode. Similarly, oil refining and distribution facilities are essential to the operation of a modern military force, yet also have extremely important civilian applications, as do railroads and bridges, which were targeted early in the campaign.

In assessing the results of the air campaign, the Pentagon notes:

> Not all the coalition's advantages enjoyed during Operation Desert Storm will be present during the next conflict. However all modern, industrial and military powers share certain universal vulnerabilities. The technological advances that make them powerful are also their great vulnerabilities. These include computer dependent C^3 systems; networked air defense systems and airfields; and easily located sources of energy. When the key nodes are destroyed, such systems suffer cascading and potentially catastrophic failure.[2]

In assessing the damage done to the various target sets, the Pentagon points out that the attacks on the Iraqi power facilities shut down their effective operation and eventually caused the national grid to collapse. They point out that this had a cascading effect, reducing or eliminating the reliable supply of electricity needed to power nuclear, biological, and chemical weapons production facilities as well as other war-supporting industries. The study also points out that in targeting the civilian telecommunication system, which was an integral part of military communications, the bombing had an immediate effect on "the ability to command military forces and secret police." In discussing the impact of the targeting of oil refining and distribution facilities, the Pentagon estimates that "for about half the bomb-load dropped on one typical refinery in Germany during World War II the coalition effectively stopped all Iraqi refined fuels production." The accuracy of modern weapons was also demonstrated in the campaign against railroads and bridges. The Pentagon's study points out "during the early years of the Vietnam War hundreds of U.S. Air Force and Navy aircraft bombed the Thanh Hoi bridge in North Vietnam. It was not seriously damaged and many aircraft were shot down. During Operation Line-backer One in 1972 the bridge was knocked down by just a few sorties using laser guided bombs."[3]

In regard to all the strategic targets in Iraq and the level of effort required to knock them out, the Pentagon list has some interesting statistics. Out of a total of 18,276 sorties only 215 were devoted to knocking out electric power and only 518 to knocking out oil.[4] Figure 9-1 shows the allied air forces level of effort against strategic targets during Desert Storm. What is of special interest in this figure is the relatively small number of sorties needed to destroy or severely damage Iraq's electrical power, oil, C^3 centers, and railroads and bridges in comparison to the much greater number needed against more specific military targets.

Advanced conventional munitions (ACMs) may be as effective against point targets as nuclear munitions and more effective than either chemical or biological weapons.[5] Thus if ACMs have assumed characteristics similar to nuclear weapons vis-à-vis certain categories of targets, it follows that there may be

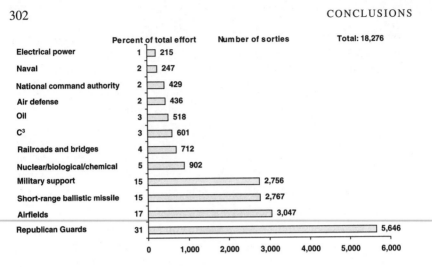

Figure 9-1. Strategic targets level of effort.
Source: U.S. Department of Defense, 1992.

similarities to certain aspects of nuclear doctrine as well. For instance, can one predict with the same degree of relative certainty the anticipated levels of damage to key infrastructure targets if an attack with ACMs is successful? Almost certainly, yes. The question next arises if, say, an adversary could assure the destruction of 70 percent (or 50 or 60 or 80 percent) of the major electrical generating and water desalination plants of a Middle East country, would that country be deterred from a certain course of belligerent action? Logically the answer would also have to be "yes." If it is possible to attack valuable economic targets with great precision, then it should also be possible to reach those fixed targets that have a direct connection with the ability of the regime to exercise control, including many components of the armed forces and the bulk of the general population. Of course, the more authoritarian and centrally controlled the regime, the more effective a "counterregime" strategy is likely to be.[6]

Whether it will be possible to "assure" such a level of destruction depends upon assumptions about improvements in missile and air defense systems and the vulnerability of the adversary's offensive missiles and aircraft. There is no doubt that over the next decade major developments in theater missile defense (TMD) will strengthen the ability of the major powers to defend against attacks from ballistic and cruise missiles. The concept being developed for an integrated TMD system includes four "pillars": active defense, passive defense, attack operations, and battlefield management/command, control, communications, computers, and intelligence (BM/C⁴I). However, the proposed TMD system being developed by the NATO countries and Russia incorporates such a wide variety of advanced systems that it is unlikely any country in the Middle

East, with the exception of Israel, will be able to deploy such a capability in the foreseeable future *independent* of cooperation with NATO or Russia.

A conclusion of this study is that the combination of advanced conventional military technology and highly vulnerable modern societies suggests a new dimension of warfare, unlike anything seen in the past, but drawing upon the theories of both early airpower advocates such as Douhet and more recent exponents of nuclear deterrence strategies. What this indicates is that new advanced conventional munitions are revolutionary and that in the context of the targeting of fixed point, soft industrial facilities, they can aptly be described as "WMDs."

Future Wars

Against this backdrop we now look at several cases in which the relationship between geography and developing technology might influence the nature of future wars in the region. In view of the number of contemporary conflicts in the greater Middle East (discussed in chapter 3), there are endless permutations for potential wars in this part of the world. We will not attempt a comprehensive review but rather focus on two basic scenarios, which we believe have the greatest potential for far-reaching international repercussions. First, the possibility of another Arab-Israeli conflict with special reference to an Israeli-Syrian war and a wider Arab-Israeli war. Second, another Persian Gulf crisis involving the United States. The Arab-Israeli case is an archetypical example of the crucial relationship between topography and technology. The U.S. Persian Gulf scenarios are illustrative of the limitations of modern power projection technologies without the cooperation of en route and local partners for access and operational support. Two other potentially major scenarios— another Iran-Iraq war and another India-Pakistan war—are discussed in detail in appendix 5.

Another Arab-Israeli War

Although it is clear to all concerned that another major military conflict between Israel and its neighbors would in all probability be an unmitigated disaster for all parties no matter who emerges tactically victorious on the battlefield, the possibility of such a conflict cannot be discounted. How such a war might come about (and who the principal participants would be) is itself a complicated question. In this chapter what we are concerned with is the influence that strategic geography will have on a new Arab-Israeli conflict. Our conclusion is that a combination of traditional geographical factors, especially topography, distance, and weather, will continue to have an impact on the battlefield along with the new capabilities generated by the revolution in military affairs (RMA) and the proliferation of weapons of mass destruction. In other words, a mixture of old and new factors will have to be carefully weighed in assessing the emerging military balance between the parties.

THE CONTINUED RELEVANCE OF TERRITORY. Israel's political debate about the wisdom of exchanging "land for peace" derives in part from different opinions as to the strategic value of the terrain it has held since the 1967 war—the Golan Heights, the West Bank, and Gaza, and, more recently, the South Lebanon security zone. It is taken for granted that the continued demilitarization of the Sinai Desert is a cornerstone of Israeli security and that no Egyptian army can ever be allowed to reoccupy the Sinai up to the 1948 border.

In the case of Gaza with its nearly 1 million Palestinians, the strategic wisdom of retaining control has shifted in favor of withdrawal. Before 1967 Gaza was seen as a "dagger" threatening vital regions in Israel along the Mediterranean coast and in the Negev Desert. From bases in Gaza, first Egyptian and later Palestinian terrorists launched attacks on Israel and caused major disruptions in both peacetime and wartime conditions. Israel occupied the strip in 1967, but since the Egypt-Israel peace treaty of 1979, the strategic utility of holding on to it has diminished significantly. Rather the political, economic, and psychological costs of the occupation became so burdensome that even the Israeli right-wing parties offered few complaints when Israel handed over day-to-day management of Gaza to Yasser Arafat in 1994. The only security issue today concerns terrorism emanating from Gaza and whether or not Israel has the right to go back into the area for counterterrorist operations. However, there is no wish on the part of Israel to become an occupier once more. In the age of CNN the images of Israeli soldiers supervising refugee camps has done great damage to Israel internally and internationally.

In the case of the West Bank the situation is more complicated. Many of the arguments about withdrawal from Gaza apply to Israel's evacuation of the Arab urban centers along the ridge line in Judea and Samaria (see map 3 in chapter 1). The strategic—as distinct from ideological—arguments for retaining some, or all, of the rest of the West Bank relate to the need to protect Israeli settlers, secure fresh water, deter terrorism, and, in extremis, to block an advance by Arab armies on the eastern front in event that they cross to the Jordan River and launch an offensive up the eastern slopes. Indeed, the introduction of foreign armies into Jordan would be treated by Israel as a major military threat and could lead to a massive preemptive attack on that force if it proceeded toward the Jordan River. Israel has insisted that Jordan be a *cordon sanitaire* and that there be no military forces from Iraq, Syria, or Saudi Arabia deployed in strength anywhere in the country. This fear reflects the reality that in conflict regions such as the Arab-Israeli theater, distances are so short that modern armies with heavy armor can make rapid progress along paved roads and change the balance of power in a particular region in a matter of hours. For this reason Israel will continue to rely on nuclear weapons and high technology to assure its ability to deter, and if necessary fight, a hostile neighbor.

The debate about the relevance of topography for military operations is most intense in the context of Arab-Israeli peace negotiations, especially concerning Israel's prospective withdrawal from the West Bank of the Jordan River and the Golan Heights. Opponents of Israeli withdrawal from the West Bank fall into two groups. The first believes that for ideological reasons the West Bank is part of Eretz Israel and should never be relinquished. Jews, not Arabs, have the first claim on the land and its resources. The second group is less concerned about ideology and most concerned about security. They argue that if Israel relinquishes control of the West Bank, it will be threatened by a Palestinian state or federation. The new state could become another Lebanon with warring factions, weapons galore, and the ability to conduct raids into Israel's most populous areas around Jerusalem and Tel Aviv. Alternatively, the Palestinian state could provide the venue for Arab armies to launch a major attack against Israel as happened in 1967. Control of the West Bank, especially the approaches to the high ground overlooking the Jordan Valley, is essential to prevent such a nightmare.

Many of the proponents of withdrawal from the West Bank accept this argument but say it is a red herring. There is no way any Israeli government will relinquish control of the approaches or permit the Palestinians to control the Jordan Valley. Thus any eventual Israeli agreement with the Palestinians would have to include iron-clad security arrangements and would never involve handing back the entire West Bank to a Palestinian authority.

A NEW ISRAEL-SYRIA WAR. In terms of possibility and current focus of defense planning, a war between Israel and Syria is considered much more likely than an Israeli war with Egypt, Jordan, or a united Arab coalition, though the latter cannot be totally discounted.

Scenarios

It is not all that easy to say what are the most likely causes or the more probable scenarios for a future Arab-Israeli war; paradoxically, one might sense that there remains a high likelihood of one more such war, the peace process notwithstanding.

One obvious war opening might involve a Syrian attack on the Golan, if further negotiations over its ultimate status should stall and if Syria thought it could nudge the negotiations along with a "seize and hold" operation. If in this scenario Syria were going it alone, it would presumably have little if any hope of prevailing in a protracted war but would hope the major powers would quickly intervene and halt the war if it could grab off most or all of the Golan. Anthony Cordesman has bruited this possibility in discussing the possible utility of Syria's ballistic missiles armed with chemical weapons, noting that "while any such use of such missiles would risk Israeli nuclear retaliation, some Israeli experts have suggested that Syria might risk limited strikes against Israeli air bases and mobilization assembly sites as part of a surprise attack on the Golan."[7] In a related vein, Cordesman notes that other Israeli experts "believe that Syria will try to use its chemically armed missiles as a deterrent to Israeli strategic strikes and to allow it to attack the Golan using its armored forces without fear of massive Israeli retaliation,"[8] in effect amounting to a form of escalation dominance for Syria, at least for a short time.

In a related scenario, one might envision a preemptive attack by Israel on Syria, if the former came to see the existence of the latter's chemical weapons and associated missiles as unacceptably dangerous. That might constitute a repeat in some sense of the Israeli preemptive knockout of the Iraqi Osirak reactor in 1981, though in the case of Syria's chemical capabilities the dispersion of its chemical stores and missiles might virtually preclude a one-shot, all-out elimination of the threat to Israel. Such a preemptive strike might, of course, occur as part of an escalating crisis or in response to strategic warning on the part of Israel that Syria was about to attack.

Still other scenarios may be envisaged, after an agreement was struck on the Golan, which might provide various options regarding mutual withdrawal of troops, demilitarized zones, and the presence of multinational peacekeeping forces. After such a "peace pact," Syria might try to launch an attack, even despite the presence of such peacekeepers, with the intention of capturing the Heights overlooking the Galilee area or of moving beyond that, perhaps in conjunction with simultaneous attacks by Egypt, Jordan, or both. Or Israel, fearing such a Syrian move, might itself attempt to preempt in retaking the Heights and precluding their takeover by Syria. Whatever the nature of a "final" outcome on the Golan, it will presumably remain, for a long time to come, a hair trigger flashpoint, for the obvious reasons having to do with the critical nature of the control of the Heights if a full-scale war should erupt. Syrian control of the Heights would put Israel back in the position it occupied in 1967, needing to assault a rugged escarpment with the high likelihood of severe casualties even in the event of a successful assault. For Syria, there would be the concern about the situation that has existed since 1967, that is, Israel's possession of the high ground only a short distance from Damascus.

Of course, some might argue that an Israel-Syria war scenario, presumably one initiated by the latter to overturn an unacceptable status quo on the Golan, is highly unlikely. They might claim that Syria would even be unlikely to start such a war with merely a "seize and hold" operation in mind, somewhat in imitation of 1973 in the sense that Damascus might bank on the major powers halting such a war, freezing some limited Syrian gains and subsequently resulting in greater pressures on Israel to be forthcoming with some kind of settlement that would abort the Israeli annexation of Golan. That view would insist that 1973 is not repeatable with Syria lacking a great power patron capable of offsetting the United States (and creating anxieties in the latter about escalation to a big power crisis as almost happened in 1973) and of providing a rapid and massive arms resupply if necessary. That view would further be affected by the impact of such a war on Syria's internal political situation (a key issue for the minority Alawites), not to mention the Syrian assessment of the balance of military power even if a mere quick "seize and hold" was the basic objective. But the 1973 and 1990–91 wars (and maybe that in 1980 between Iran and Iraq) lend caution to assessments of the possibilities for war initiation based solely or predominantly on military force balances and the presumed accuracy and "rationality" of perceptions associated with them.

In summary, a new round of fighting between Syria and Israel might involve wholly different conditions than those that existed in 1982 or before, pertaining to both weaponry and geopolitics; as a result, scenarios are difficult to draw. Seemingly Syria could not do much by itself, save for the remote chance of a "seize and hold" operation that it could make stick. Israel might, however, have a freer hand in any new conflict to the extent it would worry less about needing U.S. support to counter or alter what earlier might have been a direct Soviet intervention—although, as ever, it will be sensitive even to relatively low levels of casualties. The previous near-ritualization of such conflicts, in which everyone had to worry about relations between the superpowers, no longer applies.

In any new war pitting Israel against Syria, much will depend on which side achieves control of the Golan Heights at the outset. If Israel remains in control it might then pursue a strategic campaign while sitting still in an essentially defensive mode on the ground. This would be similar to U.S. strategy in the Gulf War and would run counter to the Israeli tradition of an early all-out preemptive ground offensive. Anthony Cordesman, in adumbrating such a strategy, refers to Israeli plans "to use a combination of air, conventional missile, and naval power to launch a crippling sequence of blows to suppress Syrian air defenses and use conventional munitions to so devastate the advanced elements of Syria's economy and infrastructure as to deprive Syria of recovery capability for several years."[9] He sees this air, missile, and naval approach as a "stand-alone" capability that would allow Israel's forces to remain in their defensive positions and launch a rapidly escalating series of air, naval, and missile attacks to force Syria out of a war while minimizing casualties.[10] Presumably, such a strategic campaign would be followed by some sort of ground offensive, but Israel, banking on the lessons of the past, might recognize that external pressures from the "international community" would likely preclude a march on Damascus, which, at any rate, might be costly in terms of casualties. In short, this is a scenario that sees Israel attack Syria in imitation of Desert Storm, it being assumed that for the near future at least, the technological equation between Israel and Syria largely mirrors that which obtained between the United States and Iraq in 1990–91.

Patrick Garrity lays it out as follows:

Assuming that Israel is not surprised, it has the opportunity to emphasize fire rather than maneuver in the early phase of a war. Traditional Israeli doctrine emphasized maneuver as a way of taking the war to enemy territory, through flanking maneuvers or breakthrough battles. This doctrine has been challenged by the depth and complexity of Syrian fortifications, because of the costs in Israeli lives and time that they might impose. With new technologies, Israeli ground forces could remain initially on the defensive during massive air and artillery strikes against Syrian armor and troop concentrations and air defenses. These strikes would prepare the battlefield for breakthrough operations.

After the battlefield was prepared, Israeli air and ground forces could begin to fight a breakthrough battle rapidly shifting and massing fire from stand-off ranges, creating gaps in Syrian deployments while minimizing direct contact with the enemy, and thereby reducing friendly losses. The maneuver portion of the breakthrough battle would be initiated once Syrian first-echelon forces had been sufficiently attrited—at least at selected points—and once second- and third-echelon forces were reduced by deep strike systems or diverted to protect vulnerable rear areas. Israeli forces could then breach Syrian defenses and defeat Syrian forces in detail through traditional means of fire and maneuver and close-in combat, again with relatively minimal losses. Operation Desert Storm validated concepts and capabilities associated with this style of warfare.[11]

Would such a campaign also be feasible in circumstances in which Syria was able to achieve initial control of the Golan Heights? It is precisely such an assumption that feeds into periodic claims that territory, and particularly territory in the Middle East, no longer matters in the sense that it did when defensive barriers were more critical.[12] An Israeli campaign would be so devastating as to render control of the Golan Heights relatively unimportant, and the Heights would fall at the end of such a campaign with an ease roughly comparable to 1967, and with the same implications for limited Israeli casualties. The role of the Israeli army, and particularly its armored and infantry units, would, in this scenario, be largely defensive, that of preventing the Syrian army from moving off the Heights down into the valleys in the Galilee area.

A number of critical caveats can be offered here that point to the difficulties that Israel might have in conducting such an operation. First, of course, is the matter of time. Unlike the United States and its Desert Storm coalition, Israel cannot afford a protracted war—and protracted by Israeli standards is a very short time. The 1973 and 1982 wars had disastrous consequences for the Israeli economy because Israel had to mobilize a substantial proportion of its male population and had its trade and tourist industry interrupted for substantial lengths of time. Hence if Israel were to attempt a strategy along the lines of the above, it would presumably be attempting a strategic campaign against Syrian infrastructure and industry and to "prepare the battlefield," in a much, much shorter time than that taken by the United States and its allies in the month-long period between the opening of the air campaign and the opening of the ground campaign. More than a week or so would be a long time for the Israelis to sit still on the ground.

Then, too, there is the question of whether the Israeli air force, for all of its vaunted skill and advanced technology, would be able to conduct a sufficient number of sorties for a decisive strategic campaign and preparation of a military technical revolution (MTR) battlefield. It would have little more than 200 or so attack aircraft to do it with. Attrition could mount in a few weeks and it would not necessarily have the assurance of U.S. resupply of replacement aircraft or spare parts. Following such a strategy would, for Israel, actually run counter to its long-held approach that takes into account the need to act quickly before the major powers and the UN can intervene and apply pressure for a stalemate on the ground; this situation would be exacerbated if Israel were to lose some territory at the outset of a conflict as Syria pursued a "seize and hold" strategy.

For Israel control of the air would be the primary requirement for success. As we have noted the small size of its air force relative to that of the United States has led some observers to question whether it could pull off a strategic campaign with even a remote resemblance to the U.S. aerial onslaught of 1991. But then it must be stressed that the distances involved are much shorter in the Israel-Syria context, and that allows the Israelis to maximize the number of sorties they can conduct even with a small air force, particularly as the requirements for tanker refueling are much less demanding than they were for the United States in 1991. That involves a paradox of sorts because the very short flight distances from Syria to Israel also make Israel's situation with regard to strategic warning a precarious one.

On the other hand, there is also the question of the extent to which Syria has been able to absorb the lessons of the Gulf War and to devise strategies to combat an Israeli attempt at a repeat of what the Americans were able to achieve. Indeed, there have been extensive analyses in Washington and presumably elsewhere of "asymmetric" strategies or responses to the type of strategic campaign conducted by the United States in 1991.[13] These asymmetric strategies can take any number of forms. They could, of course, involve terrorism, perhaps the use of chemical or biological weapons within Israel, subject, of course, to a draconian Israeli response. But they could also involve an enhanced effort at protecting communications infrastructure by moving things under ground with fiber optic cables and the like. Or they could involve relatively "simple" countermeasures: far better and more extensive use of camouflage, cover, deception, decoys, among others. Despite its lack of access to high-tech equipment, Syria might devise some hi-tech solutions amid a high-tech–low-tech mix to combat some of the key Israeli technologies, perhaps in the area of electronic warfare or in missile air defenses. Mitigating total Israeli control of the air or at least causing Israel considerable attrition of aircraft and pilots would, presumably, be at the heart of the Syrian strategy, along with the goal of taking some chunks of territory and then seeking a cease-fire in place.

The weather and the seasons might be a factor. Syria might attempt another war with Israel during the autumn or winter, when there is often inclement weather on the Golan Heights and, generally, in and around the Israel-Syria area of confrontation. Such a possibility makes the question of the extent of Israel's ability to acquire and use more new U.S. Air Force types of all-weather capabilities an important one.

A Broader Set of Arab-Israeli Scenarios

Still other scenarios may be envisioned involving the active involvement of other Arab states in various combinations. In reality here Egypt is the key. In the 1967 and 1973 wars Egypt played the biggest military role on the Arab side by far. In both of those conflicts Israel had to deploy well more than half of its forces in Sinai. It is the specter of having to fight Egypt again that presents Israel with the prospect of unacceptably high casualties (the bulk of Israel's some 2,500 mortalities in the 1973 war came in Sinai), the bane of its military planning.

Jordan, or even Jordan plus Saudi Arabia, are not likely to join Syria in an attack on Israel and, indeed, their participation might alter the military equation only marginally while rendering them both highly vulnerable to an Israeli aerial strategic campaign. Egyptian involvement, however, would present a wholly different problem for Israel. Egypt and Syria together could attempt a virtual repeat of the 1973 war, presenting Israel with the dilemma of having to fight a two-front war, albeit one that it would presumably approach, as it did in 1967, in a partially serial fashion. Or one might envision a war between Israel and Syria in which an Egyptian government, goaded by pressures from the Cairo "street," felt compelled to join, if only some time after the initiation of combat. The reverse scenario, an Egyptian-Israeli war to which Syria was added on, is also a possibility. If the peace process should break down, with or without an Islamic revolution in Cairo, a war could result from a premeditated Egyptian attack or from an Israeli preemption in response to "salami tactics" in which Egypt attempted gradually to remilitarize the Sinai by employing such strategies as moving mechanized forces into the desert and setting up SAM sites. The most likely trigger for a war involving Egypt would be a Palestinian uprising that Israel attempted to crush.

However, just as some analysts have questioned the realism of an Israel-Syria war scenario, so long as the latter is not availed of a major power supporter that could, if necessary, conduct an arms resupply operation on its behalf as well as provide political backing (as the USSR did in 1973 and 1982), so, too, have analysts questioned just how there might be a new round of fighting between Egypt and Israel. If Egypt were to undergo an Islamic revolution, it would no doubt be cut off from its U.S. source of arms, which now constitute the backbone of the Egyptian air force and its armored divisions. That would presumably render Egyptian participation in a war with Israel unlikely as the former's U.S. weapons inventory could be attrited even in a relatively short war without replacement systems and spare parts. If such a "new Egypt" were to attempt a transition back to Russian weapons, or a reliance on European systems, it would take many years, be very costly, and presumably have to depend on Saudi and other Gulf Cooperation Council (GCC) states' financing at a time of economic constraints in those countries. Egypt would, in other words, be in straits similar to those endured by Iran at the outset of the Iran-Iraq war in 1980 when the postrevolutionary Khomeini regime was largely cut off from its U.S. arms source.

In the shorter term there is some chance that an Egyptian regime could participate in an Israeli-Syria war, even if a U.S. embargo on resupply were inevitable. In such circumstances Egypt would be banking on a short war in which resupply issues would not be relevant.

There are still other possible scenarios. Iraqi, Iranian, or Libyan missile strikes against Israel, perhaps in the context of some form of a repeat of the Gulf War, could escalate into a broader Arab-Israeli or Islamic-Israeli war. Israel could, of course, retaliate via air strikes or missiles, with or without weapons of mass destruction. At the far end of a worst case scenario for Israel, it could end up in a wider war involving all three of these nations, as well as a postrevolutionary Turkey that had undergone its own Islamic revolution. On the other hand, there is the question of whether in some circumstances the current tenuous Israel-Turkey security relationship could result in Turkey siding with Israel in a wider Middle Eastern war, in which case Syria would be confronted with the major strategic problem of a two-front war.[14]

No matter what the cause of war, the key geographic factor in an Israeli-Egyptian war is control of Sinai. But what would be the contending strategies in an era of MTR? Would Egypt try to move large armored forces into the Sinai, where they would be highly vulnerable to air attacks if Israel gained control of the airspace above the desert? Or would the Egyptians lay back with their armored forces, hoping to add to whatever attrition the Syrians would be able to achieve vis-à-vis the Israeli air force, and then attempt a later move on the ground? Would the Israelis rely solely on their air force at the outset of such a conflict or would they try to move armored forces out into the Sinai, perhaps in an attempt to capture the key Gidi and Mitla passes in the low mountains of the western Sinai, which were critical to military calculations in both 1967 and 1973? Would the Israelis, instead, maintain a defensive posture along the current borders or in eastern Sinai? Depending upon what happened in the air wars, there might be the overriding question of whether the main ground forces of the two armies would slug it out in Sinai or whether the outcome would be decided largely in the air.

Of course one or more among Jordan, Saudi Arabia, Iraq, and Iran could enter the fray, which would further complicate the scenario and might further test some of the issues regarding the evolving relationship between MTR weaponry and geography. The utility of the Saudis' large U.S.-equipped air force in attenuating Israel's advantage in the air would be tested, at least to the extent of exacerbating the problem of attrition for Israel's relatively small air force. Jordan's air force could play a similar and parallel role. In the case of Iraq there is the question of whether it could move its armored forces across Jordan or into Syria in the face of Israeli dominance of the air and the fact that its tank transporters and other armored traffic would be highly vulnerable to attack in the open terrain and on the desert highways. In regard to Saudi Arabia there is the question of whether it could move some of its fighter aircraft forward to the base at Tabuk in the northwest and conduct air offensives against Israel, or at least succeed in diverting the efforts of the Israeli air force.

The mix of scenarios we have just described raises serious but unanswerable questions about ladders of escalation leading to the possible use of weapons of mass destruction. The matter has already been bruited with respect to possible Syrian use of chemical weapons at the outset of a conflict, or its use of a chemical weapon deterrent in the face of an Israeli strategic campaign aimed at destroying its infrastructure and industry or one leading toward a march on Damascus or both. But for a long time the possible Israeli use of nuclear weapons in last-resort circumstances has been considered the most likely trigger to the use of weapons of mass destruction, and that remains a specter despite Israel's seemingly greater level of security in the 1990s.

Let us assume that after weeks or even months of fighting on the ground the Israeli army, bled white by casualties in the tens of thousands, finally cracks and begins to unravel, perhaps even surprisingly quickly, as often happens to armies that lose heart after a long battle. Let us further assume that as Arab armies enter the Galilee area or envelop Beersheba or both, reports begin to come in about large-scale massacres of Israeli civilians. At such a juncture it is possible that Israel would ask the United States to intervene militarily on the ground, a request that might or might not be granted. Alternatively, Israel might choose not to make such a request.

What then might Israel do instead? Obviously, such circumstances would invoke the possibility that it would resort to nuclear weapons, bearing in mind that in 1973 some Israeli leaders seemingly panicked at the possibility that "the third Temple was coming to an end," and the country apparently went into a nuclear alert.[15] Whether Israel would eventually have used nuclear weapons if the military situation had worsened (and against what and with what types and how many weapons) is not known here; it was also clear that part of the reasoning behind the nuclear alert was the idea of forcing the U.S. hand on an immediate weapons airlift. It should be emphasized that in 1973 Israel faced the prospect of a Soviet nuclear counterthreat, which might or might not have been partially balanced out by a U.S. countercounter threat.

Presumably, in such a future scenario, given the size and flexibility of its nuclear arsenal, Israel would resort first to the tactical use of nuclear weapons. If action were taken early, it could involve the use of nuclear mines or other nuclear munitions on the Golan Heights or along the Sinai-Negev frontier. Alternatively, such a development could come later after some Israeli territory had been lost to the Arabs. Initial use of a limited number of such weapons against Arab troop concentrations might be envisioned, perhaps in combination with suggestions about a truce plus threats of further use of nuclear weapons. On the other end of the spectrum, there might be counter-cities attacks with the hope of decapitating Arab leadership cadres, or, alternatively, nuclear strikes against military headquarters, infrastructure, and follow-on forces. Enhanced radiation weapons might be a preferred choice in order to limit downwind radiation in Israeli urban areas, particularly if such weapons had to be used against Egyptian troops advancing from the southwest across the desert toward the Israeli heartland.

Presumably, Israel's decisions with regard to tactical versus strategic use of nuclear weapons and, generally, its targeting doctrine, would greatly depend on how many weapons it had, the yield of those weapons, and the accuracy of the delivery systems. There is a broad range in the estimates of the size of the Israeli arsenal. Further, these weapons presumably have not been tested for reliability or explosiveness (the possible importance of the explosion in the Indian Ocean south of South Africa notwithstanding); neither, for that matter, have the delivery systems been adequately tested with nuclear weapons. There is also the question of how many such weapons might constitute a counter-cities deterrent threat if tactical nuclear exchanges or chemical or biological warfare were initiated. If Israel had only Hiroshima-size weapons at around 20 kilotons each, those weapons might cause 100,000–200,000 casualties in large metropolitan areas such as Cairo or Teheran with populations of 10 million or more. Boosted 200-kiloton weapons or hydrogen bombs would, of course, be another story. If Israel had to target large urban centers with numerous smaller weapons, it might mean that the vague estimates of some 20–30 weapons needed for strategic deterrence are somewhat on the low side; this would in turn raise questions about how many weapons it would have left over for tactical, battlefield purposes, that is, by how much would the "effective" size of its stockpile be reduced.

However, some of the impending developments in the U.S. RMA, specifically with respect to counterproliferation, may upset current calculations about the possibilities for nuclear war in the several greater Middle East regional contexts. The new arsenal ship and land-based theater antiballistic missiles may give the United States the capability to interdict regional missile salvos, even from Israel (or India), that is, to "shut down" a nuclear war. If so even Israel (or India) could be compelled to look to new asymmetric strategies in lieu of what have been thought to be more or less invulnerable and unstoppable nuclear retaliatory capabilities.

But all in all the primary counterthreat to Israel in these circumstances, in the absence of an Arab nuclear deterrent, would be the possible use of chemical weapons by the Arabs, it being noted that at present, Egypt, Syria, and Iraq are all in possession of a militarily significant nerve gas capability. Just possibly Pakistan might represent a countering nuclear threat by transferring nuclear weapons to one or another Arab state in the midst of such a crisis.

In such a scenario Israel would clearly have escalation dominance. It could use tactical nuclear weapons on the battlefield and then threaten to use them against population centers if the Arabs should respond with chemical weapons, either on the battlefield or against Israeli population centers.

However, if one or more Arab states should acquire nuclear weapons, particularly if this is coupled with an Arab conventional superiority as hypothesized above, the entire equation changes. At that point the strategic calculations on both sides would be similar to those in the hypothetical circumstances of a possible NATO-WTO (Warsaw Treaty Organization) conflict in Central Europe, albeit in the context of an Israeli situation of last resort. A central question is whether a limited nuclear war, presumably initiated by Israel in the face of a collapse of its armies on the conventional battlefield, could be kept from expanding to an all-out counter-cities exchange. Actually, in the type of scenario outlined above, but this time assuming Arab possession of nuclear weapons, there might be some advantages on the Arab side, though much would also depend on the respective sizes of the nuclear and chemical arsenals. Israel possessing 300 atomic weapons and the Arabs 10 or 20 is one thing, but the latter possessing, say, 100 or 200 such weapons would be something else. However, if Israel were on the verge of conventional defeat, it would probably entail, as a corollary, a very constricted remaining Israeli aircraft inventory; indeed, the loss of much of its air force might well be one reason for a conventional collapse. Some of Israel's air bases might be overrun and others be within range of Arab artillery. Fixed missile launchers might also be under heavy pressure. The Arabs, with their larger spaces, might well have an advantage with a lot of room in which to maneuver mobile missile launchers armed with WMDs.

Just possibly Israel might move toward an asymmetric mix of nuclear and conventional deterrence in the face of a massive buildup of forces in an Arab grand coalition, as outlined above. It might, for instance, declare that it would not engage Egyptian forces in a conventional slugfest in Sinai, threatening Cairo with the early-on use of tactical nuclear weapons, while at the same time trying to conduct a conventional defense on the eastern front against a coalition that might include a nuclear-armed Iraq. Egypt might, of course, attempt to respond to such a threat by the threat of use of chemical weapons on the battlefield.

We have assumed that the most likely scenario in which Israeli nuclear weapons would come into play is a last-resort situation, but it is at least possible that they would feature as well even in less dire circumstances. Israel might contemplate the use or the threat of the use of nuclear weapons in response to a lengthy war of attrition in which it was being bled white, in terms of both battle deaths and damage to its economy. It will be recalled that the 1973 war was considered a virtual disaster for Israel, in that it entailed some 2,500 battlefield deaths plus numerous additional casualties and an overall net cost somewhere in the range of the equivalent of its GNP for an entire year. For Israel the 1973 war was considered to have lasted far too long, much longer than the quick victories of 1956 and 1967, which had set the standard for the vaunted Israeli preemptive *blitzkrieg* tactics. Yet, in historical perspective, twenty-two days does not constitute a long war. Far from it!

What if a future war with an Arab coalition should go on for several months, for six months, or even for a year? What if Israel should suffer tens of thousands of battle deaths in such a war, perhaps even 50,000? And in such a war Israel's economy would be devastated, with most of the male population mobilized and trade with the outside world largely diminished. At what point in such a scenario might nuclear threats or even use come into play? If Israel were the sole nuclear power in the region, threats in this context would presumably have a high degree of credibility, if not as much as in the case of last-resort circumstances. If, however, Arab states had nuclear second-strike capability, the credibility of such a threat would be much less, amounting to a threat to commit mutual suicide, that is, if counter-city strikes were anticipated. More credible would be a threat to initiate use of tactical nuclear weapons.[16]

War in the Gulf

There are any number of scenarios for future wars in the Persian Gulf, with the most significant strategic uncertainty being the role of the United States. If the United States is not directly involved in a new major Gulf War, there are only three countries capable of launching serious offensive operations one against the other in the foreseeable future—Iran, Iraq, and Saudi Arabia. While Iran and Iraq are potentially the dominant regional military powers, by the mid-1990s both countries had severe limitations on their ability to sustain, let alone modernize, their armed forces, with Iraq clearly in a worse situation than Iran. Saudi Arabia, on the other hand, has spent billions of dollars since the Gulf War in upgrading its equipment, and, on paper, now possesses significant military capabilities, especially in terms of airpower for both offensive and defensive operations. Still, over a fifteen-year time frame this balance could change significantly. If Iran and even more so Iraq can get out from under sanctions, they will be able to rebuild their armed forces and perhaps purchase the sorts of advanced arms the United States had a monopoly on during the Gulf War. Such modernization could change the regional dimension of warfare. Meanwhile the geography of the Gulf continues to constrain potential land operations by all three powers and, to a lesser extent, their air capability. These have been discussed in detail in earlier chapters but to reiterate briefly, they relate to Iran's strategic depth and long littoral, Iraq's lack of strategic depth and its tiny littoral, Saudi Arabia's deserts and the fact that it has three coasts to defend, the regional weather conditions, and various other more esoteric considerations such as the hydrography of the Gulf itself and the ease with which any of these powers could interfere with shipping through the Strait of Hormuz.

Future Scenarios Involving U.S. Forces

An approximate replay of the Gulf War has been a major focus of concern among defense planners in Washington; indeed, the possibility of the United States, with or without a coalition, having to fight again in the near future against either Iran or Iraq has probably been the most closely examined among the various possible regional war scenarios. For purposes of this study we shall examine two widely different, if also closely related, scenarios. We shall deal with limiting cases, which can be used to illuminate basic geographic issues, including in-theater access, that have important political overtones in regard to the appropriateness and likelihood of future major American intervention in the region.

On one end we can envisage a new Gulf War in which, similar to the 1990–91 crisis, the United States has a very favorable, or permissive, situation with respect to coalition allies—both within the region and outside—and broad international support, and, hence, a high degree of en route and local basing and staging access. At the other extreme is a scenario in which the United States is constrained to act virtually on a unilateral basis in a war against either Iran or Iraq (or both of them) with little if any political support either within the region or elsewhere, possibly even with political opposition from both former coalition partners and countries that had remained aloof and disengaged in 1990–91.

Consideration of these "limiting case" scenarios should not cause us to ignore myriad other possibilities for scenarios involving U.S. forces, beyond those that constitute, to one degree or another, virtual reruns of Desert Storm involving Iran or Iraq. Those possibilities could include more limited military confrontations: for instance, some kind of repeat of Earnest Will if Iran should again decide to interrupt tanker traffic in the Gulf or an action against Iraq if it should choose to challenge the "no-fly" zones imposed on it since the end of the Gulf War. The United States could get involved in suppressing or overturning internal revolts by radical Islamist or Shia groups in Saudi Arabia, Bahrain, or other GCC states. It might undertake a large-scale punitive operation against Iran or Iraq in response to a terrorist attack on U.S. territory or against a U.S.-flag airliner or U.S. forces in Europe or the Middle East (note the precedent vis-à-vis Libya after a discotheque bombing in Berlin) if it was determined that such an act was "state sponsored." The possibility of preemptive attacks against nascent or burgeoning WMD programs has been noted. These alternative scenarios involve, fundamentally, an extrapolation of the political facts and political orientations in the region that have obtained since Desert Storm on up to 1997.[17]

Gulf War in a Permissive Environment

In the first of these scenarios some of the basic "facts" of the earlier Gulf War bear repeating for emphasis. Within the region access to air and naval bases in Saudi Arabia, in some of the other GCC states, and in Egypt and Turkey was critical, both to the logistical buildup for the war and for its actual prosecution. This access allowed for a slow, deliberate five-months-long buildup of forces and materiel, in a situation in which the air and naval points of access were hardly threatened, much less interdicted. It was an ideal setup. In a related political vein, the United States benefited greatly not only from the UN mandate, but from the active support and collaboration of key Arab states, particularly Saudi Arabia and the other GCC states, Egypt, and Syria, with Jordan constituting an exception. This broad political support also resulted in easy access for the U.S. logistical effort in such en route countries as Spain and Portugal, as well as all over the rest of Western Europe and the airspace over some of the former Warsaw Pact countries. Indeed, with the exception of a few minor political problems in Thailand and India, U.S. access was practically uncontested, which is why this undertaking was hoped by many to become a model for a UN-mandated, multilateral collective security effort.[18]

There were also, as we noted earlier, some other key factors. One was the existence of the huge American force structure left over from the then recently terminated cold war in Europe, which could easily be moved to the Middle East. Another was the fact that Iraq did not have nuclear weapons, even if it did have extensive stocks of chemical and biological weapons as well as missile delivery systems.

It would be easy to say that if such a war were to be repeated with the same permissive or supportive international environment, it would be a virtual replay, with perhaps only minor modifications. But depending on how far out in time we are projecting such a scenario, inevitably some things will have changed, thus precluding such a replay, even if we assume that the U.S.-led coalition would again be fighting against just one more or less lonely regional power—Iran or Iraq—that has only the kind of local moral support and access that Iraq had from Jordan in the previous conflict.

Some of these changes could be significant. No matter what the level of global (or at least European) support for the war, there will be far fewer U.S. forces in Europe, so that moving forces to the Persian Gulf area will, by that measure alone, be a bigger job, one in which larger numbers of personnel and equipment will have to be moved from the continental United States (CONUS), Another possible change, one more likely with each passing year, is that a U.S. regional foe might have nuclear weapons and be able to deliver them some distance or have a greater ability to deliver chemical or biological weapons than Iraq had a few years ago. This factor, changing gradually as the 1990s proceed, could involve a massive shift in the overall equation, even if the United States were to retain the advantages of allies and access at the same level as before. Indeed, both Iran and Iraq would likely heed the advice of the Indian commentator we quoted earlier, to the effect that one major lesson of the previous war was that it was not wise to take on the United States or a U.S.-led coalition without having nuclear weapons.

A U.S. effort at a repeat of Desert Storm would suffer not only from the drawdown of U.S. forces in Europe, but also from the overall reduction in U.S. forces since the end of the cold war. In various major categories, for example, numbers of army divisions, active combat ships, and air force wings, the United States had, by the middle of the decade, a somewhat smaller military establishment than in 1990.[19]

On the other side, however, Iraq's forces were, by the mid-1990s, only a shadow of what they had been in 1990, and there were no indications that that force could be rebuilt to a comparable level any time soon, particularly without the kind of support that had been provided in the earlier instance by the Soviet Union. Iran's forces were reportedly in not much better shape. So it is possible that if the United States, at the head of a new coalition, were to fight either Iraq or Iran anytime in the mid- to late 1990s or even a bit beyond the turn of the millennium, both sides would start off weaker than they were in 1990, in absolute terms or at least in quantitative terms (a smaller U.S force would presumably benefit from qualitative improvements).[20] Indeed, all concerned have had ample time to study and restudy the lessons of the Gulf War and to plan for a new one with an altered game plan.

Both Iraq and Iran would, presumably, have come to the conclusion that they could not possibly match U.S. military power in another such showdown. Both will have learned that even given the U.S. point of vulnerability when it comes to casualties, another effort at trying to draw it into an attrition-type land war would probably be fruitless, with the most likely outcome being another conflict largely decided by U.S. airpower. But numerous recent analyses have pointed to the possibility that regional powers such as Iran or Iraq may be availed of what has come to be called "asymmetric strategies" to compensate for military weakness. For the most part these strategies are attempts at regaining the deterrence that would be provided by any method of increasing the chances that higher casualties could be inflicted on coalition forces. But some are also efforts to at least modify the huge technological disparities between the forces exhibited in the earlier war. We can merely summarize these measures here.

One set of asymmetric strategies has to do with improved camouflage and concealment, so as to make another such unhindered aerial assault as the one that was conducted in 1991 more difficult.[21] Whether great improvements can be made on that score, given the prevailing terrain in much of Iran and Iraq, is open to question. However, greater use could be made of dummy tanks and other systems, or such "simple" passive countermeasures as putting a lot of hot objects on the battlefield so as to fool the infrared sensors that were so successful at "tank plinking" and other tasks. Communications could be moved underground with large-scale use of fiber optic cables. Then as part of a "high-low" mix of technological responses, it is possible that countries such as Iran or Iraq might acquire some competence in electronic warfare techniques or learn to combat stealth aircraft, either of which might take them some distance beyond the utter helplessness and futility that characterized Iraqi efforts in 1991.[22] Presumably, neither Iran nor Iraq will achieve the capability to conduct antisatellite operations any time soon, certainly not without major technological assistance from the outside.

In another conflict an Iraq or Iran, noting some of the impact of the CNN coverage of the last war, might make a more determined and larger-scale effort at exploitation of collateral damage, that is, locating major military installations, command headquarters, communications nodes, and weapons of mass destruction amid civilian populations in a way that the facilities could not be destroyed without massive civilian casualties and the associated public outcry resulting from media coverage of the carnage, as happened in Baghdad.

A major concern for the United States and perhaps its putative allies would be the possibly much greater use of terrorism overseas both as a deterrent and as a rebuff to an actual strategic campaign against a regional power. Since the Gulf War the bombing of the World Trade Center in New York, the Oklahoma City bombing (not related to the greater Middle East), the Algerian fundamentalists' terror campaign against France featuring the bombings of trains and subways and the apparently planned exploding of an airliner over Paris, and the nerve gas fatalities in the Tokyo subway have all underscored the growing threat of terrorism. Perhaps the Algerian case comes closest to a possible model of what Iraq or Iran might be capable of in an attempt to deter or to combat another U.N. effort to "Iraq" one of the region's powers. A more extensive effort to use ballistic missiles, in imitation of the earlier Scud attacks on Saudi Arabia and Israel, might also be anticipated. Indeed, both of those countries might again be prime prospects for attack by either Iraq or Iran; the use of one of the latter by the other for such a purpose should not be ruled out, but what degree of deterrent effect that would have against the United States or its coalition partners is not easy to predict.

Finally, it is obvious that any regional power aspiring to take on a multilateral coalition will have looked carefully at Iraq's failed strategies in the last war and would adjust accordingly. If Iraq or Iran were to move on Kuwait, they would presumably be aware that it might not be a good idea to simply sit and wait for the inevitable onslaught, and they might be more inclined than Saddam Hussein was in 1990 to move immediately into Saudi Arabia, the ultimate consequences perhaps notwithstanding (such a move would guarantee a very strong U.S. response). Generally speaking, massing defensive forces in fixed positions behind a wall of tank ditches, berms, and minefields, all with dug-in armored forces, would presumably be seen as understandably successful in the Iran-Iraq war, but clearly unsuccessful in the Gulf War. Further, the kind of "Hail Mary," or "left hook" offensive that the United States executed in 1991 could presumably not be pulled off again as a surprise, though that does not mean that it could not work anyway.[23]

Of course if Iran were the U.S.-led coalition foe, rather than Iraq, that alone would produce a considerably altered geographic situation. Whereas the war with Iraq was triggered by the latter's invasion of Kuwait, it is not altogether clear what the trigger would be for an Iranian aggression of some sort, what that aggression would be directed at, and what would be the geography of a coalition response. Iran might also assault Kuwait, but to do so it would have to get past the Iraqi army in the Shatt-al-Arab area around Basra, raising the question of what Iraq's role would be in the subsequent struggle. Or Iran might attack Saudi Arabia or one or more of the other GCC states across the Persian Gulf, perhaps in support of an insurrection. Although at present Iran may have the capability to lift only one battalion in such an operation, that would probably provide the United States with adequate strategic warning. If Iran were to install its forces across the Gulf, that might mean that U.S. and coalition forces would have to build up again in Saudi Arabia to expel them, with a subsequent decision about whether to launch a ground attack into Iran, which would mean traversing Kuwaiti and Iraqi territory, or countering with an amphibious operation that could be restricted to areas inside the Gulf or mounted against the Iranian coast outside the Strait of Hormuz. Generally speaking, scenarios pitting a U.S.-led coalition against Iran in an only somewhat recognizable repeat of the Gulf War remain ill-defined.

Probably the main intellectual dilemma involved in assessing a new but essentially repeated version of the Gulf War has to do with the possibilities for escalation of such a war to the use of WMDs. A key decision point in such a conflict might come early on, if Iran or Iraq attacked one of the Gulf states (an attack by Iran or Iraq one against the other would presumably not engender a U.S. or UN-mandated response any more than it did in 1980), and the United

States appeared to be gearing up for another strategic campaign similar to Desert Storm. This time around the aggressor might think of attempting to use ballistic missiles (with or without chemical warheads) against the Saudi air-fields that are needed in order for the United States to mount such a campaign. Or it might threaten a ballistic missile assault against Israeli or Saudi cities so as to deter U.S. involvement. If the United States proceeded with a strategic campaign to wreck the aggressor's infrastructure as it did in 1991, it could be countered with a threat to use nuclear weapons against Israel or Saudi Arabia or against U.S. bases in Saudi Arabia and other GCC states, or against U.S. carrier battle groups offshore in the Persian Gulf or the Arabian Sea. In effect this kind of projected retaliation might fall under the somewhat novel concept of "triangular second strike," the idea that a nuclear-armed regional state could deter the United States from escalating military action against it by threatening nuclear destruction of nearby (and not necessarily contiguous) states.

If Iran or Iraq threatened such retaliation, their nuclear-armed missiles might be located in built-up civilian areas so that the United States would be in a dilemma about preempting or responding when to do so would inflict large-scale collateral damage. As gruesome as it may seem, this could conceivably require the United States to threaten counter-cities nuclear strikes in order to deter a regional foe from using nuclear weapons against its allies' cities or its own military installations. But as time goes on and the technological race proceeds, the United States may be able to circumvent some of the problems of this scenario by various counterproliferation measures, for example, boost phase interception of outgoing Iraqi or Iranian missiles, an effective antimissile theater defense, perhaps to be mounted on ships but with extensive area coverage, and the ability to detect and hit nuclear and chemical weapons storage sites, even in or around urban areas or underground, with precision conventional munitions that could avoid massive collateral damage.[24]

With regard to a "triangular second strike" and the prospects of a repeat of the 1991 Scud attacks against Israel and Saudi Arabia, it might be pointed out that five years after Desert Storm, in the words of the U.S. Army chief of staff, the United States is "a hell of a lot better off" than it was then in combatting theater ballistic missiles and WMDs.[25] To some extent that has to do with the impending deployments of the new PAC-3 missile in 1999 and, later on, the theater high-altitude area defense and medium extended area defense. But one analysis also explained the new synergism and synchronization provided by unmanned aerial vehicles (UAVs), the ability to project targets in an F-16 cockpit, and the engaging of Scuds located by UAVs with army tactical missile system air-to-surface missiles as well as the capabilities of the joint tactical ground station, a regional missile warning system.[26] Obviously the Pentagon now thinks it could do a lot better in a new Scud hunt, if that should be necessary.

The above scenario is, obviously, one of concern over a longer term. For a shorter-term perspective, the basic questions are whether Iran or Iraq can, with the use of some asymmetric strategies, alter the seemingly durable equation that emerged from the Gulf War, wherein the United States should again be able to conduct a very effective strategic campaign and then dominate the battlefield—one characterized by open desert terrain and clear weather. It is hard to see how, in the near future, a regional power can alter this basic equation, which leads one to think either that such a scenario is unlikely or that if it should occur it might readily lead to the use or threat of use of WMDs.

A Gulf War with a Very Constrained Environment

At the extreme opposite end of the spectrum one can hypothesize another Gulf War–type scenario in which the overall political context—the "facts" of alliances, basing access, and arms resupply—militate a very difficult and constrained environment for a U.S. intervention. Simply stated, the scenario is one in which the United States must go it alone, without allies or the trappings of multilateralism and UN mandates and hence without the access en route and within the theater that made Desert Storm (with the aid of hindsight) such a relatively simple and easy operation. It is also a scenario in which the techno-logical basis of the conflict has been altered by time, so that a regional foe may be assumed to be armed with nuclear weapons and their delivery systems.

Let us hypothesize the following: The United States finds itself embroiled in a growing crisis with Iraq or Iran, or possibly even both combined, if we can imagine a new constellation of forces in the region that amounts, more or less, to an all-Islamic alliance, something approximating what Samuel Huntington had in mind when he conjured up the now popular theme of a coming "clash of civilizations." Not only do we hypothesize an Iran-Iraq alliance here but also a situation in which Egypt, Turkey, and Algeria have all been taken over by Islamic fundamentalist forces. Let us assume that Saudi Arabia remains under its now durable monarchy, and that the remainder of the GCC nations are still under the control of more or less "moderate," Western-leaning regimes, albeit politically intimidated by the nearby radical regimes in Iraq and Iran to the point of a Middle Eastern version of "Finlandization." In this scenario, perhaps assumed to be taking place some time beyond the millennium, the oil equation has shifted to the advantage of the OPEC suppliers as the import needs of Asia, Europe, and the United States have grown, as concluded in our previous analysis of the future of these markets, which projected a worsening of Western dependence. The once cohesive NATO alliance has long since withered be-cause of the absence of a real outside threat, the Bosnian fiasco, and the development of a "Euro-nationalism" directed against the United States. There has been a sharp parting of the ways in regard to U.S. and European Middle Eastern policies, the former still supporting Israel even in the face of the growing oil crisis and the latter having moved further in the direction of appeasing the region's now more cohesive Islamic forces. Meanwhile Russia, the "clash of civilizations" notwithstanding, has moved markedly in the direction of a reassertion of its traditional nationalism (this would, of course, contradict the above assumption of a lack of threat to NATO) and, stung by the humiliation of the loss of the cold war and its aftermath, has begun to rebuild militarily. As part of that rebuilding process it has become a larger factor in the Middle Eastern arms markets, among other things involving tight client relationships with both Iran and Iraq, as well as with Syria and, again, Egypt. In effect Russia has tacitly signaled that it considers both Iran and Iraq to be under its extended deterrence umbrella, a warning to the United States (and to Israel) not to threaten those countries with nuclear weapons. Further, any of the countries in the region that end up in conflict with the United States are virtually assured of being resupplied with Russian arms. The resupply operation would be simple in logistical terms because Russia is so close.

The nature of the crisis could take any of several forms. It could, for instance, involve a threat by either Iraq or Iran to again take over Kuwait, and the possibility of a move beyond that into Saudi Arabia or some of the other GCC states. It could be a threat to attack the oil installations of those countries as part of a squabble over oil prices and market shares. It could involve a large coalition arrayed against Israel, with the threat of the use of WMDs. Or it could be the aftermath of an unusually serious terrorist incident directed against Europe or the United States, let us say one involving the use of chemical or biological weapons or a much deadlier version of the World Trade Center bombing.

The United States would be faced with a very unpromising situation in regard to basing access en route and within the theater of conflict (see map 33, which shows the contrasting more and less permissive basing access environments), a situation presaged in the crisis of 1996 involving Iraq, in which Jordan and Saudi Arabia denied the United States use of air bases and Turkey hinted at more limited access in the future owing to changes in its political situation. Saudi Arabia, Kuwait, Oman, and Bahrain would presumably provide the United States access in a situation in which they themselves were threatened, unless they were so cowed by the neighboring states as to be seeking a diplomatic solution that would not force them to call for American assistance. One could, of course, also hypothesize that some or all of these countries had also fallen under the sway of radical and anti-American regimes, so that the United States would have no access at all within the region and would in effect be in conflict with all of them, with oil at stake.

One might venture a prioritization of access here amid the various combinations and possibilities. Access in Saudi Arabia is, of course, the most crucial issue, as without it a major U.S. military operation in the region must become very problematic. Lack of such access could, in the case of a U.S. conflict with

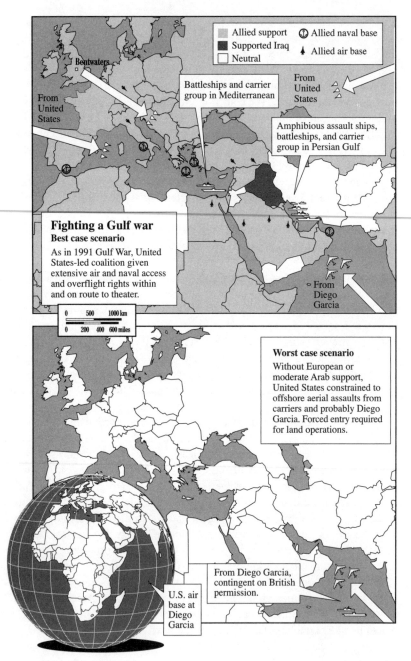

Fighting a Gulf war
Best case scenario

As in 1991 Gulf War, United States-led coalition given extensive air and naval access and overflight rights within and on route to theater.

Battleships and carrier group in Mediterranean

Amphibious assault ships, battleships, and carrier group in Persian Gulf

From United States

From United States

Bentwaters

Allied support
Supported Iraq
Neutral

Allied naval base
Allied air base

From Diego Garcia

Worst case scenario

Without European or moderate Arab support, United States constrained to offshore aerial assaults from carriers and probably Diego Garcia. Forced entry required for land operations.

From Diego Garcia, contingent on British permission.

U.S. air base at Diego Garcia

Map 33

Iran, be mitigated somewhat by access to some among the remaining GCC states, such as Kuwait, the United Arab Emirates, Qatar, Bahrain, and Oman. Next, loss of access to Egypt would pose serious problems. Israeli airfields could be used in lieu of those in Egypt for staging purposes, but closure of the Suez Canal would require routing warships and supply vessels all around the African cape. (A massive U.S. airlift over a narrow Israeli corridor via Eilat might also be threatened by combat interceptors from neighboring states, depending upon the nature of the hostile coalition.) Loss of Turkish air bases would hurt; so too would the loss of Diego Garcia for a variety of purposes, including bomber missions and basing for prepositioning ships. Loss of all of the Western European bases used in 1990–91 would also be a major problem, though in future years there will be many fewer U.S. forces and materiel to be moved from there, so that most of it would have to be moved directly from the CONUS. Loss of access to Israeli bases (granted, almost unthinkable) would, if it happened, be somewhat of a last straw, though as noted, the utility of such bases would be reduced if the Suez Canal was closed and if aerial access over Jordan and Saudi Arabia was denied. Under these negative geostrategic circumstances it is unlikely the United States would have the domestic support to conduct such risky military operations.

In a scenario that is less ominous for the United States it does have access within the region, but little if any such access en route or in adjacent zones. That would mean that all access had been denied in North Africa (Egypt and Morocco), and throughout Europe, which would mean also the Azores. One interesting question would be whether the United States would have access at Diego Garcia if Britain had moved further toward security integration within Europe and was under pressure from the other European powers to deny such access, in Ascension as well as Diego. Further, it would be assumed that the United States would have no access to major air and naval bases in the Far East and South Asia, with Singapore, Thailand, the Philippines, Brunei, and even Australia having been coerced by the oil states. In short, the United States is on its own and must, more or less, attempt to conduct a military operation out of the CONUS. Or in a modified scenario one might allow for access to Saudi Arabia, which would greatly change the equation, but might also raise the question of the extent to which Iran or Iraq or the combination of the two might succeed in knocking out some of the Saudi airfields and ports, assuming that they are then armed with better and more accurate missiles than is the case in the mid-1990s.

In such a scenario the United States might be forced, at least to a large degree, to conduct a strategic campaign on the basis of carrier-based aircraft, missiles fired from surface ships and submarines, and long-range bombers based in the CONUS (retention of Diego Garcia for basing of bombers would greatly alter this scenario). We are also assuming here that despite a lot of talk, the United States had not yet built any floating bases, that is, artificial bases that could be towed into a region, or left there on a more or less permanent basis, and be used in lieu of aircraft carriers. (Its arsenal ships might or might not be ready.)

There are some critical questions here related to future force acquisitions and readiness. If the United States had in the interim gone ahead with a large force of B-2 bombers, it could much more easily conduct a strategic campaign based in the CONUS. If it had heeded the well-publicized lessons of the Gulf War, it would have moved with dispatch to improve the capacity of the navy's carrier planes to deliver precision-guided munitions, so that if land bases were not available, the navy could conduct an efficient campaign all by itself. Quite probably an increase in the capability for launching Tomahawk missiles or their follow-ons both from surface ships and submarines would be a necessary part of this package, particularly if stealth aircraft could not be brought into the theater and if high losses in naval aircraft were anticipated. But if the regional foes had succeeded in their efforts at working toward asymmetric strategies, there might be some question about just how successful such a strategy could be, that is, whether the outcome of such a crisis could be determined by airpower alone.

On the other hand, the political stance of Saudi Arabia and the other GCC states being a big if here, there is the question of whether the United States would have to conduct a "forced entry" if it wished to carry out a large-scale land campaign. That would be the case if the countries of the Arabian Peninsula had become part of a hostile coalition or if they were merely "Finlandized" to the point of disallowing an unopposed U.S. landing of large-scale forces. If the latter were the case and if, as previously hypothesized, the United States had few forces left in Europe and no access there, it would mean that it would have to mount a large-scale amphibious operation all the way from the CONUS, a near-impossible job that would be difficult for any president to sell to the Congress and the country. Again, Diego Garcia being available as a staging point would alter the equation. There is also the question of whether the United States might be willing, despite the political cost, to use Israel as a staging point, although it would still have formidable problems, particularly if a hostile Egypt had closed the Suez Canal to U.S. warships. Even if Saudia Arabia were willing to provide access, in a case where Iran was involved as a regional foe, it would be difficult to land and mass ground troops along the coast of the Persian Gulf. In such a case, presumably, the United States would have to land ground troops on the Red Sea coast and then move them all across Saudi Arabia in order to confront the Iranian army. It must be remembered that in this scenario Egypt is no longer a U.S. ally, so the Suez Canal would presumably be closed, forcing the United States to mount a sealift all around the African horn. Even then, if Yemen or Somalia or both were opposed to American purposes, that is, were part of a broader Islamic coalition, transit via the narrow Bab el Mandeb Straits might also become a problem.

Hence in this scenario, a United States that is by far the world's preeminent military power is faced with multiple access problems, all leading to the specter of a very difficult and precarious operation that it might not really be willing to try, despite the stakes: access to Persian Gulf oil. It is likely that under these circumstances the United States would abandon any pretense of remaining the world's "oil policeman" and would leave the region to its fate. The bottom line of our analysis is that in the last resort the oil of the region can only be fought for if the United States is part of a coalition of industrial and regional powers who share similar interests. Under the "worst case" hypothesis explored here it is likely that no American president would get support for such an intervention no matter how superior U.S. forces were vis-à-vis the military adversary.

As things now stand the combination of in-theater basing arrangements and prepositioning, added to the U.S. system of basing access en route to the Persian Gulf area, would allow for an effective military campaign in a crisis seeing some degree of repetition of Desert Storm vis-à-vis Iran or Iraq. But there are possibilities rooted in the prospects for political changes in the region—Egypt, Turkey, and Russia are crucial considerations—that might result in a far more constrained environment for U.S. intervention. All of these factors coupled with the burgeoning proliferation of WMDs in the region, the drawdown of U.S. military forces in an era of budgetary stringency, and the projected expansion of other maritime powers, including China, mandates caution in assessing the future.

The reality is that despite all the advanced technologies that make power projection to the Middle East logistically so much easier than in the past, location is still a critical factor. The United States has no permanent commitment to the security of the Persian Gulf and as the above scenarios illustrate, there could be circumstances in which intervention and power projection would not be contemplated because of unacceptable political and military costs. From an American perspective, the Middle East is not the Caribbean. Furthermore, if the projections on energy demand are correct, European and Asian stakes in the energy security of the region are growing faster than American concerns. In the last resort one of the most basic elements of geography—distance—still remains a key determinant of foreign policy.

TEN

The Emerging Geopolitical Economy: The Search for Cooperative Prosperity

PRECISELY BECAUSE the greater Middle East remains so dangerous, many observers believe that the only rational way out of the cycle of violence and war is to encourage cooperative efforts among its nations to strengthen the political economy of the region and integrate it more fully into the emerging global marketplace. One of the principal underlying problems throughout the region has been the unwillingness of political leaders, until recently, to rethink the basic premises that they have used to promote economic development. Economic failure was one reason the Soviet Union imploded and why the newly independent states in the Caucasus and Central Asia are now attempting to liberalize economic activities. The Arab world for years has embraced socialist planning. Some Arab countries, notably Egypt and Jordan, are trying to break out of the old socialist mold, but it is difficult and politically risky. India and Israel have similarly been hobbled with central planning and red tape. India's economy was so unsuccessful and produced such abysmal results that the term "Hindu rate of growth" was used derisively to characterize its failure to deliver goods and services for the country's huge population. In 1991 India began to liberalize, and the results to date have been promising. Israel has also benefited from its efforts to decentralize control and revamp its Scandinavian model of socialism, yet much remains to be done.

Steven Plaut has suggested that the greater Middle East adopt what he calls the "Venice model" of development. For centuries the tiny city-state of Venice was independent, prosperous, and strong despite having no natural resources. Its success, according to Plaut, was due to a mixture of free commerce, capitalist initiative, liberalism, tolerance, and political stability.[1] This reasoning is

329

mirrored in a recent World Bank study of the current economic situation in the
Middle East, which presents the following vision for a prosperous Middle East:

> By 2010 the countries of the Middle East and North Africa have the potential to
> double incomes, increase life expectancy by close to 10 years, and cut illiteracy
> and infant mortality by almost half. They could also become full members in the
> global economy using integration with Europe and within the region as a
> stepping stone to international competitiveness. Peace, macroeconomic stabil-
> ity, and an attractive investment environment could attract billions of dollars of
> capital from nationals and foreign investors. The faster economic growth would
> reduce poverty and bring down unemployment, restoring hope to millions.[2]

The Bank argues that in the modern age poverty can no longer be attributed
solely to an absence of natural resources or poor geographic location. The
changes in the marketplace and the extraordinary improvements in communication technologies mean that the riches of the world can now be shared, at least
in theory, in a way that was never possible before. Unfortunately, in the case of
the Middle East, significant opportunities have been squandered. If the vision
of prosperity is to be fulfilled, there will have to be more far-reaching and
fundamental reforms in the economic arena than those implemented so far. The
Bank points out that during the period from the 1960s through 1985, which
coincided in large part with the oil boom, the growth rates of the Middle East
countries were very impressive and outstripped those of all other regions except
for Southeast Asia.[3] However, the oil-producing countries failed to capitalize on
their luck during this period, and as a consequence were not in good shape to
weather the storms of the late 1980s and early 1990s when a world economic
recession coincided with a precipitous fall in oil prices.

In order to prosper the countries of the Middle East must make their private
sectors more efficient, produce more skilled and flexible workers, and increase
foreign investment. Apart from economic imperatives, the political motivations
for this are very clear. The best way by far to reduce poverty is to induce
growth, and by implication this will have a critical impact on the political
situation. It is true that the necessary reforms are painful but the costs can be
minimized and partners in the region can share the burden. It has to be assumed
that oil prices will remain flat for the coming decade even though demand and
thus revenues will probably increase. However, the benefits of a rise in income
will be offset by a growing population.

The primary concern of most political leaders throughout the greater Middle
East today is the state of their economies. From the oil fields of the Caspian
Basin to the stock market of Bombay, from the Chambers of Commerce in
Egypt and Israel to the unemployment lines in Gaza and the crowded port of
Dubai, the need for economic reform and growth is now regarded as a sine qua
non for the political survival of incumbent regimes. For some leaders—Israel's
prime minister Netanyahu, for instance—the challenge is to privatize a semi-

socialist economy and develop new high-technology products for worldwide markets, drawing upon Israel's considerable intellectual capital, which has been further boosted by the influx of highly trained Russian immigrants. Netanyahu believes Israel has the potential to become one of the richest per capita countries in the world within a decade. Yet his dream, like that of his predecessors, Shimon Peres and Yitzhak Rabin, assumes a stable region in which Israel can flourish. For other leaders, for instance the PLO's Yasser Arafat, Jordan's King Hussein, and Egypt's Hosni Mubarak, the immediate goals are very different—how to provide jobs for a growing population and avoid social and political upheaval within their respective communities.

For the Gulf Cooperation Council (GCC) countries blessed with oil riches, different economic problems have arisen, namely, how to sustain a high standard of living at a time of declining relative oil prices, growing populations, and high expectations of an increasingly educated public. The GCC countries also face the problem of an overwhelming number of guest workers from Asia who have the potential to create political trouble if the social conditions in the countries deteriorate. Azerbaijan's leader, President Aliyev, believes that his country could become another Kuwait if its huge oil resources can reach the market. Similarly, Kazakstan and Turkmenistan have high hopes that oil and gas will eventually make them as prosperous as the GCC countries. Turkey faces political turmoil because chronic inflation and unemployment are the other side of the coin of the enormous profits being made in certain sectors of the Turkish economy. Successive governments have not been able to trickle wealth down evenly to the majority of the population. While Turkey is sometimes touted as a model for economic development, this view is not reflected in the recent political climate in Ankara.

A most interesting country to watch is India. If economic reforms continue and the energy and skills of the huge Indian population are tapped successfully, India's growth could parallel that of the countries of East Asia, including China. If India "takes off" it is likely to stimulate greater economic reform in Pakistan. Whatever happens, both India and Pakistan need massive capital investment if they are to achieve even minimal goals of economic growth and overcome the crippling poverty that still bedevils over 80 percent of their populations. They both have a crucial need for investment in electricity production, transportation, particularly road and rail modernization, and new communications.

Finally, there are the new pariah countries—Iraq and Iran. Both have the potential to end their current poverty and achieve higher per capita incomes. However, so long as Saddam Hussein remains in power in Baghdad and the mullahs in Teheran are cut off from many foreign investment opportunities because of pressure from the United States, their problems, including lack of investment in new infrastructure, population growth, and corruption assure that they will sink lower and lower on the economic scale. Both Saddam Hussein and the mullahs realize that in the last resort, poverty will weaken their power base, even though in the short run they can both capitalize domestically on their pariah status.

Against this background of concern about economic growth and prosperity in the region, the more global geopolitical issues discussed in the previous chapters take on special importance. Several are worthy of recapitulation or further elaboration. We examine four: energy security, water scarcity, joint infrastructure projects, and the communications revolution.

Energy Security

It is worth repeating a basic statistic: the Persian Gulf and Caspian Basin contain over two-thirds of the world's proven oil reserves and possibly forty percent of the world's natural gas reserves. The Persian Gulf alone is capable of meeting much of the growing demand for energy in the industrial world *provided* that intra-Gulf conflicts do not escalate to war and destroy oil fields as happened in Kuwait in 1991. Additional capital investment is necessary to modernize and develop the existing fields in the region, especially in Iraq and Iran. This raises a catch-22 situation. Without conflict resolution major investment may be difficult to obtain; but without investment the poverty of countries such as Iran and Iraq will not be relieved, which will have a serious impact on their political and social processes. In view of the dangers in the region, and yet the need to export more energy, it is logical and natural that additional pipelines to transport energy to market should be laid. Iran, for instance, has a major strategic stake in developing land routes to Asia rather than continuing to rely on the highly vulnerable Strait of Hormuz. But to develop these access routes requires cooperation and money, which is unlikely to be forthcoming as long as the United States and Iran remain enemies.

Aside from the threats to the Gulf states posed by Iraq and Iran, the internal situation on the Arabian Peninsula also poses threats to energy security. While the United States presently has the military capability to deter high-level threats from Iran and Iraq, it has little control over the internal security problems that the Gulf countries will face in the coming decade, which have to do with social and religious factors, including autocracy, corruption, asymmetric income distribution, migration, human rights abuses, and Islamic extremism. The potentially negative political impact of large numbers of foreign workers is demonstrated by the fact that the most serious recent internal security threat to the United Arab Emirates occurred when Hindu and Muslim expatriates began rioting following the destruction of the mosque at Ayodhya in December 1992. Citizens of the United Arab Emirates had never heard of Ayodhya, but were terrified by visions of foreigners rampaging in downtown Dubai. With over 80 percent of its population being made up of foreigners and half of these coming from South Asia, the United Arab Emirates and other Gulf countries cannot remain insulated from the long-standing conflict between India and Pakistan.

Indeed, the Indian subcontinent, with over 1 billion people, is virtually on the doorstep of the Gulf. Bombay and Karachi are only 981 miles and 530 miles, respectively, from Musqat. Although India has traditionally been on friendly terms with most of the Gulf states, its long-standing rivalry with Pakistan, the lack of resolution on the Kashmir issue, and the periodic outbursts of incendiary right-wing Hindu fundamentalism (seen as threatening to India's 130-million strong Muslim minority) are serious stumbling blocks to a truly close relationship between India and the Gulf states.

One of the key factors in this relationship is the constant movement of Indian migrant labor to the region. At the time of the Gulf crisis, the six GCC states—Bahrain, Kuwait, Oman, Qatar, Saudi Arabia, and the United Arab Emirates—had a combined total of over 5 million migrant workers, comprising around 68 percent of their total labor force.[4] According to pre Gulf War estimates, up to 30 percent or some 1.5 million of these foreign workers were Indians.[5] In addition to laborers the Indian communities in the Gulf included technicians, builders, nurses, traders, and, later, domestic workers. A highly organized recruitment system developed rapidly in response to this need for labor. South Asians (primarily Indians and Pakistanis) initially became popular as workers because they were willing to perform unskilled labor in harsh conditions and their wage levels were far lower than those of the non-Gulf Arab migrant workers—the Yemenis, Egyptians, Jordanians, and Lebanese. For example, in Saudi Arabia in 1987, Asians could be hired for only half of what Arab workers commanded and by 1989 the ratio had dropped to one-third.[6] If the subcontinent's 1 billion inhabitants fail to make further economic gains because of domestic incompetence, war, or international indifference, the inevitable political earthquake would have repercussions far beyond its borders. (For more on India's energy needs and economic dilemmas, see appendix 3.)

In the case of the Caspian Basin, the geographical dilemma facing the three key countries, Azerbaijan, Kazakstan, and Turkmenistan, is even more acute. As we noted in chapter 3, they cannot get their energy to market without crossing someone else's territory. In the past virtually all the great oil and gas exporting countries have had direct access to the world's shipping lanes. Given the complicated nature of the geopolitics of the Caspian region, the key question is whether the major oil and gas importing countries will resort to traditional competition to secure access or whether will they be persuaded that a more cooperative approach is both sensible and necessary. For this reason the routes discussed to transport the energy have taken on political and strategic as well as economic and environmental overtones. The choices for oil and gas routes involve many complicated decisions. Many different options have been proposed, and each has costs and benefits. The stakes for control of the Caspian resources are very high. Presently the scramble for the riches of the region is unregulated.

It is clear that it would be difficult to insulate disputes in the Caspian from the three adjacent areas—the Caucasus, the Persian Gulf, and South Asia. But it is also true that if some way can be found to solve the access problems amicably and give all the regional countries a stake in Caspian stability, a new era of cooperation would begin, which could help to diffuse some of the more intense regional conflicts. This is not going to be easy. As long as the United States and Iran are bitter enemies, no progress can be made in developing southern routes for oil and gas. As long as Russia adopts the hegemonic approach to the control of energy and uses military force to suppress dissent within its borders as in Chechnya, chances are that chaos, sabotage, and violence will continue. On the other hand, unless Azerbaijan, Kazakstan, Georgia, and Turkey pay attention to Russia's sensitivities and the growing nationalism in Moscow, they could find themselves subject to Russian reprisal and intimidation.

In sum the question of Caspian energy access is one of the most important facing the international community today. It is very much in the interests of all the parties to work out cooperative arrangements to develop these resources, otherwise Caspian Basin oil and gas simply will not reach market and there will be greater pressures than ever on the Persian Gulf and the entire oil market.

Water Scarcity

Another vital strategic geographical issue discussed in chapter 3 concerns water scarcity throughout the greater Middle East. Given population growth and industrial development in the region, the demand for fresh water could reach critical levels in decades to come. While there are many unilateral ways for each country to improve its situation with regard to water—developing more efficient agricultural techniques or even abandoning agriculture, for example—cooperation among neighbor states to develop new sources of water may well be essential. From an economic standpoint desalinization may be the most efficient route, but new pipelines cannot be ruled out, provided the political and security objections can be overcome.

Water remains a key strategic issue in the region particularly in regard to where it is to be found. Israel will never relinquish control of the Golan Heights until it is assured that the water supplies from that area are under either its own or international control and cannot be tampered with by Syria. Similarly, until the Palestinians and Israelis can reach some agreement on the distribution of water on the West Bank, this will remain one of the most critical unresolved problems of the Arab-Israeli conflict.

Over the years Israel has proposed several canal projects to resolve the water crisis caused by the shrinking of the Dead Sea, including a Med-Dead canal from Gaza; a Red-Dead canal from Eilat-Aqaba—both to the southern Dead Sea and potash works area—and a Med-Dead canal from Atlit (under Mount Carmel via Beit She'an and down the lower end of the Jordan Valley) to the north end of the Dead Sea. However, because it is highly unlikely that Jordan will agree to any of these proposals other than the Red-Dead canal, this is the project that is probably the most politically realistic at this time, though whether it can ever be financed is another matter.

The proposal for a canal between the Red Sea and the Dead Sea was first made in 1851,[7] but it is especially timely today, both politically and environmentally, first, because Israel and Jordan are now at peace and, second, because the Dead Sea has shrunk by 25 percent over the past thirty years, primarily because of a reduction of inflow from the Jordan River. The Red-Dead plan proposes that Red Sea water be pumped from the port of Aqaba and conveyed uphill northward along the Israeli-Jordanian boundary running through the Wadi al-Araba (known in Israel as the Arava Valley). Sixty-two miles north of Aqaba, at Mount Edom, the canal will reach an elevation of 660 feet above sea level and then commence its 2,000-foot descent to the Dead Sea, the lowest point on earth at 1,335 feet below sea level. On its way the annual 35 billion cubic feet of water, flowing down the valley at 21,200 cubic feet a second, will snake back and forth across the border, creating boating lakes and fish farms along the way, and will generate power to drive the largest reverse osmosis desalination plant in the world.[8]

According to the plan's advocates the inflow of Red Sea water will decrease the salinity of the Dead Sea and, equally important, reduce its absorption of fresh water from the surrounding land, freeing it for agricultural development. Hydroelectric stations along the canal will generate 600 megawatts of electricity, which will be used to power desalination plants. Preparations for feasibility studies began after the Casablanca conference in November 1994. The studies will take an estimated three and a half years to complete (that is, until June 1997).[9]

Finding financing for the construction of a Red-Dead canal is likely to prove difficult. The European Investment Bank is not allowed to fund more than 50 percent of any project and has so far never loaned more than 20 percent. The World Bank has announced that it will not loan more than $200 million to the project. International Finance Corporation (IFC) analysts predict that bilateral sources will provide up to $1 billion. Private sector investment will obviously be crucial and it is not at all clear that the project will survive financial and environmental scrutiny.[10]

Conservationists fear that the natural balance of the entire area could be destabilized, and that sea water could leak into the aquifer under the Rift Valley should tremors cause cracks and leaks in the canal's construction. Critics also

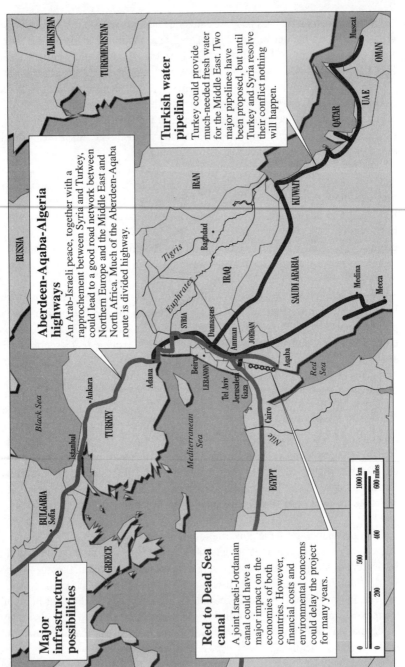

Major infrastructure possibilities

Aberdeen-Aqaba-Algeria highways

An Arab-Israeli peace, together with a rapprochement between Syria and Turkey, could lead to a good road network between Northern Europe and the Middle East and North Africa. Much of the Aberdeen-Aqaba route is divided highway.

Turkish water pipeline

Turkey could provide much-needed fresh water for the Middle East. Two major pipelines have been proposed, but until Turkey and Syria resolve their conflict nothing will happen.

Red to Dead Sea canal

A joint Israeli-Jordanian canal could have a major impact on the economies of both countries. However, financial costs and environmental concerns could delay the project for many years.

Map 34

argue that the effect of combining brine with the Dead Sea salt water cannot be predicted, and wonder what will happen to the potash and magnesium production at the Dead Sea.[11]

In 1987 Turkey's then prime minister, later president, Turgut Ozal proposed an even more complicated, high-risk project: a water "Peace Pipeline" to channel fresh water from Turkey's Ceyhan and Seihun rivers to the Levant and the Arabian Peninsula. The two rivers originating in the central Anatolian mountains dump 39 million cubic meters of water into the Mediterranean Sea daily. The Turkish proposal involved two major pipelines. The 1,700-mile "Western pipeline," with a projected cost of $8.5 billion, would pump 3.5 million cubic meters of water daily from Adana, Turkey, and transport it through Damascus, the West Bank, Amman, Medina, Mecca, and Jidda. The second, "Gulf pipeline," which would cost $11 billion to construct, would carry 88 million cubic feet a day and transport it 2,400 miles through Kuwait, Saudi Arabia, Bahrain, Qatar, the United Arab Emirates, and Oman.[12]

However, the project has been postponed indefinitely. The Arab countries have not been keen on the idea of dependence on Turkey for so vital a commodity as water. Iraq and Syria are already uncomfortable with their dependence on Turkish goodwill for the waters of the Euphrates, and Saudi Arabia considers that there is more security (perhaps at a higher financial cost) to be gained from desalination, in which the Saudis have already invested heavily. Turkey's ambassador to Iraq before the Gulf War, Nacati Utkan, argued that the Peace Pipeline was a "confidence-building project" that might help to break the circle of suspicion in the region.[13] Whatever happens politically in the next decade, the simple equation remains compelling: Turkey continues to have a surplus of fresh water while many countries to its south face an increasing water deficit (see map 34). Without such a pipeline the development of massive new fresh water facilities along the Arabian littoral is essential given the growth rate of the economies and the population increase in that region. Yet as was realized during the Gulf War, water desalinization plants are at risk of being destroyed or damaged during an armed confrontation. The reality is that as more and more fresh water production plants are built, they become more vulnerable in the event of conflict.

Other Cooperative Infrastructure Projects

In the past few years, as well as joint water projects, numerous other proposals have surfaced for regional cooperation in infrastructure. While there is a long history of such proposals, in the past most of them were advocated by external players hoping to use economic incentives to help resolve the Arab-Israeli conflict.[14] But they did not work because the parties were not ready for peace. What is new is that the proposals that have surfaced in the past few years have been made by the regional countries themselves. Some have been suggested by regional leaders and others were formally presented at two Middle East economic summits organized by the World Economic Forum and the Council on Foreign Relations held in Casablanca in 1994 and Amman in 1995. The basic premise behind all these proposals is that the region as a whole would benefit from economic cooperation and this, itself, would be a deterrent to further conflict.

There have been many studies describing the possibilities for new or improved roads; railways; canals; pipelines for oil, gas, and water; airports; electricity grids; power stations; ports; and communication facilities. On the face of it the possible benefits of such projects seem considerable. For instance, the potential for Middle East tourism would clearly be greater given shared facilities such as airports and roads. However, it must be stressed that the meetings in Casablanca and Amman brought together the political elite and business leaders, two groups in the Middle East most identified with the peace process. In the Arab world many of the intelligentsia and much of the public remain cynical and hostile toward this type of cooperation.

The belief that economic opportunity and development can determine the political agenda for a country and reduce the risks of conflict is not new. In 1910 Norman Angell published his book, *The Great Illusion,* arguing that war between the major capitalist countries would not happen because it made no economic sense. Rational man—a creature influenced by economic motives—would prevail. Two years later World War I began. On this count Lenin, who argued war between the imperialist powers was inevitable because of the inherently competitive nature of capitalism, was more right than Angell. The presumed economic benefits have frequently been paralleled by suspicion over the strategic risks of cooperation. As we discussed in chapter 2, the most famous joint infrastructure project of the nineteenth century, the Suez Canal, generated great friction among the leading powers, especially Britain, France, and Austria, and among the two key regional players, Turkey and its vassal, Egypt.

In contrast, in the aftermath of World War II France and Germany, bitter rivals with hatreds going back over the centuries, forged new ties. Together with other West European partners, they developed one of the most successful regional cooperative systems in recent history. The European Coal and Steel Community originally proposed by Jean Monnet and Robert Schuman as a venture between West Germany and France to regulate coal and steel production provided the foundation for the European Union. The hope was that if the two countries could cooperate on basic industrial requirements, they could begin a new, less conflictual and competitive, relationship. The idea found acceptance and in 1951 Belgium, the Netherlands, Luxembourg, and Italy joined to sign the Treaty of Paris. Although Britain was suspicious of the supranational nature of the organization and did not join initially, the concept was so successful that it formed the basis for the Treaty of Rome and the European Union, of which Britain is now a member. What happened in Europe had a profound influence on Shimon Peres' vision of a future Middle East. Peres has argued that the Arabs and the Israelis can overcome their hatreds and forge a new Middle East. He has persistently made the case that economic cooperation is an essential ingredient of this vision and has consistently been optimistic about the long-term relevance of his ideas. His optimism has been shared at times by other regional leaders including former president of Turkey Turgut Ozal and Crown Prince Hassan of Jordan.[15]

The views of these leaders contrast with the more pessimistic outlook of those who argue that fundamental asymmetries in the region, especially between Israel and its Arab neighbors, make talk of mutually beneficial economic cooperation just that—talk. One problem is that Israel's workforce is skilled and its most important markets are in Europe and Asia, not with its Arab neighbors. Furthermore, the Arab economies, including those of the oil-rich Gulf states, are backward, socialist, and inefficient. (In fact, much like the Israeli economy before the reforms in the mid-1980s, which began to show real growth in 1989.) In differing ways but with some common problems, the economies of Saudi Arabia, Egypt, Syria, and Jordan are all in trouble. Peace will not bring major economic dividends without much needed but politically dangerous reforms.[16]

Despite such pessimism it is worth reviewing some of the infrastructure proposals if only to show how far the rhetoric of peace and cooperation has advanced and why economic geography plays such a prominent role in contemporary analyses of the problems and prospects for the greater Middle East region.

In preparation for the Casablanca economic summit in November 1994, the government of Jordan prepared a large brochure outlining a number of projects that could benefit from peace. While coy about specific deals with Israel (the brochure was prepared over a year before the Jordanian-Israeli peace treaty), the list of possibilities was impressive. The brochure argued that the port of Aqaba "can serve the needs of Jordan, Israel, and the Palestinians for years to come."[17] While this is a rather obvious statement, it is one with which the Israelis certainly concur. Among other suggestions, oil and gas transit from the Gulf could use a pipeline to Haifa for distribution via the Mediterranean. The potash and chemical production prospects of the Dead Sea could be expanded provided there was no harmful environmental impact. New bridges and roads would open up access to more tourism following peace.

Israel's proposals have been more extensive and more specific. As we noted earlier, water cooperation is seen as the key and the proposals include cooperative measures in conservation and prevention of pollution in shared water systems. Regional monitoring of hydrological activities along with water management of the lower Jordan River has been suggested. This would include catchment of overflow from the Yarmuk River and the Sea of Galilee and the use of winter rain flood waters in the Arava. There could be joint development of water resources around the Dead Sea, including desalinization projects jointly administered in the Dead Sea, Arava, Eilat, Aqaba, and Sinai areas.

In regard to tourism Israel is interested in the creation of a regional tourism organization that would help develop a Red Sea riviera at Eilat and Aqaba and the Lowest Place on Earth Park near the Dead Sea. The Israelis also envisage cross desert tourism in the Arava and along ancient spice and pilgrim routes overland from Egypt to Mecca via Israel and Jordan.

As for transportation they have delineated land, marine, and air projects. For land access, among other ideas, they discuss the central corridor along the Great Rift Valley from Syria and Lebanon through Israel and Jordan to Egypt; a highway connecting Cairo to Saudi Arabia via Eilat and Aqaba; and east-west road connections between the Mediterranean and the Gulf via the Rift Valley. They also propose east-west rail links among Israel, Jordan, and Syria connecting Haifa to Amman and Damascus, including a restoration of the Hejaz railway from Medina to Damascus. Marine projects include intraregional maritime shipping and transportation among Nuweiba (Egypt), Aqaba, Eilat, a Saudi Arabian port on the Gulf, and a joint seaport on the north shore of Eilat along the border.

Air traffic would benefit if there was agreement to universalize air traffic systems in the region. Both Jordan and Israel support a joint Israeli-Jordanian airport (Ein Eurona/Aqaba). Also mentioned is the establishment of a meteorological forecasting center.

Energy projects are some of the most practical and potentially important. They include the interconnection of electricity grids between Israel and its neighbors. Plans have been proposed to reopen the oil Tapline from Saudi Arabia, across Jordan and the Golan Heights to Sidon, Lebanon. The Eilat to Ashkelon pipeline (built during the time that the Suez Canal was closed, 1967–75) could be reopened, as could the Iraqi Petroleum Company line from Iraqi oil fields to Haifa via Jordan. Natural gas lines are proposed from Egypt and the Persian Gulf to Jordan, Israel, and the Palestinian Authority. In addition Israel has suggested cooperation in earth sciences, particularly in the Rift Valley, including the Dead Sea; joint oil shale exploitation between Jordan and Israel; solar energy projects across the Sinai, Negev, Jordanian, and Saudi deserts; and the need to establish a regional energy information center.[18]

Egypt's contributions to the infrastructure proposals have been most recently highlighted in several studies published by the American Chamber of Commerce in Egypt. Of special interest is a 1995 publication on the prospects for cooperative road development linking Egypt and North Africa to Europe via Israel, Syria, and Turkey, a so-called "eastern Mediterranean corridor." The proposal discusses the need to "restore the land links between Europe, Asia, and Africa."[19]

Ultimately three developments will probably do more to open up the traditional Middle East to new economic growth than anything else. First, a peace treaty between Israel and Syria; second, a resolution of Syrian-Turkish conflict over water and terrorism; and third, Turkish membership in the European Union. If these three things happen a new land bridge from Europe to Asia and Africa could become a reality.[20]

Good roads and peaceful coexistence from Turkey to Arabia would herald a revolution in social economics and thus in political relationships. Suppose similar developments occurred across the region from Iraq and Iran through Pakistan and India? India and Pakistan are notorious for the lack of good land communications. Though there are paper plans to build dedicated superhighways on the subcontinent, there are none in India today. In Pakistan progress on building the first divided highway has been far slower than was hoped. Yet both countries, but especially India, may be on the verge of significant economic growth. India's ground transportation, as that of the other Asian giant, China, is hopelessly outdated. While some may find the archaic railways quaint or even romantic if one is not in a hurry, the roads are terrible. The vast bulk of Indian commerce travels by road, yet there are fewer than 100 miles of divided roadway throughout India.[21]

A Note of Caution

While all these projects seem exciting and symbolic of a new Middle East, there are precedents that suggest that pushing too hard and too strong for joint projects can raise anxiety levels, especially among the weaker or less sophisticated partners in a joint venture. Also such proposals can generate a negative response from third parties who either feel threatened by certain projects or believe they will not benefit materially in any way. There are also many questions in regard to various economic factors that must be taken into account as calculations are made regarding the opportunity costs of joint projects and what they mean for investors. For instance, are joint projects sensible if the market alone determines profitability? If not, how great a subsidy may be needed to launch the projects? Where will the funds come from?

Perhaps no project symbolizes the considerable benefits and costs of new ventures better than Iran's efforts to open up a rail link between Central Asia and the Indian Ocean and the Mediterranean of which the final section between Turkmenistan and Iran was completed in the spring of 1996. It is now theoretically possible to travel by train from Beijing to Istanbul without touching Russian soil. On the face of it this option has great geopolitical benefits for the landlocked countries of Central Asia, which are otherwise dependent on Russia and China for rail access to the world. It is far too early to judge whether the project will be a success or a "white elephant" (see map 35). However, if U.S.

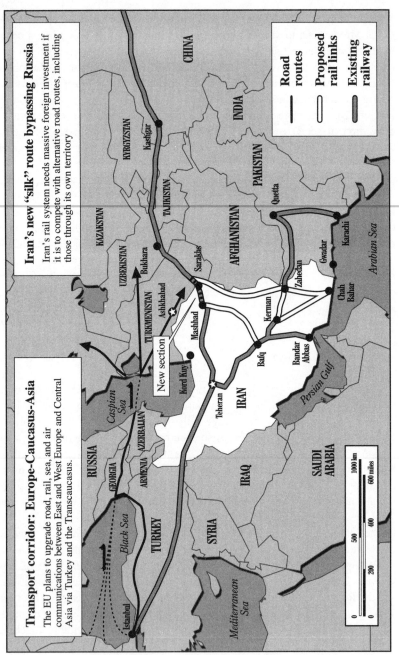

Transport corridor: Europe-Caucasus-Asia

The EU plans to upgrade road, rail, sea, and air communications between East and West Europe and Central Asia via Turkey and the Transcaucasus.

Iran's new "silk" route bypassing Russia

Iran's rail system needs massive foreign investment if it is to compete with alternative road routes, including those through its own territory

Map 35

relations with Iran were to improve and the United States was prepared to invest there again, the rail link could become an important new route and a hedge against chaos and anarchy in Russia.

There is a chicken and egg issue here. Until the political climate for cooperation between adversaries or former adversaries improves, the risks of investing huge amounts of capital in new schemes is likely to be too great for private groups. Nevertheless the economic benefits of greater cooperation make such sense that if the peace process is sustained and expanded, some of the wish list of projects will almost certainly materialize. If they do, they will radically change the map of the greater Middle East.

The Communications Revolution

Less discussed but potentially of great importance in the future are the expanding communication links between the greater Middle East and the outside world. No one knows what impact the information and communications revolution will have on political behavior, except that it will likely be profound. The most basic reality is that the proliferation of communications systems will make it increasingly difficult for closed societies, which at the same time wish to become part of the global economy, to deny their citizens access to uncensored information. There is already a heated debate in several countries—most notably Iran—about the "dangerous" impact of satellite television on Islamic society. Two pioneers of the new geography of communications quote the distributor of *Baywatch,* the world's most widely viewed program: "Once a show is on satellite, it is like the rain. It falls on the rich and poor alike—and both watch."[22]

Not only are the means of communication making stunning technological progress, but the economic costs of using the systems are falling. Four related developments are of particular relevance to the greater Middle East, which includes a large number of countries generally classified as "not free," especially in their efforts to control the information that reaches their citizens from the outside world. First, the dramatic expansion of fiber optic submarine cables from Europe and East Asia to the greater Middle East will facilitate much easier and cheaper communications with the rest of the world (see map 36). Second, this will allow the use of telephones and the Internet system to expand in the region. Third, the proposed Iridium system, which will put hundreds of nongeostationary low- and medium-earth-orbit satellites into space, will provide a virtual global service for handheld satellite mobile phones. Fourth, the expansion of direct broadcast TV satellites is attracting millions of new viewers each year, including most of the greater Middle East, which is now covered by Star TV's Asia Sat I.[23]

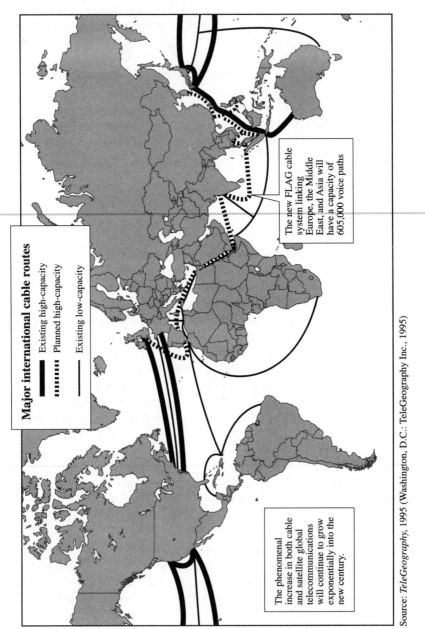

Major international cable routes

Existing high-capacity
Planned high-capacity
Existing low-capacity

The new FLAG cable system linking Europe, the Middle East, and Asia will have a capacity of 605,000 voice paths

The phenomenal increase in both cable and satellite global telecommunications will continue to grow exponentially into the new century.

Source: *TeleGeography*, 1995 (Washington, D.C.: TeleGeography Inc., 1995)

Map 36

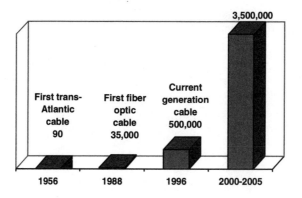

Figure 10-1. Long-distance submarine telephone cable capacity (simultaneous calls).
Source: Derived from *Telecommunications Map of the World* (London and Washington, D.C.: Petroleum Economist Ltd. and Telegeography Inc., 1996).

It is difficult to comprehend the enormity of the changes that these new technologies will bring to the world and to regions such as the Middle East, so it is useful to describe some of the trends in simple quantitative language. For instance, when the first trans-Atlantic telephone cables came into service in 1956, they could transmit ninety telephone calls simultaneously. Progress was slow until 1988, when the first fiber optic cable was introduced. At that time—and after more than thirty years in development—35,000 telephone calls could be made simultaneously. Only eight years later, in 1996, the number jumped to 500,000. The new fiber optic cables and their supporting technologies, which are in the planning stage and should be put into service between 2000 and 2005, will probably be able to carry 3.5 million telephone calls simultaneously or—perhaps even more revolutionary—several hundred thousand channels of compressed video services (see figure 10-1).[24]

Table 10-1 shows that the number of outgoing telephone calls from most Middle East countries is increasing at a significant rate. Interestingly, Iran's proclivity for contact with the outside world grew at an extraordinary rate of 33 percent between 1993 and 1994. (Iran, it must be added, has a relatively modern telecommunications system that was installed during the rule of the last Shah.)

Since economic modernization and growth rely on increasing access to information processing it will become even more costly for authoritarian regimes to control access to global information systems. Shutting down computer bulletin boards or satellite transmissions to prevent the open exchange of ideas via e-mail or access to western movies will soon become almost impossible for all but the most repressive dictatorships, such as North Korea and Iraq.

Table 10-1. International Telephone Traffic
Selected Middle East countries

Country	Outgoing millions of minutes of telecommunications traffic		
	1993	1994	Percent change 1993–94
Algeria	121	118	–2.5
Bahrain	77	87	13
Egypt	80	. . .	. . .
India	284	314	11
Iran	157	208	33
Israel	176	. . .	. . .
Jordan	50	57	14
Morocco	125	130	4
Pakistan	57	61	8
Qatar	58	63	8
Saudi Arabia	455	499	10
Syria	37	40	9
Tunisia	67	64	–5
Turkey	265	284	8
United Arab Emirates	342	428	25
United Kingdom	3,130	3,507	12
United States	7,500	8,911	19

Source: Derived from Gregory Staple and Zachary Schrag, *TeleGeography 1995: Global Telecommunications Traffic Statistics and Commentary* (Washington, D.C.: Telegeography, 1995), pp. 184–87.

The wild free-for-all that is taking place on the information superhighway is analogous to those of earlier days when dramatic new technologies came into the marketplace within a relatively short period of time. The railroads, automobiles, radio and television, and later the airlines, though greeted skeptically by some, were seen by most as potentially revolutionary in their impact on society. But few could predict exactly how and why. Initially people were overwhelmed with the new technologies and regarded the technology itself as the wonder. Over time, however, they lost interest in how the railroads worked, or why radio signals could be beamed from one end of the world to another, or what made an airplane fly, and became more concerned with practical, commercial considerations. In the case of the railways the question soon was not that you could take a train from Boston to Chicago but when could you take it, how comfortable would it be, how long would the journey take, and how much would the trip cost? Exactly the same sort of considerations applied to the other technologies. What were important were the applications and their effect on day-to-day life.

Similarly, most people no longer wonder about computers, fiber optics, and the Internet but want to know how useful they are, how easy to operate, and at what cost. Thus while there are dramatic breakthroughs in the development of hardware, it is the software and how it will help people communicate that claims the attention of specialists thinking about the future. Indeed, mapping the Internet has become a major challenge since the maps not only have to cover the simple hardware factors as shown on map 36, but also the numerous destinations to which information can be sent and from which it can be retrieved.[25]

Summary

In the final analysis the case for a new map of economic cooperation in the Middle East is enlightened self-interest. While rapid economic development invariably causes a certain amount of disruption and pain within a society, the alternatives of slow or zero growth are even more intolerable in the long run. The irony is that the Middle East could be on the verge of an economic takeoff that could transform the map to show new bridges, roads, railways, airports, pipelines, ports, hotels, electric grids, communications systems, desalinization plants, canals, and dams. However, if history teaches us any lessons, it is that structures that go up can come down. The great trade routes served to benefit east and west for hundreds of years and then suddenly were closed because of political rather than economic turmoil. The same could happen in the greater Middle East.

As we have emphasized the development of weapons technology is moving at such a fast pace that the region could find itself exceptionally vulnerable to war. Indeed, we believe this is one of the most important conclusions of this study, namely that the changing interface of geography and technology in the greater Middle East region has the potential for new, spectacular outcomes both positive and negative. Technology is overriding former physical barriers that kept countries apart or ignorant of each other and the wider world, and with this new contact could come cooperation and prosperity. Yet it is the new technology of missiles, long-range aircraft, and weapons of mass destruction that also reduces the protection offered by distance and terrain and raises the specter of regional war on a scale of destruction unthinkable a decade ago.

ELEVEN

A New Strategic Map

FOR THOUSANDS of years the geography of the Middle East has been a central factor in determining military and economic access to the region and its resources. Yet the obstacles and opportunities provided by geographic features for both military and economic activity have changed over time as a result of improvements in technology and shifts in economic trends. In addition at different times, but especially in recent years, political and demographic changes have dramatically affected the human geography and the political parameters of the region. The thesis of this study is that with the end of the cold war, the increased demand for Middle East energy and radical changes in military and civilian technology, including the revolution in military affairs and the proliferation of weapons of mass destruction, the strategic geography of the region has once again assumed a different configuration. A new strategic map is being drawn that assures that the Middle East will remain a prize in an emerging international system whose future contours are not yet clear.

The outlines of the new map can be ascribed to two distinct but equally relevant and mutually reinforcing categories—the political economy and the military balance. The political and economic dimensions refer to the impact of recent changes in the conduct of regional diplomacy, including efforts to resolve centuries-old conflicts and the expectations and the search for economic prosperity and development. The military dimensions relate to the impact of geography and new technology on scenarios for conflict and war and the possible ways that new wars will be fought.

On the diplomatic front the greater Middle East region has witnessed some of the most dramatic political changes of any part of the world over the past five years. The breakup of the Soviet Union and the American victory in the Gulf War were followed by the 1991 Madrid conference, which reinvigorated the Arab-Israeli peace process. In 1993 the breakthrough in negotiations between the Israeli government and the Palestinian Liberation Organization culminated with the famous handshake between Prime Minister Yitzhak Rabin and Yasser Arafat on the White House lawn on September 13, 1993. Since that time Israel has made peace with Jordan, completed major withdrawals from several cities in the West Bank and Gaza, and opened up, or reestablished, diplomatic relations with forty additional countries throughout the world. In parallel the American policy of building up its military capabilities in the region and isolating Iran and Iraq, while controversial, has successfully prevented any major recurrence of interstate war in the Persian Gulf.

However, in the past two years there have been some countervailing trends. The assassination of Yitzhak Rabin in November 1995, the Hamas suicide bombings in Israel in February and March 1996, and the defeat of Shimon Peres, Rabin's successor, in the May 1996 elections in Israel have put the Arab-Israeli peace process on a slower track. Although one of the authors has recently completed another study that suggests that the peace process has reached a point of no return, this does not ensure a new era of peace, as the bitter fighting between Israelis and Palestinians in September 1996 demonstrated so vividly. The Middle East remains a powder keg and trends reflecting both optimistic and pessimistic outlooks are frequently noted by analysts.

What is clear is that the greater Middle East is at a critical juncture. The violence in the Caucasus, continued conflict in parts of Central Asia, and the unresolved India-Pakistan dispute, together with the uncertainties about the Persian Gulf and the Arab-Israeli peace process, make it clear that progress toward comprehensive regional peace, or even peace in one or more of the subregions, is unlikely to come anytime soon. The political orientations and the nature of future regimes in Turkey, Egypt, Saudi Arabia, and Algeria, as well as in Iran, Iraq, and Pakistan, remain in doubt. For this reason it is possible to imagine pessimistic scenarios, while at the same time endorsing the logic of the key regional leaders who all argue that with peace and stability could come renewed economic prosperity, which would, in turn, stabilize the local politics.

In a broader context, the nature of the evolving international system and the role of the major powers remain uncertain. The United States may or may not choose or be able to play the kind of dominant role in the region that it seemed destined to assume following the Gulf War. It will depend not only on Washington's traditional political oscillation between the poles of isolationism and interventionism, but also, as we have noted, on limitations of the declining U.S. defense budget, the future of basing access within and en route to the Middle East, and the technological possibilities of long-range projection of force deriving from the ongoing revolution in military affairs. American and European regional policies could substantially diverge in the future with regard to Israel, the Arabs, oil, and the dual containment of the pariahs in Baghdad and Teheran. Russia's ambitions along its new southern border reflect traditional regional prerogatives, involving overall influence, a major role as arms supplier, and a major say in the distribution of Caspian Basin energy. If India's economy continues to grow, it will become a more major player in determining the region's security policies. It will become more closely linked to the Persian Gulf in regard to trade, oil imports, the provision of guest workers, and perhaps eventually the transfer of military technology. China, if it makes its expected bid for superpower status (now ominously referred to as "hegemonic China" in Pentagon planning circles), and as it becomes more dependent on Middle Eastern oil, is likely to increase its transfers of conventional armament and weapons of mass destruction technology, and eventually may project military power into the region, which would be a throwback to the era of Zheng He's navy.

Overall, as we have noted, the ongoing trends in economics and military technology have produced a greater Middle East that may either move toward a more cooperative and peaceful future or further devolve into violence. Some see it through the lenses of Pollyanna, some of Cassandra. We have laid out the possibilities for a more idealistic, prosperous, and venturesome future for the region in line with the visions of Shimon Peres and some of his interlocutors. This could involve a host of possible cooperative economic projects (including canals, highways, water pipelines and dams, oil and gas pipelines, communications grids, and regional airports). If some or most of these were to come to fruition, it would militate toward a more prosperous greater Middle East, which might nearly come to resemble the modern regions of Europe or East Asia. Much of the modernization could be underwritten by oil and gas revenues, which are predicted to increase substantially in the wake of projected global shortages, as well as by external investments, although significant increases in per capita income are difficult to achieve in the face of rapid population growth. If nevertheless such an increase in regional per capita income could be paralleled by democratization—more often than not an expected correlate—then that combination of wealth and democratization might allow the region to become part of a "zone of peace," an appellation now applied only to North America, Europe, East Asia, and the Antipodes.

On the darker side the region may remain a cauldron of instability and tension. That could entail the continued unresolvability of its three major conflict pairings (Arabs versus Israel, Iran versus Iraq, and India versus Pakistan), various and endemic ethnic and religious conflicts both within and across national boundaries (now made more volatile by the new geographic relationships resulting from the breakup of the USSR), confrontations over oil and water, and the struggle between modernity and religious fundamentalism in countries as diverse as Pakistan, India, Israel, Iran, Egypt, and Turkey. Now introduced into this volatile mix is the prospect of nuclear weapons in each of the three major conflict pairings and the acquisition by all of the region's major players of ballistic missiles, cruise missiles, and bombers that can reach targets throughout the region. The short distances involved, in some cases outright contiguity, produce hair-trigger preemptive possibilities in connection with what some see as primordial hatreds and the prospect of irrational decision-making under duress. The open terrain and normally clear weather in much of the region give an advantage to those who control the air and have access to satellite reconnaissance data, and the newer weapons of the "system of systems" associated with the revolution in military affairs, as demonstrated in the Gulf War, promise to overturn traditional assumptions about the value of massed armies as well as the kind of linear fixed defenses that Iraq was able to use in the earlier Iran-Iraq war.

As the countries of the region modernize and develop the economic and technological infrastructure of modern states, they will be increasingly vulnerable to strategic campaigns that could wreck their industries, electrical grids, oil installations, desalinization plants, and other vital facilities, a specter previewed by the missile exchanges and use of chemical warfare in the Iran-Iraq war, the Israeli air attacks on Syrian industrial targets in 1982, and the U.S. takedown of Iraq's infrastructure in 1991. The weapons now becoming part of regional inventories will allow for far worse destruction, not to mention the prospects for terrorism with weapons of mass destruction, both within the region and far afield. The prospect of increased competition among the United States, Europe, Russia, China, India, and Japan for the region's oil and gas resources, in the absence of alternatives elsewhere, raises all manner of possibilities for revival of big-power military rivalries.

Between the nether poles of optimism and pessimism, the jury is still out. What is clear is that the strategic importance of the greater Middle East will continue to hold center stage in world politics, even as its traditional configuration and alignments change.

APPENDIXES

Alternatives to Persian Gulf and Caspian Basin Oil

COST-EFFECTIVE alternatives to greater Middle East oil could reduce the importance of the region's energy reserves and its strategic significance to the industrialized powers. There are two ways in which this could happen: a sizable increase in developed oil reserves outside the region or the expanded use of new fuel technologies that replace oil and reduce its centrality to the world economy.

Oil from Other Regions

Many other regions of the world contain potential oil deposits that could provide alternatives to greater Middle East resources. The most important are in Russia, the Americas, China, Southeast Asia, and Africa.

Russia

By World Bank estimates Russia holds 10 percent of the world's oil reserves and 40 percent of its natural gas.[1] The International Energy Agency (IEA) projects that from the late 1990s onward, oil production in the former Soviet Union, which rests largely on Russian reserves and capacity, will increase to 10 million barrels a day (mb/d)—a substantial increase but still below peak production of 12.5 mb/d in 1987–88.[2] The exploration and development of its undoubted oil reserves requires political stability as well as economic and technological capacity, all of which Russia presently lacks.

355

There is also the question of how much oil Russia can spare for export given its pressing domestic demand for energy. Despite its position as one of the world's largest energy producers, cities from Kaliningrad to Khabarovsk were underheated during two consecutive winters because they did not receive promised oil supplies. The former Soviet Union has suffered an energy shock "more traumatic than what the West endured 20 years ago."[3]

Indicative of some of the problems associated with developing Russia's oil resources is the example of Sakhalin Island, located in the far eastern part of the country. In June 1994 Prime Minister Viktor Chernomyrdin signed the Sakhalin-2 agreement in Washington, which called for a local Sakhalin company and three foreign oil firms to develop the gas and oil reserves off the northeast coast of the island. The investment for exploration and development of Sakhalin's resources could reach $20 billion or more over the next thirty years. However, with a weak economy and facing high oil transport costs, the population of the island—including necessary labor—is leaving in search of better living conditions. There are legal obstacles as well: the agreement signed in Washington for Sakhalin's exploration will bear no fruit unless new legislation defining property and export taxes is implemented.

North America

According to the IEA's *World Energy Outlook 1995,* U.S. production will continue to decline through the year 2000 with a possible recovery in production thereafter if costs fall and discoveries are made to supplement current reserves. Other estimates suggest U.S. production stabilizing briefly in 1997–98 and then declining sharply after 2002.[4] During this period production in Prudhoe Bay is expected to drop 6 percent a year. Increased production in the Gulf of Mexico could turn U.S. output levels upward after 2000. Indeed, new discoveries in the Gulf, made possible with new 3-D seismic sensing technology, have generated a much more optimistic forecast about the region's potential. There is talk of a new American oil frontier, and wells that are now producing 3,000 barrels a day (b/d) could produce as much as 30,000 b/d.[5] Overall, North American production has been and will continue to be suppressed if the low price of oil on the world market continues to hold. Particularly affected are the stripper wells, with individual yields of less than 10 b/d, but which together account for approximately 1 mb/d of U.S. production.[6]

Compared to fields in other oil-producing regions, oil fields in the United States have been extensively explored. Yet according to the Office of Technology Assessment, sizable quantities of "conventionally mobile oil remain to be recovered," mainly in existing fields. The report recommends "sustaining exploratory and development drilling activities in known fields, horizontal drilling, accelerating enhanced oil recovery, bringing shut-in or marginal oil fields back into production, and limiting the premature abandonment of existing wells."[7]

As far as Canadian oil production is concerned, the offshore fields of eastern Canada are especially important; if developed they could add 200,000 b/d by the year 2000.[8] In addition Canada has enormous potential for expanding oil production from tar sands, estimated at 200 to 300 billion barrels. As in the United States, however, prevailing low oil prices inhibit the development of Canada's resources.[9] Surprisingly, drilling for natural gas and oil in Canada has boomed; in 1994 drilling activity reached its highest level in a decade, exceeding all predictions. At the same time the drilling boom has exposed some serious flaws in the Canadian oil and gas industry—including shortages of equipment and qualified oil field workers—which have caused prices to skyrocket. Comparable drilling rigs cost twice as much to run during the summer of 1994 in Canada as in the United States, where drilling may soon intensify.[10]

Mexico is one of several non-OPEC countries in Latin America with significant oil resources. The U.S. Geological Survey estimates Mexico's remaining identified oil reserves at nearly 50 billion barrels as of January 1, 1993. Unfortunately, the North American Free Trade Agreement (NAFTA) has not brought the opening of the country's oil industry to foreign investors. Mexico's oil production is projected to remain stable until 2005 at 3 mb/d, increasing thereafter to reach 3.6 mb/d by 2010. Rising real crude oil prices over the next decade could, however, result in a more hasty dismantling of current political and financial constraints, which could lead to significant increases in investment in the oil industry and the expansion of output to 3.5 mb/d by 2000 and 3.9 mb/d by 2010.[11]

South America

In Colombia, two new oil fields currently under development are expected to begin operation in the late 1990s, bringing total production up to 1 mb/d. Brazil has oil in its deep-water Campos basin and in some remote regions of the Amazon. Initial exploration in the Santos River basin has also proven promising. In spite of high projected development costs, these deep-water basins are expected to add to Brazil's production, which could reach 1.3 mb/d by 2001 and 1.5 mb/d by 2010.[12] Peru could expand output to 200,000 b/d by 2000.[13]

Among the Latin American countries Venezuela's oil output and exports are particularly important to the United States, its second largest source of supply after Saudi Arabia, delivering 956,000 b/d.[14] According to the IEA 1994 global energy forecast, however, Venezuela, like the Middle East, will have to more than double its exports of oil by 2010 to meet rising global demand.[15]

Latin American gas reserves are estimated to be about 6 percent of the world's total, and they are located primarily in Venezuela, Mexico, and Argentina. The expansion of the natural gas industry has been rapid in certain areas of Latin America, but there are several factors that could hamper it, not only in Latin America but in all developing countries. Gas reserves are usually developed in conjunction with oil production, since the gas has been used primarily for reinjection for oil extraction. Many of the reserves are far away from major urban and industrial areas, which makes it impossible to exploit them without huge investments to build gas pipelines. The debt-burdened Latin American economies, which seem to be in constant flux, are unlikely to be able to afford a large expansion in gas production. Overall, even if the domestic use of gas increases sharply, net exports are expected to remain virtually nil.

China and Southeast Asia

Whether or not China will become a major oil importer in the next fifteen years will depend upon its success in developing two potentially lucrative areas: the Tarim Basin in the Takla Makan Desert in northwest China and the offshore resources of the South China Sea.

Some observers have great expectations for the Tarim Basin, as it is believed to be the largest unexplored oil basin in the world. However, considering the area's temperature range—from 50°C in summer to –40°C in winter—together with shifting sand dunes and vicious winds, as well as the depth of the oil-saturated layers, the Tarim will likely be a very costly source of oil. Rough estimates put the cost of building a pipeline at about $10 billion.[16]

Further exploration and development of oil reserves in the South China Sea will also be difficult. Governments of the six surrounding countries would all stake claims—in part or in full—over the South Sea's Spratly Islands: China, Vietnam, Malaysia, the Philippines, Taiwan, and Brunei all covet the islands' potential riches. The Philippine government gave permission to Vaalco Energy of the United States and its Philippine subsidiary, Alcorn Petroleum, to explore a part of the disputed Spratlys. In a reaction similar to Russia's in regard to the Caspian Sea, China immediately protested, claiming territorial sovereignty over the islands. The diplomatic impasse was temporarily solved by ignoring the sovereignty dispute and endorsing the principle of "common exploration and development of the Spratlys."[17] Nobody is sure how long the settlement will last. In the past China has used force to resolve territorial disputes with Vietnam in the South China Sea.

Brunei is Southeast Asia's third largest oil producer, which has made the small autocratic sultanate (its population in 1995 was only 292,000) one of the richest countries in the world, boasting a per capita income of $14,518.[18] *World Oil* estimates liquid reserves in Brunei at 1.2 billion barrels, but production is controlled for purposes of conservation, leaving total output at around 170,000 b/d of crude and condensate. If current production levels are maintained, Brunei may continue to produce oil until 2015, at which time its wells will dry up.[19] In nearby Indonesia predictions are even grimmer because of the added burden of domestic demand. Indonesia is the fourth largest country in the world in terms of population, with about 205 million people in 1995 and an annual growth rate of 1.56 percent.[20] Indonesian crude and condensate production was down 1.6 percent in 1994 to 1,505,419 b/d and the trend is for slowly falling rates to continue. In 1995 exports were around 750,000 b/d.[21] Analysts predict that Indonesia will likely become a net importer within a decade.[22]

Europe

In 1996 North Sea production peaked at more than 6 mb/d. Future exploitation of the North Sea, which only twenty-five years ago was considered one of the world's most promising sources of oil, is hampered by the low-price policy OPEC has adopted to recapture lost markets. However, the region still has huge potential and new field developments, and technology could compensate for the declining production of the older fields if the price is right.[23]

Africa

The main oil-producing countries of Africa include Angola, Algeria, Egypt, Gabon, Libya, and Nigeria. Egypt, Libya, and Algeria, together with Nigeria, hold nearly 90 percent of Africa's proven oil reserves of 60 billion barrels: over 30 billion in Libya, 17 billion in Nigeria, 9 billion in Algeria, and 5 billion in Egypt. Political instability may hamper production expansion plans in Nigeria and Algeria. Recent exploration has shown some further promise in Egypt and, in addition to drilling, the Egyptian government has been working to create favorable conditions for investment and free up oil for export by replacing oil with gas for domestic consumption. In order for oil production to continue at the currrent level, however, new oil will have to be found.[24] Algeria's gas reserves are more abundant than its oil reserves, and the decline of oil prices in the mid-1980s gives Algeria little motivation for oil exploration, but the government is trying to create economic incentives to increase foreign investment and expand exploration activities.

Increasing the oil output of most of the North African countries will be difficult owing to general economic weakness and financial constraints. Both Algeria and Egypt have, to varying degrees, been in a state of internal war with militant fundamentalists for the last few years. Under such circumstances it is unlikely that many foreign investors will risk making significant investments in either country.

Libya is totally dependent on oil exports. However, as a result of the Pan Am aircraft bombing over Scotland in 1988 Libya was placed under a UN embargo that, to no one's surprise, wreaked havoc on the economy. The embargo is sometimes breached, as it was by the Norwegian-owned Saga Petroleum company, which intends to invest in an oil field in Libya.[25] Overall, however, Libya's potential for exporting oil while under the embargo is limited.

Alternatives to Oil

In view of the global dependence on Middle East and, potentially, Caspian Basin oil and the political and economic problems associated with expanding production in other areas, it is not surprising that alternative ways to generate electricity and operate transportation systems have been sought. Several potentially important technical developments are under way, especially regarding alternatives to the gasoline-fueled transportation system. However, the degree to which these alternatives will seriously affect expected demand for petroleum imports is a function of a number of factors apart from technology. These factors revolve around government and private industry, including legislation and incentives to encourage increased use of alternative energy sources. We do not believe that such alternatives will seriously diminish worldwide demand for oil within the next ten to fifteen years.

Alternative Methods for Electricity Generation

The key alternative methods for electricity generation are coal, nuclear power, hydroelectric power, and natural gas.

COAL. As far as solid fuels are concerned, the IEA projects an overall increase in world consumption for the next fifteen years at an average annual rate of 2.1 percent; the figure for the Organization for Economic Cooperation and Development (OECD) is 1.3 percent a year. Solid fuels, especially coal, are expected to maintain their current share in the base load generation, especially in light of the expected decline in nuclear capacity expansion. Hard coal is expected to continue to be important in power generation—especially in North America—even beyond the turn of the century. Primary solid fuel demand is projected to increase most significantly in the ROW (Rest of the World, which excludes the United States, Japan, Western Europe, and Eurasia).[26] By 2010 the ROW could consume over 50 percent of the world's solid fuel.

As a result of these developments coal consumption is expected to almost double by the year 2010 in the ROW countries. This increase in coal use poses very serious environmental concerns; air pollution and the greenhouse effect will probably require policies to restrict the use of solid fuels, especially in developing countries. As we have already pointed out, however, solid fuel's share in the total energy mix will expand but is not projected to exceed that of oil.

NUCLEAR POWER. The future of nuclear power is highly uncertain, in both industrialized and developing countries. In China increasing energy needs are presently satisfied mostly by coal. However, domestically designed nuclear power technology has already been developed and, according to ambitious government plans, new nuclear plants will be constructed. Nonetheless, nuclear power in China is still in its early stages, and, with the exception of Japan, South Korea, and Taiwan, the same is true of the rest of Asia.[27]

In India, for instance, proponents of nuclear power argue that if the Indian government was prepared to accept new conditions for safeguarding nuclear material in power plants that used nuclear technology, the U.S. government might eventually be prepared to modify its current policies of embargoing all nuclear sales to non-NPT (nonproliferation treaty) signatories. However, this seems unlikely, and as long as India is denied this technology, no foreign investor will put money into its nuclear program. This leaves India no alternative but to invest heavily in oil and liquefied natural gas power generators, as well as to try and attract foreign investment in electric power.

In both the developing and the industrialized countries, the public is highly concerned with nuclear safety and problems of nuclear waste storage, and the 1994 IEA global forecast predicts only a slight expansion of nuclear power generation, with the two exceptions being France and Japan. Most of the developed countries, fearing another Chernobyl, are unlikely to build new nuclear power plants. The overall OECD nuclear capacity is expected to grow by 1 percent until 2000 and then decrease by 0.4 percent until 2010.[28]

Nuclear power is used mainly for generating electricity. In the United States the percentage of electricity generated by nuclear energy was 19.8 percent in 1992, much less than in Japan (27.7 percent) or Western Europe (41.2 percent) (see figure A1-1). It is predicted that the share of nuclear power in total U.S. electricity production will decline slightly by the year 2010. There are no plans for significant nuclear construction, and many of the nuclear power plants presently operating are expected to reach maturity and be taken out of service in the next fifteen years.[29]

HYDROELECTRIC POWER. Increasing hydroelectric capacity is also problematic. In most of the OECD countries, the possibilities for building hydroelectric plants have been virtually exhausted, and where they have not, environmental and population concerns preclude most potential large-scale projects. Among the OECD countries Turkey is the only exception—it plans to almost triple its hydroelectric capacity by 2010. Overall, hydroelectric capacity for the OECD is projected to increase only by 1 percent annually.[30]

Both in the developing countries and in the industrialized world, it is increasingly difficult to build large hydroelectric dams owing to pressure from environmental and human rights organizations. In Canada, for example, the huge Hydro-Quebec project in James Bay has encountered strong opposition from environmentalists and also from indigenous peoples, who will be dis-

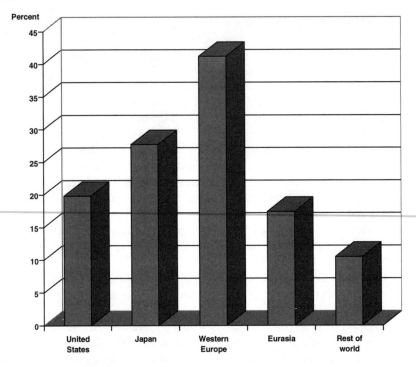

Figure A1-1. 1992 Nuclear share of total electricity.
Source: Energy Information Administration, World Nuclear Capacity and Fuel
Cycle Requirements, 1993.

placed by the flooding of the area. The Three Gorges Dam project on the
Yangtze River, which was finally approved by the Chinese government, has been
the subject of numerous protests by domestic and international environmental
organizations. The dam is the largest public works project in China today, probably
comparable to the Great Wall; its cost is estimated at about $10 billion, and it is
expected to supply 17,000 megawatts of electricity. A second project, which
includes the construction of an 860-mile aqueduct, will cause even more contention
than the Three Gorges project, since it will displace about 225,000 people.[31]

NATURAL GAS. Natural gas is an increasingly attractive fuel alternative for
much of the world and for this reason requires a closer look. Not only does it
produce less carbon dioxide than coal, but it is low-cost and efficient, and
gas-fueled power stations can be built comparatively quickly.[32] Coal emits
67 percent more carbon dioxide than gas to produce the same amount of heat.

In the years 1973–89, the increase in natural gas consumption in the former
Soviet Union, Eastern Europe, and the developing countries, including the

Middle East, constituted the most significant change in the non-OECD energy balance; demand for gas almost tripled. This trend is expected to continue, and gas consumption is projected to increase until 2005 at an annual rate of 4.9 percent. In those same years the use of gas in the OECD countries is expected to increase by only 2.3 percent a year, as the demand for gas as a fuel is greater in the developing countries than it is in the industrial world. About 45 percent of the increase in gas consumption is expected to come from the Middle East, which will try to free as much oil as possible for export. Another 21 percent of the increase will come from Latin America, and 21 percent will come from the Asia-Pacific region.[33]

Japan is Asia's biggest consumer of liquefied natural gas, but other booming Asian economies also show a preference for gas. Thailand, for example, is considering long-term agreements to import natural gas from Gulf Arab states as part of its plan to switch from coal to gas in its power stations. Industry sources add that natural gas will play an increasing part in the region by the year 2000, owing mainly to environmental concerns.[34]

Natural gas is also increasingly important for the European Union, but reserves in the European continental shelf account for only 3 percent of the world's total. The Middle East, in comparison, holds 33 percent, with Iran, the United Arab Emirates, Saudi Arabia, and Qatar all controlling significant fields. Russia, which holds approximately 41 percent of the world's reserves of natural gas, intends to expand its gas exports to Europe. It currently supplies over 20 percent of Europe's gas requirements, and it is likely to increase this figure in the future.[35]

Alternative Fuels for Transportation

Unless a cost-effective substitute for the gasoline engine is found, world demand for oil for transportation will continue to rise. Proponents of alternative fuel vehicles (AFVs) argue that they are more energy-efficient, environmentally clean, and economically viable in the long run than their gasoline-fueled counterparts. They argue that the government should intervene in the market to foster technological progress and curb the petroleum companies' monopoly. Whether governments are likely to take such measures when oil is comparatively inexpensive and the supply is steady is questionable, but it certainly seems plausible that in the event of another oil crisis, AFVs will make much faster progress and capture new markets more easily.

ELECTRIC VEHICLES. All these advantages, however, do not provide an answer to the most important question of all: When will electric vehicles become a viable alternative? The battery technology that could make electric cars generally affordable is not yet available. In spite of progress, say some specialists, reducing the cost of electric batteries is unlikely to be possible until 1998 at the earliest.[36] One major constraint is charging time; completely discharged batteries for a 1993 Chevrolet Geo Metro EV can be recharged for four to five hours of use in a quick-recharge mode, and up to ten hours in the regular mode.[37] Most EV producers suggest that, as such, recharging should be done overnight. Another obstacle is that the lead-acid batteries that EVs presently use are quickly exhausted if all the systems in the car (e.g., air conditioners, blowers, lights) are working at full capacity. The current-generation EVs are estimated to be able to go from 50 to 100 miles on a charge, depending on the battery and the driver's habits on the road. Because of these drawbacks even the proponents of EVs acknowledge that for the present, they are practical only as a second car for commuting and urban driving.[38] Nevertheless, the General Motors electric coupe, the EVI, went on sale in California in the fall of 1996. The debate about when EVs will become a marketable, viable alternative continues. Estimates of the timetable for developing the new technology vary from five to ten years. Here the impact of the law can be important. California remains committed to a requirement that 10 percent of all cars sold in the state have zero-emission by 2003.

Some of the major technological problems concerning EVs could probably be solved by the use of the advanced flywheel, a new technology that raises high hopes for the future of EVs. (Flywheels—which store and produce energy—are not new. The potter's wheel was a flywheel, and these devices were in much demand during the Industrial Revolution.) Flywheel batteries use electric motors with a large attached wheel that stores mechanical energy. Flywheels can take in household electric current, which sets them spinning; the motion produces mechanical energy, which is stored in the spinning wheel. The flywheel then acts as a generator, turning the stored mechanical energy into electricity.

One of the most serious disadvantages of current EV batteries is that they have a limited range and wear out quickly, especially if a lot of power is needed in a short period of time. In contrast to lead-acid batteries the flywheel has a very high power density—a term defining the ability to absorb and deliver power at a high rate. It provides energy efficiently and quickly when the vehicle has to accelerate. The flywheel also recaptures the extra energy released at braking and stores it, a feature not found with lead-acid batteries. The flywheel is not affected by cold weather as lead-acid batteries are, which is an important advantage in the northeastern United States.[39] Using the flywheel not only improves the performance of EVs but it also provides a longer life for the chemical battery.

There are companies that consider the flywheel a potential stand-alone source of energy for EVs. American Flywheel Systems, Inc., in Seattle plans to produce an EV fueled by six to twelve flywheels; such a car would go from 0 to 60 miles an hour in 7 seconds and have a range of 400 miles.[40] Others, however, such as United Technologies Corporation—a multinational conglomerate—consider flywheel batteries appropriate only for complementary use in conjunction with existing chemical batteries.

As with many new technologies, the flywheel concept raises questions about cost, durability, and safety. These issues are crucial in making predictions about whether and when the flywheel will appear on the market. According to most alternative transportation specialists, more demonstrations of flywheel performance, cost, and safety are necessary before marketing is possible.[41]

The efficiency and power of flywheel energy storage have been demonstrated in many laboratory experiments, but much remains to be done before the technology is complete. The Department of Energy recommends that "a complete prototype system" be engineered, built, and tested.[42] The existing components of the flywheel energy storage technology have yet to be integrated into a "mass producible energy storage system."[43] Some observers say that when the EV is parked, the flywheel will slow down and lose energy. Others maintain that the stored energy will last for days, but nobody knows exactly how long.[44]

Chrysler has been working on the Patriot, a hybrid EV that uses both a flywheel and a clean-burning natural gas turbine engine. The car can accelerate very quickly and can reach a speed of 200 miles an hour. The flywheel adds more energy when the vehicle needs to accelerate and stores the extra energy that is released. The Patriot was displayed at an auto show in Detroit in the winter of 1994 but is still a "concept" vehicle.[45] Two brothers from California, the Rosens, have developed their own design for this concept, which they believe will be revolutionary. The Rosen design marries two very different technologies: a high-power jet turbine and an advanced flywheel suspended in a vacuum in a 12 7-inch cylinder that can spin at 55,000 rpm.[46]

HYDROGEN GAS FUEL CELLS. Technological difficulties currently inhibit the development not only of electric cars, but of all kinds of vehicles utilizing alternative energy sources. Hydrogen technology, a new innovation with great fuel efficiency potential, is still too expensive for the average consumer. The technology uses hydrogen extracted from natural gas and oxygen from the air to produce energy, releasing only water and heat as by-products. This virtually emissionless system, according to some optimistic forecasts, is between five and six years from realization.[47] Developing a technology does not, however, mean having the finished product ready for the market.

NATURAL GAS, METHANOL, AND PROPANE FUEL. According to one expert, "natural gas could become the world's most popular energy source if it can replace petroleum products in vehicles." Right now, however, there are only 700,000 compressed natural gas (CNG)-powered vehicles in existence. Most CNG users are found in Brazil, the Netherlands, New Zealand, Italy, and Japan. The United States is ranked tenth, with 497 CNG stations for gas-fueled vehicles.[48]

A Canadian campaign—begun in the late 1980s—to expand the market for natural gas– and methanol-powered vehicles has progressed at a much slower pace than expected. Further cooperation between the government and producers of methanol, petroleum, and gas will be required for the project to be successful.[49] In the United States a company that spent $15 million on natural gas vehicles (NGVs) over a period of three years has not yet realized expected revenues.[50] Methanol vehicles, projected to eventually be compatible with their gasoline-fueled counterparts, presently have certain disadvantages, including high levels of iron and other metals in the vehicles' oil. Natural gas– and propane-fueled vehicles appear to be among the most promising new technologies. The Department of Energy has reported that in the United States, there are 3,297 propane refueling stations, 497 CNG stations, and 52 alcohol stations.[51] Proponents of methanol and CNG forecast a 45 percent share on the AFV market for each of them by the year 2010.

At the moment any gasoline-fueled vehicle can be adapted to run on CNG in one of several different modes: a vehicle running on CNG only, one that can use either CNG or gasoline, or one that runs on a combination of CNG and diesel fuel. The conversion of a gasoline-fueled vehicle into an NGV costs between $2,000 and $3,500 and takes a day or two. There are CNG home dispensers that can be connected directly to a home's domestic natural gas line. An important advantage of NGVs is their emissions record, which compares favorably to gas and diesel. Natural gas produces up to 97 percent less carbon monoxide, 39 percent less nitrogen oxides, and 72 percent fewer reactive hydrocarbons.[52]

The development of NGVs in the United States was stimulated by the Clean Air Act amendments of 1990 and the Energy Policy Act of 1992, which created significant market opportunities for all AFVs, including NGVs. The executive order signed by President Clinton in April 1993 further enhanced NGV production by calling for a 50 percent increase in government purchases above those mandated in 1992. The main constraint on NGVs is lack of supply infrastructure, but proponents of natural gas say that the system of refueling stations is expanding rapidly.

Ethanol, which is also considered a possible future substitute for gasoline, is still twice as expensive, but new biomass conversion technologies are currently being developed that will lower the price. At present the United States provides domestically for most of its ethanol consumption—only 8 percent of the ethanol consumed in 1991 was imported. Ethanol has disadvantages, however; it produces approximately the same amounts of nitrogen oxides, carbon monoxide, and nonmethane hydrocarbon emissions as gasoline.[53]

After examining the various AFVs, the proponents of oil as a fuel point out these disadvantages: batteries for today's electric cars cannot deliver the energy of even four gallons of gasoline, they are twice as expensive, and they need to be recharged at least every 100 miles. Ethanol provides only two-thirds the energy of gasoline, its production cost is twice as high, and its use increases emissions of nitrogen oxides. Thus, according to the president of the American Petroleum Institute, the only alternatives that may play a significant role will be other petroleum fuels.

Table A1-1. Alternative Fuels for Transportation

Fuel	Advantages	Disadvantages
Gasoline	The infrastructure—refineries, pipelines, and service stations—is in place; gas at the moment is cheap and abundant; new technologies for "reformulated," environmentally cleaner gasoline are being developed.	Produces significant carbon dioxide and carbon monoxide emissions; oil resources are unevenly distributed all over the globe; there is a potential danger of Western dependence on the Persian Gulf.
Compressed natural gas (CNG)	Natural gas is abundant and the Western world has huge natural gas reserves; gasoline-fueled vehicles can be turned into NGVs easily and quickly (for $2,000–3,000); refueling is quick, like gas engines; better emission record than gasoline or diesel fuel; less of a health hazard than gasoline or diesel fuel.	Lack of infrastructure; CNG vehicles require heavy, bulky tanks that have limited range; not as clean as EVs.
Electricity	EVs weigh less than gasoline counterparts, cost less to operate; no oil changes and decreased maintenance; significantly lower emissions of hydrocarbons, carbon monoxide, and nitrogen oxides.	EVs need a completely new technology; electric batteries are still too expensive and need recharging very often (approximately every 100 miles) especially if all systems in the car are on full blast; the power plants that produce electric batteries cause even worse emissions than gasoline.
Methanol	Variety of sources for production: natural gas, oil, coal, and biomass; lower flammability, which is a good safety indicator.	Higher accumulation rates of iron and other metals in the oil of the vehicles; carbon monoxide and nitrogen oxides are the same or higher than gasoline; such vehicles require more maintenance.
Ethanol or mixed ethanol and gas	The United States can satisfy most of its need through domestic production.	Ethanol costs twice as much as gasoline; it produces the same nitrogen oxides, carbon monoxide, and nonmethane hydrocarbon emissions as gasoline; it provides only two-thirds the energy of gasoline.
Hydrogen	Uses oxygen and hydrogen; the technology is emissionless and the resources are unlimited.	The hydrogen technology is still too expensive and underdeveloped.

REFORMULATED GASOLINE. Given this analysis of alternative fuels it seems justifiable to give considerable credence to the proponents of oil as a fuel, who predict that gasoline will remain the world's primary transport fuel (see table A1-1). The president of the American Petroleum Institute suggests that the "fuel of the future" will be a "descendant of the fuels of the past and present": it will be a new, clean-burning, "reformulated" gasoline. The reasons for such a statement are several: the infrastructure for gasoline (refineries, pipelines, service stations) already exists, oil discoveries continue to outpace oil use, and there is no technologically, economically, and environmentally viable alternative.[54]

Other proponents of petroleum also regard "reformulated gasoline" as holding great future promise. A "reformulated" gasoline in current-model cars will allegedly be environmentally cleaner and economically viable. Such a view is supported by the fact that today's automobiles produce emissions that are 96 percent lower than those of cars from the 1960s.

Military Geography:
Standard Generalizations about
Terrain and Weather

THE FIRST generalization about the role of terrain and weather in regard to
military geography is that there are differences in the effect on conventional
and unconventional warfare and on the gray areas between them. In some
circumstances that divide is blurred, as for example in the Lebanon war of 1982
or the Soviet invasion of Afghanistan.

On the conventional side, flat and dry terrain (but not necessarily desert) is
deemed favorable for mobility, for offensive maneuver warfare featuring deep
penetration and encirclement by armor, conjuring up the familiar images of
Guderian, Patton, Sharon, and other legendary tank commanders. (Northern
Europe, for instance, is also considered "good tank country.") On the other
hand, wet and rugged terrain (or just rugged terrain with cover or with craggy
defiles) is considered unfavorable for rapid offensive movement for mechanized warfare. Flat jungle terrain or land dominated by rivers, swamps, or
marshes is said to provide equally tough going for large mechanized conventional forces. These are crude generalizations, of course, and any given theater
of war does have terrain and weather specifics, idiosyncrasies, sometimes
man-made features (the *bocage* of massive hedgerows in Normandy was tough
for allied armored forces in 1944, despite the overall flat nature of the terrain).

Similar sets of common generalizations verging upon cliché can be made
regarding unconventional or guerilla warfare, as well. Areas with lots of cover,
for example, jungle canopy or rain forests, and with rugged terrain are good for
hiding guerrillas, for providing bases, hideouts, and arms caches that are
difficult to find on the ground or to see from the air. Such terrain—all the more
so combined with rainy weather and habitual cloud cover—renders airpower

(including the use of attack helicopters) relatively useless. Altogether such terrain conditions tend to favor insurgents over counterinsurgency forces and to increase the ratio required for the latter's victory, as well as lead to a more protracted, stalemated conflict that, more often than not, will be to the political detriment of an incumbent regime on the defensive. Conversely, flat and cover-less terrain, particularly if combined with clear weather, will be advantageous to counterinsurgency operations, allowing technological advantage to come into play, particularly in regard to airpower. Again these are ideal-type or hypothetical pure cases, and in the history of real-world conflict there have often been surprises and, of course, more complex conditions that defy such easy generalizations.

One problem with easy generalizations is that the theaters of war in many areas, perhaps in most wars, have a complex mix of terrain and weather, rendered all the more complex if the wars last long enough to span the various seasons.[1] There are also many other "complexities." There is, of course, the question of just how human settlement patterns and road networks are juxta-posed to terrain features, going beyond the simple and obvious generalization that inhospitable terrain and climate will normally correlate with low popula-tion densities. There are also factors such as the location of borders and major cities vis-à-vis, for example, mountains and rivers, here ramifying further into issues such as the defense in depth with which some countries are provided or not provided, or the proximity of guerilla hideouts to major core areas. The relationships of lines of mountain ranges to axes of advance and to key core areas are also crucial, as is the relationship of terrain features to the size of countries and battle areas. What all this adds up to, obviously, is the require-ment that military geography be put in specific contexts for specific wars. What is further implied is that there are very rigid limits on really comparative analyses here. Each war, each country is, of course, *sui generis,* in terms of its military geography, and generalizations can be ventured only with caution.

Some examples from the Middle East will serve to illustrate the importance of context. Take, for example, the matter of mixes of terrain for given countries and wars. The Iran-Iraq frontier is mountainous in the north, with the Zagros range making offensives by either side very difficult. In the south, however, there are some areas favorable to offensive warfare, though much of the Tigris-Euphrates frontier features rivers, marshes, and floodplains whose ame-nability to mechanized warfare and movement of troops will vary greatly according to the season. The conflict in Afghanistan was fought in snow-covered mountains, in cities, and on rough, hard-scrabble deserts. The Spanish Sahara war was fought on sandy desert but also in dry, rugged terrain featuring wadis good for shielding guerilla fighters. The Arabs and Israelis, in 1973, fought on the flat, sandy deserts and dry mountain passes of Sinai, on the rugged plateaus and "tels" of the Golan Heights, and on the snowy slopes of Mount Hermon. India and Pakistan have fought in the mountainous regions of

Kashmir, on the desert terrain of the Rajasthan Desert, and on the salt flats of the Rann of Kutch. Numerous insurgencies have been fought, variously, in mountains, cities, and forests. Generally speaking the smaller the theater of war, the more specific the terrain, but even this generalization has been belied, for example, in Lebanon and Sri Lanka.

There are various combinations of ruggedness and cover. The mountains of Yemen, Afghanistan, and Iran have little cover; those of Lebanon have much more. None of these mountainous areas would be good tank country, but those that are covered make aerial surveillance and interdiction very difficult.

Few of the wars of the third world have been fought in heavily populated areas, unlike the cold war scenario that was anticipated in Central Europe in regard to a possible World War III. Patrick O'Sullivan points to Lebanon's Tyre, Sidon, and nearby refugee camps as exceptions during the Lebanon war.[2] The early 1980 fighting in Abadan and Khorramshahr in Iran's Khuzestan Province was another exception. There was fighting in 1977 in some of the Ethiopian towns along the Somali border. The contenders in Afghanistan have fought in Kabul and Kandahar. The Gulf War saw some brief fighting within Kuwait City. But mostly the region's wars have been fought out in open terrain, away from urban populations. Some analysts, however, see a new trend whereby, increasingly, third world conflicts, particularly those in the low-intensity range, will be fought out in cities, a point underscored by the battles in Grozny between the Russian army and Chechnyan insurgents.

The relationship of mountains and rivers to borders and the axes of advance and retreat of armies is critical and, again, this can modify broader generalizations about the nature of terrain. The Zagros range runs parallel to and near the Iran-Iraq border, making offensives in either direction difficult. In effect this serves to protect Baghdad from invasion, counterbalancing Iraq's lack of defense in depth because of the relative nearness of its capital city to the border. Ethiopia's mountains are also mostly perpendicular to a potential Somali line of attack. The low mountains of western Sinai provide a similar barrier to advance in either direction by Israel or Egypt. On the other hand, the mountains of Lebanon ran parallel to Israel's line of advance in 1982, channeling the Israeli advance into the valleys between mountain ranges or onto roads along their lower slopes. It is hard to say here where the advantages and disadvantages lay. Contrary to the usual assumptions, mountains parallel to a line of advance can still allow for a penetration and breakthrough to easier terrain beyond. But, as Israel learned in 1982, fighting in the valleys between mountain lines can be tough, particularly because a major breakthrough is difficult where the opponent can fight endless delaying actions in falling back

The contextual placement of major terrain features and borders can be important. The fact of the Hindu Kush's relationship to the Soviet-Afghan border is an example. The distances of Teheran and Addis Ababa, respectively, from the borders with attacking Iraq and Somalia largely explain their attackers' failure to win decisive victories at the outset of their wars. Pakistan's extreme vulnerability to an Indian offensive against its major population centers is obvious. One could as easily point to Israel's vulnerability (but also the reciprocal vulnerability of Amman and Damascus) or, conversely, to the defensive depth provided in the case of the Libyan and Chadian capitals, Tripoli and N'Djamena.

India's Energy Needs:
The Gulf Connection

THE GULF WAR proved to be critical in shaping future relations between India and the Gulf states. First came the shock of increased fuel bills owing to the short-run hike in world oil prices, as India depended on Iraq and Kuwait for 40 percent of its oil imports.[1] More important, however, was the loss of remittances from Indian migrant workers in the Gulf, which constituted a major source of India's foreign currency. By time the Gulf War broke out these remittances came to approximately $2.2 billion annually.[2] In addition the trade embargo on Iraq led to a loss of Indian export markets. Added to this were the costs of repatriating and resettling the returning workers, all of which had to be borne by the Indian government. Some states that had contributed large numbers of workers were hit particularly hard, among them Kerala. It is estimated that one Indian Gulf worker could support twenty people in the home state. With approximately 150,000 Indian nationals stranded in the area of hostilities, the Indian government had to airlift most of them from Kuwait shortly after the invasion in August, and then run special evacuation flights from the region.[3]

Favorable relations with the Gulf states will become more important to India in the coming decade. Early 1995 saw a frenzy of ministerial visits and bilateral economic agreements between India and various Gulf states. This is particularly true in the energy sector. In April India's Oil and Natural Gas Corp. and the National Iran Oil Co. agreed to launch a joint venture to explore and produce oil and natural gas. The two also reviewed progress on a $5 billion deal arranged in 1993 to build a 1,367-mile-long pipeline from Iran to India.[4] Deals have been signed with Qatar, and India's private Essar Group has signed a memorandum of understanding for 2.5 million tonnes of Qatari liquefied natural gas for the next twenty-five years.[5] The Bahraini Development and Industry Ministry is also working on an oil and petrochemical agreement with India.[6]

In April 1995 the Indian external affairs ministry released its annual report, in which it stated that relations with the Gulf states had "acquired a significant socio-economic dimension," as bilateral trade was on the upswing.[7] Clearly, as both India and Pakistan attempt to redefine the post–cold war geopolitics of the region, India has cast its net westward in search of new economic and political allies, leading to what some have called "Indo-Pakistani competition in public diplomacy" in the Muslim world.[8] Political and military ties between Pakistan and Southwest Asia have always been solid, and India has everything to gain by trying to make a dent in Pakistan's traditional influence in the region. As the pace of consumption and industrialization in India grows, no issue reflects the challenges more clearly than the energy crisis and the debate about how to solve it. Well before economic reform started, daily power cuts (brownouts) were common in most major urban areas, especially in the sweltering heat of summer. For many years this discomfort was tolerated by the majority of Indians. It was part and parcel of a national ideology that called for India to be self-sufficient in virtually all primary staples of life, even if this meant inefficiency.

Today the Indian middle class not only consumes more electricity but has become much more dependent on electrical systems for day-to-day living. A brownout today can mean that the air-conditioning stops, the water filtration system backs up, the computer with advanced programming crashes, and the fax and telephone recorders stop. The Indian elite today is no different from its Western counterparts in that it wants the perks and comforts of modern living, most of which are based on high per capita energy consumption.

India's national energy programs have failed to meet the increased demand for electricity and gasoline for industry, agriculture, and the domestic consumption of the growing middle class. Electrical generation capacity has grown fiftyfold since independence, from 1,700 to 81,000 megawatts, 69 percent being generated by thermal power (coal) and hydrocarbons (oil and natural gas), 28 percent by hydroelectricity, and 3 percent by nuclear power, but will have to double in the next decade if it is to keep up with projected demand.[9] According to the *Economist* it will cost India more than $200 billion over the next twelve years to provide an extra 9 percent of power capacity a year.[10] To meet this challenge India will have to address both the demand and the supply side of the equation.

At present 14 percent of India's primary commercial energy supply is met by imports in the form of crude oil or petroleum products. This percentage will have to increase in order to meet the greater demand, leading to a further balance of payments deficit in an already struggling economy. There are also other dilemmas. While new projects relying on petroleum and natural gas make eminent sense from an engineering standpoint, they pose strategic as well as economic risks. For instance, the Enron thermal power station just outside Bombay was designed to rely on oil, which will have to come from production rigs off Bombay's shore, and could be subject to terrorist attack, or liquefied natural gas, which would come from the Gulf on tankers that would be vulnerable to interdiction in a crisis. India has also investigated the possibility of laying a gas pipeline under the Arabian Sea from Oman to Bombay, although the costs appear to be prohibitive.

There has also been talk of cooperation between Pakistan and India on bringing natural gas by pipeline from Iran, which has huge resources and is anxious to sell them. Such a proposal would have benefits and costs for both countries. Since they would be dependent one upon the other for energy supplies, such an arrangement could be a deterrent to further conflict over Kashmir. Equally, though, both countries could be held hostage to Iranian control of the supply. In any event India and Pakistan are going to have to import more oil from the Persian Gulf in the coming decade, and this will make them even more vulnerable than in the past to the effects of another Gulf crisis.

Middle Class Aspirations

Although there is controversy about the precise number, the Indian middle class is estimated to be between 150 to 250 million, out of a total population of 900 million.[11] Regardless of what numbers we choose to believe, it is clear that this new class of Indian consumers is very large in absolute terms. Most analysts define Indian households with an annual income of between $1,200 and $3,000 (Rs. 36,000 and 90,000) as middle class.[12] Annual incomes of $2,500 (Rs. 78,000) and above are increasingly seen as upper middle class. Since these numbers seem deceptively low by Western standards, it is easier to think of them in terms of simple purchasing power: in other words, the type of consumer goods a family can afford on its base income is one of the indicators of class in India. With standard washing machines priced at about $166 and color TV sets at $466, these items are now within the reach of many Indians, especially given the high national savings rate of about 20 percent and newly instituted credit procedures.

This purchasing power makes the Indian middle class a significant factor in Indian economic and political life. Economic reforms have produced an orgy of consumer spending at the same time that lower customs duties for imported goods have forced domestic producers to reduce prices of consumer items, while improving quality and choice. Middle-class Indians, long deprived of the luxury goods enjoyed by their Western counterparts, are on an extended buying spree.

This has been paralleled by another important phenomenon: the slow but steady erosion of India's socialist ethos, long the pillar of the country's economic philosophy. For years socialism in India had stigmatized consumerism and wealth, while crippling income tax rates were a further incentive to hide wealth. The arrival of private satellite and cable television channels (both foreign and domestic) has exposed middle-class Indians to previously unavailable international programming. Expectations have risen, in tandem with the reduction of interest rates, an increase in credit and leasing (or hire-purchase) facilities, and an increasing number of dual- or multiple-income households. Meanwhile income levels have risen as more and more Indians are hired by foreign companies on foreign salaries. All of this has converged to produce a radical change in Indian attitudes and aspirations—moneymaking is now a major preoccupation, with the acquisition of material goods a close second.

Equally significant, over half of India's population—521 million people—is under the age of 24.[13] Born after the fiery *swadeshi* (self-reliance) rhetoric of Gandhi brought India its independence in 1947, and long after the subsequent wars with Pakistan and China, this new breed of moneyed, ambitious, and mostly young Indians has no philosophical ties to socialism, and it is the major force in the retreat from economic isolationism toward notions of the free market and entrepreneurship. Furthermore, they believe India is destined to be a great power and are not shy about saying so. Traditional professions such as medicine and the civil service, which were the preferred route up the socioeconomic ladder, are now being spurned in favor of careers in business and finance.

The most striking aspect of the new middle-class phenomenon, however, is that its ranks are increasingly being swollen by India's 750-million-strong "rural underclass," which is also modernizing as it experiences an increase in income. Alarmists have long argued that the economic reform process in India is far from irreversible because the wealthy consumers remain a minority and will inevitably be outvoted by the underclasses, who are increasingly disaffected by how hard life has become under liberalization. The claim has been that the rural poor, traditionally the bastion of socialist support, have been hit hardest by the steep increases in basic food prices, caused by runaway inflation and the end of certain subsidy programs. It is clear, they have said, that it is impossible to distribute the gains of liberalization equally in a society already suffering from deep economic cleavages, and that while the burgeoning middle class has frantically acquired its cars, cable TV, and satellite dishes, the alienated masses have battled inflation and unemployment.

In fact most indications are to the contrary. There are signs that the trickledown effect of consumerism seems to have reached the villages, where 70 percent of India's population resides. The "need to climb up the ladder," according to an *India Today* survey, "is increasingly cutting across the rural-urban divide."[14] The only difference seems to be that the rural masses still consume basic, less technologically advanced goods. One survey by the *Los Angeles Times* found that more than 75 percent of the bicycles and portable radios sold in India and more than 60 percent of the table fans, sewing machines, bath soap, and wristwatches are bought by the rural population, though many of these are still locally produced and sold at fixed prices.[15]

Thus while the Westernized urban Indians may be busy acquiring their fax machines, air conditioners, and cellular phones, the rural underclasses are buying black-and-white television sets and VCRs, albeit one to a village rather than one to a household. In its economic survey *India Today* identified a burgeoning "aspiring class" of about 670 million people, mostly poor and rural.[16] It is this group that is constantly mobilizing upward and feeding into the middle class. Thus a two-tier society that was once divided simply into a small class of "haves" and a huge one of "have-nots" has gradually been redefined into a multiple-tier one of "haves," "have lesses," "aspiring to haves," and "have-nots."

The Energy Crisis

Most energy supplies in India remain in the hands of the public sector monopolies, operated and heavily regulated by the state. Individual State Electricity Boards run government-owned power plants, where overstaffing, inefficient operation, and artificially controlled prices have created huge distortions in the energy sector. Government subsidies have kept prices low in order to mollify rural voters—farmers in some states receive virtually free electricity. Power plants operate at about 50 percent of their earning capacity, with the other half being lost in transmission and distribution owing to shoddy equipment, stolen through "jumper" cables, foregone through inadequate billing procedures that result in nonpayment, or given away in ad hoc subsidies by corrupt local officials. It is clearly possible to do much better than this—power plants owned by private industries routinely operate at 85 percent.[17]

At present India consumes about 382 kilowatts of electricity per person per year, compared with 11,000 in the United States.[18] Even so in the 1980s alone, long before any economic reforms imposed greater consumer demands on the energy supply, demand increased by 5.7 percent.[19] Now, as middle-class consumption rises, so too will the pressure on the energy supply. This is especially true because India's economy—like those of all fast-growing developing countries—is extremely energy-intensive.[20] One survey estimates that the demand for energy will double over the next ten years, while another forecasts the need for a tripling in energy capacity over the next fifteen years. The *Economist* provides the more conservative estimate of a 9 percent annual growth in power capacity, assuming a 6 percent growth in GDP.[21]

In theory India could be self-sufficient in the basic requirements for electricity production. It has abundant quantities of coal, but it is of poor quality and creates serious pollution and huge amounts of slag as it burns. In addition, transporting it makes increasing demands on the already overused and undercapitalized railway system. So if coal use is to grow commensurate with the anticipated demand for fuel, there will need to be major new investments in both transportation and environmental protection.

The other principal alternatives for the production of electricity are natural gas, oil, and hydroelectric and nuclear power. However, India has very little cheaply available oil and gas, and importing these fuels from the Gulf adds dramatically to the balance of payments deficit and poses strategic risks. Hydroelectric schemes have run afoul of environmental complaints and mismanagement and conflicts with Nepal and Bangladesh over the control of waters from the north. The future of nuclear power is equally controversial. Proponents argue that if the Indian government was prepared to accept new conditions for safeguarding nuclear material in power plants that used nuclear technology, the U.S. government might eventually be prepared to modify its current policies of embargoing all nuclear sales to non-NPT (nonproliferation treaty) signatories. However, this seems unlikely and as long as India is denied this access, no foreign investor will put money into India's nuclear program. This leaves India no alternative but to invest heavily in oil and liquefied natural gas power generators.

In a bold move Prime Minister Rao threw the sector open to private investment in 1992, permitting 100 percent foreign ownership and wooing investors by waiving the bidding system and negotiating contracts rather than putting them out to open tender. Project proposals for energy development from foreign firms came pouring in, and seven of them were hastily accepted on a "fast track" for clearance and given generous terms that make the investments almost risk-free. Among these is the controversial $2.8 billion power station project in Dabhol, Maharashtra, run jointly by the U.S.-based companies Enron Development, Bechtel, and General Electric.[22]

No matter how individual energy cases are resolved, what is abundantly clear is that India's energy crisis can only get worse with burgeoning consumer demand from the middle class. Another indicator of increased energy consumption is the growth of sales of automobiles, trucks, and two-wheeled vehicles (motorcycles and scooters). Car sales in India have grown at an annual rate of approximately 25 percent in the last two years since reforms broke up auto manufacturing and assembly monopolies, while the average annual growth rate between 1970 and 1988 was in the neighborhood of 7 percent.[23] Spurred by a lowering of excise taxes on cars from 55 to 40 percent, as well as a relaxation of restrictions on consumer financing, Indian consumers have demonstrated an appetite for vehicles so great that Indian auto manufacturers have even begun diverting supplies from exports to the domestic market.[24]

Two-wheelers, however, remain the vehicle of choice for most Indians, with sales four times higher than that of cars at 821,600.[25] More affordable and well within the reach of the lower middle classes, motorcycles and scooters show a similar annual sales growth of 24 to 26 percent. Meanwhile, trucks sales have increased by about 35 percent.[26]

The continuing increase in the number of vehicles in use since the 1991 reforms has exacerbated the strain on India's road transport system, which is laboring under an archaic, inefficient, and poorly maintained infrastructure, long seen as a major hurdle to industrial growth. The Indian government has now thrown the entire infrastructure sector open to private and foreign investment, introducing a five-year tax holiday in the latest budget for any enterprises that build, maintain, or operate roads, highways, expressways, or bridges, and other such transport infrastructure. It has delineated $25 billion worth of expressways and $8 billion worth of highways and bridges as priority projects.[27] However, none has been built so far, and India has yet to open a modern divided highway on trunk routes.

In addition to stimulating a boom in foreign investment and in the construction industry, this added demand for cars will certainly translate into increased demand for gasoline. After electric power generation, transportation is the other major area of hydrocarbon use, and increased car ownership is directly related to aggregate demand for gasoline. Demand for gasoline is also inversely related to fuel efficiency, and Indian cars tend to rank rather low in this respect. All of this points inevitably to even more pressure on the country's already limited hydrocarbon supply, as well as to more demand for gasoline imports from the Persian Gulf.

The Status of Nuclear, Chemical, and Biological Weapons Programs in the Greater Middle East

BY THE MID-1990s there was a widespread assumption that Israel, India, and Pakistan all had acquired operational nuclear weapons. Depending upon the source Israel was reported to possess anywhere from 50 to 300 nuclear weapons, perhaps comprising a mix of Pu-239 and U-235 fission weapons, fusion weapons, and related neutron weapons for tactical purposes. India and Pakistan were widely believed to possess much smaller but growing nuclear arsenals, perhaps in the range of 10–20 weapons, though some experts suspected that India, which is now more than twenty years beyond its test of a nuclear device and is known to possess several unsafeguarded plutonium production facilities, had a much larger arsenal. India has refused to go along with a nuclear test ban; Israel may or may not have tested in the ocean below South Africa (but no doubt has conducted computer simulation tests minus the bomb material); and Pakistan may have had its weapons tested by China.

Iraq and Iran are both assumed to be moving toward a nuclear capability. The former may have been only a year or two away at the time of the Gulf War, after which its nuclear infrastructure was largely—but presumably not totally—dismantled by the UN. But the knowledge and much of the wherewithal remain. Iran appears to be striving mightily to acquire nuclear weapons and the amount of time that may require is estimated variously from five to ten to fifteen years. It will if it can (or cannot be stopped) and may be getting crucial help from sources in Russia, China, and Pakistan. Algeria and Libya have given indications of wanting atomic bombs, but in neither case is such an eventuality feared for the short run. Egypt, Syria, and maybe Saudi Arabia and Turkey loom as other long-run regional possibilities for nuclear status.

Virtually all of the major actors in the region have been reported developing either or both chemical warfare (CW) and biological warfare (BW) weapons: Syria, Egypt, Iraq, Iran, Israel, and Libya primarily, but also India and Pakistan. CW weapons have already been used in Yemen, Sudan, Afghanistan, and in the Iran-Iraq war, and in the hands of Iraq, Syria, and Egypt vis-à-vis Israel are commonly discussed in the context of a "poor man's nuclear weapon," that is, an asymmetrically weak but still potentially potent counterdeterrent. BW weapons may, however, have the potential of causing widespread casualties roughly on a par with nuclear weapons. It is clear that large-scale inventories of anthrax and other BW materials have been acquired by Iran, Iraq, and perhaps other regional nations—weapons that can be delivered by aircraft or missile.

Israel's Nuclear Program

If one can rely on the recent works of Seymour Hersh and of William Burrows and Robert Windrem, Israel now deploys a large and diverse nuclear force, with perhaps as many as 200 or even 300 nuclear weapons.[1] According to these sources, backed up by the now perhaps largely accurate revelations of the London *Sunday Times* stemming from the Vanunu case, Israel may possess both Pu-239- and U-235-based weapons, produced by the Dimona reactor's output of weapons-grade plutonium in the first instance, and by the possible existence of a gas centrifuge operation, if not also some laser isotope separation capability, in the second.[2] It may also have produced thermonuclear weapons, an assumption implied in the Vanunu revelations about lithium operations in one section of Dimona. Finally, it appears that Israel may also have produced a number of neutron weapons (also based on the thermonuclear technology); indeed it has been speculated that it was this type of weapon that was tested over the ocean in 1979 some 400 miles south of the southern tip of Africa.[3]

Some analysts, however, are more cautious with regard to these estimates. They might question, for instance, how Israel might have constructed and operated a centrifuge facility, with its large electrical consumption and large profile, seemingly without detection. But one way or another, the size of the Israeli arsenal and the yield of its weapons may have important implications.

The perhaps formidable size of the Israeli arsenal has taken some experts by surprise. According to Burrows and Windrem the Israelis first deployed nuclear weapons prior to the 1967 war. The 1967 crisis now seems to have been initiated by Egyptian president Nasser because he feared that Egypt and its allies were running out of time if they wished to obliterate the Israeli nuclear program, if not the State of Israel itself.[4] With Israel's need for nuclear weapons further impelled by the massive Soviet arms resupply to Egypt and Syria after the 1967 war, and by the involvement of Soviet forces in the combat along the Suez Canal zone in 1969–70, Israel is believed to have deployed some twenty weapons by the time of the 1973 war. During that war Moshe Dayan and others reportedly panicked about the impending "fall of the Third Temple" and undertook a not very well disguised nuclear alert which, in turn, prompted the massive U.S. arms resupply effort.[5]

Afterward, for many years, U.S. and other intelligence services appear to have assumed that the Israeli arsenal had not climbed much higher than that magic number of twenty. According to Hersh and others, it was not until the early 1980s, indeed, after the "flash over the ocean" and the Vanunu revelations, that the CIA came to realize that the Israeli arsenal was much larger and much more diverse than had been thought earlier.[6] Indeed, it appears that it was only with these revelations that the intelligence community realized that the output of the Dimona reactor had long been much greater than what would have been produced by a 26-megawatt facility, and that Dimona had long operated at levels anywhere from three to six times that power output level. By now it is widely speculated, partly on the basis of the Vanunu revelations, that Israel has put into operation one or more of the physical separation methods for producing weapons-grade U-235, that is, either a gas centrifuge or a laser isotope separation operation or both, though these technologies might require a large energy consumption that might be difficult to hide. If so then the size of its arsenal may not fully be tied to the plutonium output of Dimona.[7] One recent study claimed that Israel had a stockpile of at least 60–80 plutonium weapons, but maybe up to 200–300 weapons, some with yields of over 100 kilotons, and even possibly enhanced radiation variants or variable yields. These weapons are backed by a formidable and diverse array of possible delivery systems: Jericho I and II ballistic missiles, Lance missiles, MAR-290 rockets, MAR-350 surface-to-surface missiles, and an array of nuclear attack aircraft such as F-16s and F-15s backed by extensive aerial refueling capability.[8]

Pakistan's Nuclear Program

As is the case with Israel, Pakistan's nuclear program has long constituted a major mystery for analysts, both in and out of government. Similarly, there was a long period during which analysts of various stripes, aware that Pakistan had both the intention and capability to go nuclear, spoke vaguely in terms of a "bomb in the basement" or of a "virtual" program that could be brought to fruition in somewhat of a hurry if the strategic circumstances so dictated.[9] The Pakistani program was often referred to in the context of an "Islamic bomb," partly because it was talked about in those terms by Pakistanis close to the program. Also mentioned were the long-time close relationship between Pakistan and Saudi Arabia, the latter's financial assistance to Islamabad (some of it in the context of the Afghanistan war; hence also involving the broader nexus with the United States and Egypt), and also some media reports about raw uranium shipments originating in Niger moving through Libya en route to Pakistan. During the 1970s and 1980s there were frequent reports about possible diversions of plutonium from the Canada-supplied 40-megawatt KANUPP reactor and, even more important, about the development of a gas centrifuge operation based on the near-amazing covert operation mounted in Europe, specifically at the multinational centrifuge facility at Almelo in the Netherlands, under the direction of Pakistani scientist A. Q. Khan, now widely credited as being the "father" of the Pakistani bomb program.[10] Despite fitfully and sporadically applied U.S. pressure intended to forestall Pakistan's move toward nuclear weapons possession, it was widely recognized that that pressure had had to be suppressed during the crucial period in the 1980s when the United States directed a covert effort to drive the Soviet Union out of Afghanistan. Pakistan was a key player in that operation.

At any rate by about 1989 the Pakistani nuclear weapons program was described as follows by Burrows and Windrem:

> Kahuta, the most important of the many facilities in the foothills of the Himalayas, could churn out enough bomb-grade uranium in gas form to make a dozen bombs a year. PINSTECH the site of a reactor that had been provided by Eisenhower in the Atoms for Peace program, was at that moment being upgraded and modernized by Chinese technicians. Spent fuel rods from the U.S. reactor and the KANUPP power reactor in Karachi could be reprocessed into bomb-grade plutonium at the nearby, and aptly named, New Labs. There were two research and development facilities for centrifuges, a tritium production plant capable of making tritium gas for boosted weapons, heavy water plants, fuel fabrication plants, a weapons design facility, and more, all dispersed within a 100-mile radius of Islamabad.[11]

There is also, according to this same source, a top-secret plutonium production reactor that Pakistan started building in the 1980s along the Indus River, rated at 70 megawatts, and hence capable of producing enough plutonium for another five bombs a year. This facility, at Khushab, southwest of Islamabad, is expected eventually to be able to produce both plutonium and tritium boosters for larger weapons.[12]

But what this all adds up to, now and for the near-term and longer-term future, is not entirely clear. Burrows and Windrem assert that Pakistan "was finally able to machine seven nuclear warheads during the spring and summer of 1990," otherwise noted as a period in which tensions between it and India may have brought the two South Asian nations close to the brink of war. With four to five years since having intervened, this analysis clearly implies that Pakistan has a considerably larger deployed arsenal than the seven weapons alleged to have been completed in 1990. If one can assume not only an ongoing centrifuge operation but also a fairly significant impending plutonium production facility outside the purview of the International Atomic Energy Agency

(IAEA) or other safeguards, that would further imply a weapons program of very significant magnitude. Indeed, one recent report claims that Pakistan has produced enough highly enriched uranium since the mid-1980s to manufacture six to fifteen weapons, each with double the destructive force of the Hiroshima bomb. Whether there has been a tacit U.S. wink at these developments in the context of a quiet promise by Pakistan not to "Islamicize" its weapons, that is, to give some of them to allied Islamic states such as Iran or Saudi Arabia or Libya, is not here known. It has been rumored, but sporadically and without real evidence, that Pakistan has been giving some help to the Iranian bomb program, which it is now assumed is being pushed along at full tilt. The future of that connection remains unclear.

India's Nuclear Program

India exploded its first nuclear device in 1974. Its program, however, has been referenced to China all along, particularly in the dual contexts of China's own acquisition of nuclear weapons (it became a nuclear power in 1964, and a thermonuclear power shortly thereafter in 1967), and the brief (and for India, disastrous) war between the two Asian powers in the Himalayas in 1962. Factoring out China, one might have claimed that—status issues aside—it would have been to India's advantage to keep the South Asian subcontinent denuclearized, so as to maintain the strategic edge it has by virtue of its much larger population and economy, as well as its obvious superiority in conventional weapons and warfighting capability, which was underscored in the war of 1971.

But although the Indian program was covert for many years after the test in 1974 (few if any publications in the twenty years that followed speculated on the possibility of a large and operational Indian nuclear force), in retrospect it appears that New Delhi was quietly moving forward with a large-scale nuclear armaments program. Thus, according to Burrows and Windrem, "while they loudly proclaimed their peace-loving nature throughout the 1980s, the Indians were in fact embarked on an all-out armament binge of astounding proportions. And since they neither did it openly nor bragged about it, the West remained largely ignorant of the program's true dimension." Summing it up,

> During that decade, India designed and built—not imported—five power reactors, each larger than 200 megawatts and each a potential plutonium producer. It constructed a uranium purification and conversion plant and a laboratory needed to enrich the metal to bomb-grade. It also set up three heavy water facilities, a fuel fabrication plant, and three separate research reactors. At 100 megawatts, one of the research reactors was the largest such plant in the world. The Indians even set up their own fast-breeder in preparation for a much larger one they ordered from the Russians.[13]

With such an infrastructure in place, and with more than twenty years having elapsed since the first test, the questions are how large an arsenal India has acquired by now and how much larger an arsenal will it amass in the future. Burrows and Windrem refer merely to "the dozen or so nuclear weapons" that "were the crown jewels of their national defense system."[14] But it is not likely that India would have allowed Pakistan to acquire a relatively larger arsenal, not to mention the still looming problem of deterring China. The CIA and others clearly underestimated the scope and pace of the nuclear weapons developments in Israel, Iraq, Pakistan, and maybe also North Korea. Is it not also possible that India has amassed a far larger arsenal than is assumed in Western publications, as those assumptions are based mostly on wishful thinking of interested parties who might wish that India's arsenal not exceed ten to twenty nuclear weapons? One recent report claims that according to American intelligence estimates, India has put aside enough weapons-grade plutonium from its extensive nuclear facilities to assemble twenty to fifty Hiroshima-size bombs.[15] In the opinion of Burrows and Windrem, "ultimately, the Indian military establishment intends to put hydrogen warheads on intercontinental ballistic missiles with ranges of up to 5,000 miles."[16] Whether India will ever be able to afford such an effort remains to be seen.

Potential Nuclear Powers (Iraq, Iran, and Libya)

There is the question of which nations apart from Israel, Pakistan, and India will be able to produce the bomb in the near future. In regard to the greater Middle East region, there is a near-universal view that Iran and Iraq want to acquire nuclear weapons, and that in the long run and without binding arms control regimes, both will be very difficult to stop, the current UN effort at dismantling the Iraqi program notwithstanding. Libya's Qadaffi is known to covet nuclear weapons. Egypt is a wild card, having taken few steps so far in the direction of nuclear status and seeming to rely on chemical weapons as a "poor man's" nuclear deterrent. However, it possesses a scientific and manufacturing infrastructure equal to those in Iran and Iraq, albeit without the oil wealth to finance such an expensive activity. Saudi Arabia has the money for such an effort, but does not have the scientific critical mass, so that if it were to acquire nuclear weapons, it would most likely be on the basis of purchase from existing nuclear powers, perhaps China or Pakistan, whose nuclear programs have long been tied into a broader set of alignment links. Syria is still another possibility, but there appears to have been little movement toward a nuclear program, as Damascus has come to rely on its chemical weapons arsenal as a poor man's deterrent.

Iraq's massive effort to acquire nuclear weapons has received widespread attention. Some of its nuclear facilities were damaged or destroyed by the coalition air forces during the Gulf War and most of the remainder were destroyed under the aegis of the United Nations in line with the terms of the agreement ending hostilities. Now, of course, it is known that by 1990 Iraq had come very close to producing nuclear weapons, having constructed an elaborate and diverse nuclear infrastructure, which, to a surprising degree, had escaped detection by the major nations' intelligence services. It also escaped the notice of the IAEA, which, given the fact that Iraq was a signer of the nonproliferation treaty (NPT), had the responsibility for monitoring its nuclear facilities and applying safeguards to them.[17]

In 1981 the Israelis had destroyed the French-built Osirak reactor, a plutonium-producing facility that would have required a plutonium separation facility to pull bomb-grade material out of the reactor wastes.[18] After that bombing, however, the Iraqis apparently made a decision to acquire the bomb via the U-235 route. To that end, aided by a massive and sophisticated global procurement network under the direction of one Jafar Jafar, they made significant strides over a period of a decade, focusing on two complementary methods for acquiring bomb-grade U-235—calutrons (called baghdatrons in Iraq) and the gas centrifuge. This involved a massive complex, much of it hidden underground, of what Burrows and Windrem refer to as "a vast, heavily redundant superweapon network for the development and production of chemical and biological agents, ballistic missiles and atomic bombs."[19] The core of the Iraqi nuclear program, which, it is claimed, might well have had the capability to produce some twenty atomic bombs a year, is described as follows:

> The facility most critical to the nuclear weapons program was at Al Furat, south of Baghdad. There, Iraqi scientists had built, with the help of Interatom, a Siemens subsidiary, a workshop for the design and fabrication of centrifuges, which offer the most efficient, least expensive way to enrich uranium until it is bomb grade. The baghdatrons developed by Jafar were indeed ingenious. But they would limit Iraq to a paltry three bombs a year. Thousands of centrifuges operated in tandem arrangements called cascades would lead to an atomic bomb assembly line just like Israel's, which can produce twenty bombs or warheads a year. More important, the calutrons and centrifuges would work, not in parallel, but in sequence. The calutrons would do the hard work, enriching natural uranium from 0.7 percent to 20 percent, a process requiring brute force. The centrifuges would then take it the rest of the way to 90 percent, a relatively easy task.[20]

Still another, more recent report claimed that UN inspection teams found stockpiles of enriched material and major production and research and development facilities and equipment, including calutron enrichment equipment. There was also apparently evidence of two viable weapons designs, neutron initiators, relevant explosives and triggering devices, and plutonium reprocessing and centrifuge technology. One apparent core of the nuclear system was a centrifuge complex at Rashidya.[21]

There appears to be some disagreement among experts about the degree to which the Iraqi program has been dismantled, and about how difficult it would be to rebuild it. Presumably most of Iraq's uranium has been seized and most of its infrastructure destroyed, though the job done by the UN is not likely to have been as thorough as would have been the case if the coalition armies had marched north to Baghdad, captured the nuclear-related plants, and destroyed them before the defeated regime had the opportunity to remove and hide a presumably unknown proportion of its materials and equipment.

What Baghdad will do if and when its sovereignty over its territory is completely restored remains to be seen. Surely, alerted Western intelligence services are not likely to be as asleep at the switch as they apparently were in the years preceding the Gulf War. In addition Western corporations, in Germany and elsewhere in Europe, are not likely ever again to have the opportunity of circumventing export restrictions almost at will, mostly involving dual-use technology, as was the case earlier. That being said, a key point is that Iraq surely knows how to make nuclear weapons, so that if it can get its hands on some weapons-grade U-235 or Pu-239, it will only be a matter of time before it can deploy the bomb. Finally, there is the question of the possibility of either Israeli or Iranian preemptive strikes if Baghdad should be caught rebuilding its nuclear infrastructure.

As of this writing it is not easy to predict just how soon Iraq will manage to get itself out from under the UN-mandated arms embargo (not to mention the associated embargo on the oil exports that are needed to finance arms acquisitions), and once it manages to do that, just what the subsequent progress of its weapons acquisitions would be. Also unclear is the question of just how completely the UN has managed to unearth and destroy Iraq's variegated capabilities in the areas of chemical, biological, and nuclear weapons and ballistic missiles. There is clearly a wide band of disagreement on these questions among experts writing in the public domain, with some saying that those capabilities have been largely destroyed, others claiming that Iraq has managed to hide and conserve a substantial part of its previous capability, and still others, while leaning toward the former, who claim that because technological knowledge cannot be obliterated, Iraq would be able to reconstitute some of these capabilities in fairly short order once it got out from under the embargo. Apparently, Iraq has a very advanced procurement system for importing missile parts, high-tech furnaces, and guidance systems, and that pertains to the current situation, long after the end of the Gulf War. But the time line involved, relative to Iran's capabilities (which are not constrained by an embargo, but which may be somewhat constrained by U.S. efforts to dissuade countries such as Russia and China from supplying Iran with capabilities for weapons of mass destruction) is important, not only with respect to the fundamental question of which side might be inclined to begin a war, but with respect to possible ladders of escalation and escalation dominance if a war should erupt—something that might arise from escalation of a minor crisis or "accidental" event as well as from premeditation.

Iran now appears to be moving hastily and deliberately toward the acquisition of nuclear weapons, while not laboring under the kinds of restrictions that inhibit Iraq owing to the consequences of its military defeat.[22] As it happens there were earlier fears, back in the mid-1970s, about the nuclear ambitions of the Shah, seemingly reflected in the drive of his government to acquire some twenty nuclear power reactors and to acquire a minority share of one of the multilateral, gaseous diffusion uranium enrichment consortia in Europe. But after the revolution in 1979, it appeared that the Khomeini regime had abandoned the Shah's putative nuclear ambitions. But then the defeat in 1988 and, in particular, the effective use by Iraq of both Scud missiles against Iranian cities and chemical weapons against the Iranian ground forces, seems to have caused a change of heart in Teheran; indeed, the Shah's old nuclear program appears to have been revived in 1984. Since the end of the Gulf War it has been reported that Iran has moved full tilt toward acquisition of nuclear weapons and related delivery systems. It apparently shopped for large reactors in Germany, for enriched uranium in France, and for other types of nuclear technology in Brazil and Argentina, but was turned down. It was reported to have attempted to purchase fissile material from Kazakstan.

However, Iran has apparently had success in acquiring nuclear technology from China in particular, but also from Russia and Pakistan, with some reports claiming that Khan has been actively involved in assisting Iran. That assistance may have involved centrifuge technology. China and Iran, meanwhile, have signed a nuclear cooperation agreement that is expected to result in the building of research reactors and calutrons and perhaps a couple of 300-megawatt pressurized water reactors as well. Both Russia and Pakistan are training Iranian nuclear scientists and technicians. Russia has apparently agreed to sell four reactors (two in the range of 1,000–1,200 megawatts, two others around 465 megawatts). One source reports that there is already a growing nuclear cadre of 3,000 personnel at work at Iran's main nuclear center at Isfahan, and that there is a second secret facility near the Caspian Sea at Moallem Kelaieh.[23] The bottom line here appears to be that, as Iraq before it, Iran is developing a diverse nuclear program, but one centered on the U-235 route, involving both centrifuge and calutron technology. One estimate is that if Iran must learn to separate Pu-239 or produce weapons-grade U-235, then it might take five to ten years, but only one to two years if it can acquire such fissile material from the outside.[24]

The similarity to earlier developments in Iraq is inescapable, and no one doubts that the Iranians have at least equivalent scientific and industrial capability. But with the Western nations having been burned by the Iraqi case and hence forewarned of the possibility of repeat situations, Iran may have more trouble than Iraq did in devising a clandestine international procurement network, that is, there is a focus on the Iranian program that Iraq did not have to contend with. Iran also appears to lack the financial resources that Iraq had during the 1980s and up to the outset of the Gulf War. On the other hand, the Western nations, whether or not aided by Russia, may have only limited influence over Pakistan and China in stanching the flow of crucial assistance to the Iranian bomb program. If Iran does not provoke destruction of its program as Iraq did and if the United States does not try to forestall it as it is doing in the case of North Korea (although the effectiveness of that effort is not yet clear), it is hard to see Iran being prevented from moving eventually to operational nuclear weapons. Some observers have suggested that a worried Israel might try to preempt the Iranian program, in a repeat of the Osirak operation, presumably involving conventional methods only. But the factor of distance, the dispersion of the Iranian infrastructure, and the mere fact that an Osirak raid is likely to work only once because of the element of surprise, would appear to render this possibility an unlikely one.

Libya is not expected to acquire the capability to build nuclear weapons any time soon, even with the aid of foreign scientists.[25] It was earlier reported to have attempted to purchase nuclear weapons from China and perhaps also on the black market, presumably to no avail. Further west, however, there has been increased concern about the possibility of Algeria making a move toward nuclear weapons status. Around 1991 reports emerged about the building of a nuclear facility at Ain Oussera, south of Algiers in the foothills of the Atlas Mountains. This facility is said to involve a Chinese-supplied reactor rated at around 40 megawatts, which is larger than a research reactor and would have the capacity to produce enough bomb-grade material for two bombs a year.[26] A related reprocessing plant is suspected, and Algeria has low-grade uranium resources that render it autarkic in terms of nuclear fuel if it can develop a fuel fabrication capability.

Chemical and Biological Weapons

Although the prospects for the possession and even use of nuclear weapons is commonly considered the grimmest specter haunting the future of the greater Middle East region, there is the accompanying, daunting problem of the proliferation of CW and BW weapons, not only to the region's nation-states, but perhaps to terrorist groups as well. As measured by the seeming facts of existing or expected deployed weapons arsenals and the recent history of actual combat, CW weapons would now appear to be a much larger looming threat than BW weapons, even though many commentators continue to view the latter as ultimately more ominous in terms of the possibilities for massive casualties.

Unlike nuclear weapons, whose actual use has been restricted to the original instances of Hiroshima and Nagasaki, CW weapons have been used of late. In the 1960s Egypt apparently used gas warfare against Yemeni tribesmen,[27] and in 1967 there were unsubstantiated rumors about Israel having captured Egyptian CW weapons materials in the Sinai Desert, which, at any rate, were not used.[28]

The Soviets were accused of using "yellow rain" toxins in Afghanistan, as were their allies in Laos during the long Vietnam conflict.[29] Iraq clearly made extensive and effective use of CW weapons in the latter stages of the war against Iran and against the Kurds, producing the gruesome photos of the massacre at Halabja, as well as numerous other reports of CW use against Iranian troops and indigenous Kurds.[30] Some reports, whether accurate or not, asserted that Iran also used CW weapons in that war. Libya appears to have used CW weapons, dropped from aircraft, on several occasions in its war against Chad in 1986–88. It or its allied Sudanese government or both are further alleged to have used such weapons against antigovernment and non-Islamic rebels in the south of Sudan in 1988.[31]

The matter of CW weapons came to prominence during the second Gulf War, when it was feared that Iraq might fire Scud missiles armed with nerve gas warheads against Israeli cities (something that had openly been threatened by Saddam Hussein when he ranted on about consuming half of Israel by fire) or against coalition forces during the ground phase of Desert Storm. Indeed, it is now increasingly clear that captured CW weapons destroyed by U.S. forces were responsible for the mysterious illnesses acquired by some of the coalition forces. Adding all of this up, it was clear that the inhibitions against CW weapons use, perhaps long described in terms of international legal "norms," are eroding. There is now a complex linkage between nuclear and chemical weapons in this region, with the latter in some cases, as we have noted, constituting a poor man's deterrent against the former, but where the former have also been developed as a deterrent against the latter.

Which countries then are now considered to deploy significant and operational CW weapons, and which others may do so in the near future? One recent major work on this subject concludes that at least six Middle Eastern states —Iraq, Egypt, Syria, Libya, Israel, and Iran—probably have or are about to have significant CW offensive capability, if not actually deployed and ready-to-go weapons.[32] For reasons not entirely clear, neither India nor Pakistan, both of which have the easy capability of producing such weapons is strongly suspected of having offensive CW capability or even of significant development programs in this area. They, presumably, similar to the case of Israel, chose to go straight to the development of nuclear weapons rather than to the poor man's alternative apparently chosen by Iraq, Iran, Syria, Libya, and Egypt, at least pending the possible later development of the bomb. On the other hand, Israel apparently developed CW weapons long after having acquired the bomb, for reasons not entirely clear. One source speculates on the possibility that Israel may wish to retain the option of something short of nuclear weapons up the ladder of escalation and if faced with a CW attack.[33] It may also be that it wishes to possess CW weapons as a hedge against the possibility that it may be forced to abandon its nuclear weapons by international or American pressures.

Israel's CW program, such as it is beyond the obviously big effort at devising defensive measures against CW attack, is reported to involve a mix of nerve gas and blister agents, which it is claimed are produced in a facility that is part of the Dimona complex. There is little real open-source information, though one report at least points to mustard and nerve gas production near Dimona. Much more out in the open are the Iraqi and Syrian CW arsenals. Iraq's extensive use of such weapons during its war against Iran has been well documented, with Teheran having claimed that some 50,000 of its soldiers were killed, wounded, or disfigured by those weapons, and then there was the gruesome massacre of Kurdish civilians at Halabja. Since then the post–Gulf War investigations by UN teams of the Iraqi CW infrastructure have produced an almost astonishing picture of a massive program. One report claims that the CW production facility at Al Multhanna "turns out an estimated 3,000 tons a year of mustard gas and Sarin and Tabun, as well as trying to develop a gas called VX, a persistent nerve gas, and BZ, a hallucinogen."[34] Another referred to binary Sarin-filled artillery shells, 122-mm rockets and bombs, enough precursors to produce 490 tons of VX nerve gas, and a large VX production program. Prior to the Gulf War there had been flight tests of Scud missiles with chemical warheads.[35]

Emerging from the Gulf War was a report that when U.S. F-117A stealth aircraft blasted some Iraqi Tu-16 and B6D Badger bombers at an air base at Al Taqqadum, they were in the process of being loaded with CW weapons.[36] It is believed that CW-armed Scuds aimed at Israel were at the ready, albeit never fired. At the time of that war Saddam Hussein is reported to have had no less than 150,000 bombs, rockets, missile warheads, artillery rounds, and landmines as a stockpile of CW weapons.[37] Whether that stockpile has largely been destroyed or not or how quickly Iraq can reconstitute it in the coming years are interesting points of speculation. Furthermore, according to a Center for Strategic and International Studies report, Iraq's BW capabilities appear to have been successfully evacuated at the war's end and kept intact. Another source claims that Iraq had assembled "at least 191 biological missile warheads and aerial bombs," and that "17 tons of missing growth media were used to produce biological agents involving anthrax, botulinum, and enterovirus."[38]

Syria too is known to have long been stockpiling CW weapons. One earlier report back in 1983, apparently based on a U.S. Special National Intelligence

Estimate, stated that Syria has probably "the most advanced chemical-warfare capability in the Arab world," with the possible exception of Egypt.[39] In 1989 CIA director William Webster stated unequivocally that Syria was "stockpiling a variety of chemical warfare agents for various battlefield missions and producing and amassing a variety of munitions that can be used as delivery systems for chemical agents. . . . Syria has nerve agents in some weapons systems."[40] Some information has been published on stockpile locations, pointing to Sarin or Tabun nerve gas warheads deployed in caves and shelters near Damascus. One report says that experts think that Syria has stockpiled 500 to 1,000 metric tons of CW agents.[41] It may also have production capability for anthrax and botulism, BW agents. It has also been reported that some years ago, Israel may have considered a preemptive strike against the then burgeoning Syrian CW infrastructure.[42]

Earlier Syrian CW weapons may have been acquired from the USSR, but it is now apparent that Syria has indigenous development and production capability, and may even have transferred some CW weapons to Iran.[43] Generally the CW literature points to a three-way nexus of Iran, Libya, and Syria in this connection, paralleling these nations' long-held alliance ties dating back to the Iran-Iraq war.

Whether Libya now has deployable and operational CW capability is not entirely clear, but the evidence seems to indicate that it does. Its earlier use of CW weapons—presumably acquired from the USSR, but possibly from Syria—against Chad has been noted. More recently there was the *cause célèbre* related to the Rabta facility—dubbed by pundits such as William Safire as "Auschwitz in the sand" because of the strong role of German assistance—resulting in the famous "fire" that is now believed to have been set by the Libyans themselves as a ruse.[44] Rabta was uncovered and exposed, but it is still there and operating under (presumably) Libyan military control. In 1989–90 reports began to emerge about other possible CW weapons facilities such as one at Matan as Sarra on one of the roads connecting Chad and Libya, not far from the border.[45]

In 1996 newer and unambiguous reports emerged about a giant new Libyan nerve gas production plant at Tarhuna southeast of Tripoli, sometimes referred to as Rabta II. It was described as a giant underground plant carved into the side of a mountain where no spy-satellite eyes could see the factory inside and no American jets could destroy it. It was further deemed invulnerable to commando raids, and probably only a direct hit by a nuclear warhead on top of the mountain could take it out. There were numerous analyses in the press about if and how the United States might preemptively destroy this facility.

Egypt is long considered to have had a militarily significant CW arsenal and actually put such weapons to use some thirty years ago in Yemen. One report claimed that Egypt had a production plant for the nerve agent Tabun way back in the early 1970s, while others have claimed that Egypt and Iraq collaborated on CW weapons development over many years, perhaps in parallel with the now aborted development of the Condor missile.[46] Other reports are more ambiguous, and the Egyptian government has steadfastly denied building CW plants. Many observers assume that almost by inference, the Egyptians' abjuring of a serious nuclear weapons program points to their reliance on CW weapons as a poor man's atomic bomb, to counter Israeli nuclear weapons and Libyan chemical capability.

Then there is Iran, apparently caught at a disadvantage in dealing with Iraq's use of CW weapons during the latter phases of the Iran-Iraq war, but clearly determined to catch up, both with respect to atomic weapons and CW. According to Burrows and Windrem, Teheran has produced limited but significant quantities of chemical weapons—mustard gas, nerve gas, and hydrogen cyanide—at secret locations at Qazun, Al Razi, Bahshwir, and Damghan, east of Teheran.[47] The last is reported to be producing about five tons of nerve gas a month, which can be fitted into Scud-B warheads. Webster stated in 1989 that Iran had begun CW production in the mid-1980s, and that it was "stockpiling a variety of chemical warfare agents for various battlefield missions and producing and amassing a variety of munitions that can be used as delivery systems for chemical agents" (afterward defined as mustard, blood agents, and nerve agents).[48] Earlier, Anthony Cordesman had estimated that Iran had significant amounts of stockpiled mustard gas plus a stockpile of cyanide by around 1988.[49] A subsequent report claimed that nerve gas production began no later than 1994.[50] Since the end of the Iran-Iraq war it can be assumed that Iran has been moving full tilt in this area, with the lessons in mind of the havoc wreaked on its forces by Iraq, considered by many to have been one of the final precursors of the Iranian defeat in 1988.

Iran has also done extensive work on BW weapons, which perhaps will be as lethal as small nuclear weapons. One report claims that it may have the facilities for dry-storable weapons that could be mated to corresponding missile warheads and bombs.[51]

There is an increasingly looming threat of terrorist activities in this area. The nerve gas attacks on the Tokyo subways underscored the possibilities. Indeed, writers such as Laurie Mylroie have suggested that Iraq and Saddam Hussein may have been behind the World Trade Center bombings, raising the specter of a possible CW version of such an attack at a later date in another possible crisis between the United States and Iraq.[52]

Future Greater Middle East War Scenarios: Iran-Iraq and India-Pakistan

OTHER THAN the two major scenarios with which we have dealt in detail—Arabs versus Israel and a possible reprise of Desert Storm—the two most likely warfare scenarios in this region are those involving the other main conflict pairings: Iran versus Iraq and India versus Pakistan. Iran and Iraq fought before, in 1980–88, while India and Pakistan fought major wars in 1965 and 1971 in addition to that in the late 1940s.

A New Iran-Iraq War

Conjuring up scenarios for a future Iran-Iraq war, and placing them in a geographic context, is not a simple matter. On the surface it might be surmised that if such a conflict were to break out any time soon it might well be fought essentially as an instant replay of the lengthy and brutal war of 1980–88. That would amount to the prediction that such a war would take on the same character of stalemated attrition along a fairly stable front not much at variance with the present borders, and that it would feature the inability on both sides to use combined arms tactics that would allow for sustained offensives. But there may be reasons to question whether a new Iran-Iraq war would indeed be a repeat of the earlier one, particularly given the fact that both sides have had a lengthy period in which to rethink strategies and tactics and, of course, there has been the intervening Desert Storm war, which will have had an impact on the planning on both sides.

What appear to be the major changes that have already shifted (or may yet shift) the balance of military power or alter the potential nature of such a war, as measured by duration, casualties, the use of certain weapons, or the possibly increased fluidity and movement of the battlefield? First, of course, Iraq's defeat in 1991 and its difficulties in finding replacements for the weapons it lost would appear to have weakened its position vis-à-vis Iran. However, if Iraq succeeds in getting the embargo against it lifted and recovers, that situation might change. Iran, on the other hand, which in the previous war was largely reliant on the U.S. weapons acquired during the reign of the Shah, has shifted dramatically since 1988 to reliance on weapons from the former Soviet bloc nations, a trend that actually had been begun some years earlier when China, North Korea, and Libya were sources of arms. There are also some signs that the training, leadership, cohesion, and morale of Iran's military forces may have improved since 1988, if for no other reason than that the social divisions within the Iranian military will by now presumably have been moderated. In short, the huge advantage for Iran in GNP and population may now be a more accurate indicator of relative strength than it was in the 1980s.[1]

However, it may be important to recall that among the reasons for Iran's partial collapse in 1988 were Iraq's use of chemical weapons and its relatively more effective exploitation of Scud missiles against Iranian cities. On the other hand, as noted by James Thomas, Iran had no significant chemical weapons or delivery systems in that war and was thus unable to deter Iraq from such use.[2] With Iraq having been largely disarmed in these areas (to what degree is a matter of argument) and with Iran having obviously worked hard to acquire additional capabilities, it is also possible that there has been a considerable shift in the balance in these areas. Further, unlike the situation that obtained during the mid- to late 1980s, U.S. assistance to Iraq in the form of satellite reconnaissance photos would, presumably, no longer be available. That assistance was apparently invaluable to Iraq in allowing it to know in advance just where along the long battlefront the next offensive would be, so that it could position its defensive forces accordingly.

The political relationships of both Iran and Iraq have changed since 1988, which has important ramifications for the crucial matter of arms transfers. Iraq may have less access to Russian military systems than it had in the 1980s (though that was never a consistent access and it had to rely on Egypt's Soviet-origin weapons, other Warsaw Pact sources, France, and Brazil, among others), while Iran had a variety of clandestine sources including Israel, China, North Korea, and Vietnam. It is not easy to say what would happen if another war should erupt, but clearly Iraq's political position with respect to arms sources appears to have worsened. One critical question would be whether Iraq, if it were threatened by Iran, could repair its relations with Egypt and Saudi Arabia, in particular, so as to obtain some of the former's Soviet-bloc stores and rely on the latter's political clout and money in order to acquire arms.

Neither Iran nor Iraq appears to have moved forward substantially in the principal areas of the emerging revolution in military affairs. The two key questions are, first, whether either side has developed the capability to conduct a combined arms offensive that would take a lot of territory and could transcend to some degree the factors of strategic depth mentioned earlier (Iraq is more vulnerable in this sense), and, second, whether either side could use missile bombardments or weapons of mass destruction or both to force a relatively early decision in a war.

Iraq's forces were, of course, decimated during the Gulf War, though the extent of that may have been exaggerated at the time. On paper the remnants still constitute a formidable force, though one of the lessons of the Gulf War was to interpret paper force structures with caution. Of course it is not easy to say just when or to what extent the UN-mandated embargo on Iraq will be ended, and if it is, just how quickly and to what extent Iraq will be able to rebuild its forces back to the levels that existed in 1990. For that reason short-run scenarios would be predicated on an Iraq badly weakened vis-à-vis Iran, but in the longer run—particularly given Iraq's always strong potential as an oil exporter and the possibility of using the proceeds from such activity to make large arms purchases—projection is more difficult.

By the mid-1990s neither Iran nor Iraq was in very good shape, at least to the extent that either might contemplate a major new round of fighting, but the relevant question is still how their respective strengths compare one with the other. All in all, particularly because it still had access to outside sources of weaponry, Iran appeared to be in far better shape. But neither appears to have moved very far toward acquiring the capability to conduct combined arms operations. Iraq has been badly decimated, quantitatively and qualitatively, in those areas that gave it a slight advantage in the fighting toward the end of the previous war. Iran's more recent acquisitions of advanced Soviet MiG-29s and Su-24s may give it the potential to fight more effectively in and from the air.[3]

It is hard to say what might trigger another war between these two countries. Iran is, of course, psychologically speaking, a revisionist power in this context, still stewing in humiliation over the loss of a war to a smaller power that it had long thought its military inferior. But Iran's religious fervor seems to have peaked and dwindled, and to the extent that the Islamic regime must be concerned about internal discontent and the possibility of a counterrevolution, it might be very reluctant to engage in another major war. Or could a beleaguered regime think about such a war so as to divert public attention from its internal problems? Iraq too would be thought of as exhausted, broke, and rent with internal political fissures to the extent that Saddam Hussein might not any time soon contemplate a new round, one in which (as he won the previous round) he might have little to gain and much to lose.

The diplomatic or political context of a new round between Iraq and Iran is not easily predictable. Unlike the Arab-Israel context there appear to be few possibilities for the addition of other regional actors to one or both sides of the conflict. Pakistan might or might not back Iran. Syria, with its newly found ties to the United States, might or might not support Iran again to the extent that it did in the previous war, nor might Libya. A big question mark is whether or to what extent Saudia Arabia and Egypt, both having fought against Iraq in the Gulf War but both still worried about the implications of an Iraqi defeat by Iran, would weigh in again on the side of Baghdad, with implications involving weapons supply, financial help, and pressures on the United States to be something other than neutral. Interestingly both Iraq and Iran now field military arsenals that are primarily Soviet-Russian in origin, with Iran having made many recent acquisitions from Moscow, and with Iraq relying on a largely Soviet-supplied arsenal that dates back into the period before the Gulf War. If a war should erupt, would Russia mount a resupply operation on behalf of Iran only, or would it be under pressure from the United States not to do so, and would that pressure be effective? Would Iraq also get some Russian spares, ammunition, etc., perhaps only because the Russians need the cash, or would it be forced again to try to obtain such equipment from other former Soviet client states and allies that still have sizable arsenals of older Soviet equipment, countries such as those in Eastern Europe, Egypt, Vietnam, or North Korea?

With regard to the possibility of arms resupply operations, particularly if mounted from Russia, either for Iran alone or for both sides, there are some interesting geographic problems that arise in view of the fact that Russia no longer has a contiguous border with Iran, much less Iraq. In order to mount an airlift to Iran, Russia would presumably require overflight permission from one or more of the Central Asian states, including Turkmenistan, which, to the extent that it identifies with a fellow Muslim state, might be inclined to deny it. In the past the Soviet Union apparently ignored such restrictions and went ahead with unauthorized overflights, as it was rumored to have done during the Horn war of 1977–78.[4] Otherwise, if overland and overhead access were denied, one might even imagine a Soviet sealift of sorts to Iran through the Caspian Sea.

The geography of arms resupply to Iraq might also involve some interesting new problems, depending upon whether there is a repeat of the situation in the 1980s, when Baghdad was availed of access by Jordan, in particular, and also by Saudi Arabia. Whether in a new Iran-Iraq war, begging the question of where the United States might stand, Iraq would be availed of overland routes of access for arms through the Jordanian port of Aqaba is not clear, nor is the prospective position of Saudi Arabia in such an eventuality. Also unclear are the probable stances of Syria and Turkey. Syria sided with Iran in the previous conflict and then with the U.S.-led coalition against Iraq. In the future it is possible that Arab solidarity against Iran might drive Syria and Iraq together under altered political circumstances, in which case Baghdad's problem of access routes for arms resupply, whether overland or overhead, would be substantially eased.

Quite probably a new confrontation between these two countries would be another static war of attrition. But it would appear that there is some chance that Iran might, in the near future, hold somewhat of an advantage over the decimated Iraqi forces, based on its more extensive recent acquisitions of arms from Russia. Surely, there would be an extensive alteration of the situation that existed in the previous war. Ephraim Karsh states that from the Iraqi perspective, that war had been "conducted under ideal circumstances, with superior firepower and complete mastery of the air," and was one in which the sustained air campaign against civilian targets gradually led to the collapse of Iranian national morale in 1988.[5] Again the former advantage given to Iraq in conducting its mobile defensive operations by U.S. reconnaissance satellite data would presumably not be repeated, almost certainly not as long as Saddam Hussein's regime remained in power. On that basis might Iran be able to conduct an offensive into the Iraqi heartland? If so it would immediately bring into play Iraq's vulnerability in strategic depth. As we pointed out earlier, Baghdad is not far from the Iranian border, and the lines of communication between the capital city and the key southern area around Basra could easily be cut by a short Iranian offensive thrust, which, if successful, could quickly cut the Iraqi forces in half, not to mention the implications for continued Iraqi control of its Shi'a populations in the south. But even such an offensive by Iran might require an improved capability for combined arms offensive operations; specifically the use of air cover for a sustained armored thrust.

One other interesting geographic problem is whether there would be an altered impact in connection with the long frontiers between the two nations, the matter of frontal width. In the previous war, with both sides alternating meat-grinder offensives, the main theater of battle moved back and forth from the southern to northern to central fronts, with each side probing for a weak spot and shifting its defensive forces in response to anticipated offensive moves by the other. But neither side could make a really decisive breakthrough, at least not until the Iranian forces began to crack in 1988. However, there is a question now as to whether with both Iran and Iraq fielding smaller forces than they did during the earlier war, might that result, in view of the length of the frontier, in a more fluid battlefront, where at least limited offensive breakthroughs might be possible, the lack of armor on both sides notwithstanding. In a related vein there is the question of whether either side might be able to achieve strategic surprise through a quick mobilization and a sudden offensive.

If neither side were availed of satellite reconnaissance assistance (the United States would probably not be willing to help Iraq again, but there is the question of whether Russia might extend such assistance to Iran and if so would the United States be inclined to react, perhaps under pressure from Saudi Arabia and Egypt), it might be easier, say, for Iran to launch a massive offensive with some degree of surprise, so that Iraq would not be able to shift its defenses in time. Further, if either side were availed of a major power's satellite reconnaissance information during normal times, it would then be able to anticipate and deal with a strategic surprise mobilization and attack by the other. However, there is also the question of whether chemical weapons might be used earlier on and with more frequency in another war, and that too might increase the chances that one side or the other might be able to achieve a breakthrough and a decisive victory early on, hence avoiding a lengthy war of attrition.

That brings us to a discussion of weapons of mass destruction (WMD), which involves some very complex questions having to do with various war scenarios and, particularly, escalation dominance. It also involves the matter of time lines, specifically a complex arms race in which both sides would presumably be attempting to acquire and make operational various kinds of chemical and biological weapons, nuclear weapons, and associated delivery systems, most important, perhaps, those in the ballistic missile category but also strike aircraft and cruise missiles.

At present it is not easy to predict how soon Iraq will manage to get itself out from under the UN-mandated arms embargo (and the associated embargo on the oil exports that are needed to finance arms purchases), and once it manages to do that, just how rapidly it would acquire weapons. Also unclear is just how completely the UN has managed to unearth and destroy Iraq's chemical, biological, and nuclear weapons and its ballistic missiles. One recent article reported on the opinion of Rolf Ekeus, the Swedish official in charge of the UN effort to strip Iraq of its weapons of mass destruction, that Iraq has a very advanced procurement system for importing missile parts, high-tech furnaces and guidance systems, and that referred to the current situation, long after the end of the Gulf War.[6]

Indeed, it is even possible that an attempt at preempting nascent nuclear facilities might be made in this context. An effort by Israel to preempt Iranian nuclear infrastructure cannot be entirely ruled out, nor can an effort by Israel to repeat the raid of 1981 if Iraq should again make a strong move toward nuclear acquisitions. But for reasons noted earlier, that is, the difficulty of achieving surprise a second time and the obvious efforts by both Iran and Iraq to move their WMD programs underground and to disperse them as well as to conduct redundant programs using multiple avenues of approach particularly to nuclear weapons, such a preemptive move seems somewhat unlikely. But if Iran should acquire nuclear weapons ahead of Iraq, it might be tempted to preempt the latter's nuclear facilities, to the extent it deemed such an operation technically feasible. Or a nuclear-armed Iran, fearing that Iraq would develop matching capability, might be sorely tempted to mount an all-out conventional assault in an attempt to capture Iraq's nuclear and other WMD facilities on the ground. Most are located in a periphery around Baghdad, and as we have noted, Iraq does not have much strategic depth in front of its capital city.

If a conventional war should begin, and if Iran should appear to be on the road to some degree of victory, Iraq would presumably be very tempted to move to the use of chemical weapons. Indeed, its use of such weapons in the earlier war might constitute somewhat of a deterrent against Iran's initiating such a conflict, all the more so as Saddam Hussein has threatened to use such weapons, as pointed out by Karsh with reference to his televised interview with Peter Arnett during the Gulf War.[7] But this time Iran might be equally if not more heavily endowed with chemical weapons. If this were mutually recognized it might result in some level of escalation dominance in favor of Iran, to the extent that rational decisionmaking was in evidence on the Iraqi side, something not necessarily to be counted upon, particularly if the regime's political survival was at stake. In addition, the battlefield implications of extensive use of chemical weapons by both sides may not be easily predictable, that is, whether it might precipitate breakthroughs and routs or lead to a continuing stalemate with very high casualties. If nuclear weapons were available on one side or the other, there would also be the question of whether there might be an escalation to that level, particularly by the side that was being disadvantaged on the conventional battlefield by the use of chemical weapons.

At present neither Iran nor Iraq would appear to have the number of conventionally armed ballistic missiles that would allow for an all-out decisive counter-cities terror attack. If, however, either or both could acquire, let us say, a missile inventory that was equal, at least quantitatively, to what we now know Iraq had at the beginning of the Gulf War (1,000 missiles and several hundred launchers), and if the accuracy of these missiles could be improved to the point of even modest CEPs (circular error probabilities) that would allow for reliable targeting of the rival capital and other major cities, that could be a major factor in a new conflict. Iraq's missile barrage against Teheran in 1988 took a big toll on Iranian morale. Attacks in both directions, perhaps on a much larger scale, would be not only devastating, but might also trigger escalation of the war, perhaps to the level of chemical weapons. As previously bruited, escalation to that level might occur early on in a new conflict, rather than in a very gradual manner determined by the technological shortcomings of both sides as in the 1980s. Escalation to chemical weapons could begin as a response to conventional terror attacks on cities. It could also be part of an escalation ladder driven by defeats on the conventional battlefield and fears of the collapse of military forces on one side or the other. With the improved ranges of the missiles on both sides, the distance factor has ceased to be important. In addition, unlike the United States or Israel, neither Iran nor Iraq would appear to have the capability in the foreseeable future to use its air force for the purpose of finding and destroying the other side's missiles. At the bottom line are the questions of the extent to which either or both sides can produce relatively accurate missiles in quantity (or acquire them from the outside) and what would be the relative timing of such acquisitions in relation to other relative developments of military capabilities.

War between India and Pakistan

The long-run specter of large-scale warfare between India and Pakistan remains and must be viewed with the assumption that both countries have acquired arsenals of operational nuclear weapons. True they have not fought for nearly a generation, but there have been some serious scares in recent years in which both sides anticipated the eruption of full-scale combat. Most important, there is the continuing and unresolved problem of Kashmir. Both sides continue to rearm, and that arms race has long included a nuclear component. The situation between the two nations is tense and is likely to remain so as long as the issue of Kashmir is unresolved.

Whether and to what extent both sides still plan for another round of conflict is in some dispute. Indian forces are reported by one source to be short on readiness and maintain only a few weeks' worth of ammunition and other materiel necessary to prosecute a war with Pakistan, because the Indian government reputedly recognizes that any new war would quickly be brought to an end by the major powers.[8] However, the old Soviet Union that once served as a mediator in this conflict no longer exists, and as India moved away from its former effective alliance with Moscow that entailed, among other things, extensive Indian reliance on Soviet weapons, the new Russia would appear to be much less of a stabilizing force in this context. There are few indications that the United States would act if Pakistan were on the verge of dismemberment, though the specter of the use of nuclear weapons might change some of those assumptions. Generally speaking, however, the end of the cold war and the end of the connection of the superpowers' rivalry to the South Asian conflict may have lessened the chances for major power pressure to bring an end to a new war.

The possible constellation of alignments surrounding a renewed conflict in South Asia seem largely repeatable, stable, and easily extrapolated from the past. Such a conflict would probably not draw in other nations as active combatants. Iran would not likely be able or willing to weigh in on the side of Pakistan, even if its sympathies were on that side. India has no foreseeable allies, unless in a wildly hypothetical scenario, Iran were to join Pakistan and Iraq were to weigh in on the side of India, Muslim solidarity notwithstanding. The United States and Russia would probably remain more or less neutral, although they might apply some pressure to foreshorten a conflict, which would likely take the form of an embargo on arms resupply. The wild card here would be China. Relations between India and China have improved in recent years, but they are still rivals and China remains a major supplier of conventional weapons to Pakistan as well as a significant source of technology for its nuclear and ballistic missile programs. There is some chance that China might come to the aid of a Pakistan on the verge of military defeat by India, hence triggering off a far more dangerous conflict the ramifications of which could not be easily predicted.

The most likely scenario for a another full-scale India-Pakistan war would not be a repeat of the 1965 and 1971 conflicts since Bangladesh, a.k.a. East Bengal, has been eliminated from the equation and with it India's problem of planning and mobilizing for a two-front war. But so too has Pakistan's logistical nightmare of having to supply forces across India. In both of those wars there was heavy fighting in the Punjab near Lahore, particularly in the area around Sialkot and Chhamb, and some more limited fighting further south in the Great Indian Desert and in the salt flats and marshes of the Rann of Kutch, as well as in the air and in waters immediately offshore. In neither war, at least on India's western front, was there a successful, sustained offensive well into the others's territory, despite India's large advantage in size of forces. Perhaps another war would repeat this pattern, resulting in a limited Indian victory, some Pakistani loss of territory, and the cessation of the war after a few weeks because of exhaustion and possible outside pressures and, now, because both sides might fear escalation to nuclear weapons. Such a war could be triggered not by a premeditated strategy, but rather because of crisis escalation, perhaps involving Kashmir, or internal upheavals within India, or Indian allegations of Pakistani involvement in insurrections in the Punjab or elsewhere, or the reverse regarding internal problems within Pakistan. Hence partial mobilizations on both sides might trigger more major ones and then perhaps a conflict caused by one or the other side deciding to teach a lesson to its rival or to mollify internal political pressures with a limited war.

However, a premeditative war cannot be ruled out. Given the massive force imbalances involved, it is difficult to imagine under what circumstances Pakistan would launch a major, premeditated war. Bangladesh is gone, and Pakistan does not covet Indian territory or control over Indian populations. It would also have little hope of getting at India's nuclear infrastructure with a conventional assault. India, on the other hand, has long been accused of wanting to dismember multiethnic Pakistan into its several constituent parts, so as to completely eliminate Pakistan as a major player in the region. But it is not likely that it would want, on a permanent basis, to control even larger Muslim populations. However, an India frightened by the possible implications of Pakistani nuclear weapons, particularly if more radical Islamic political forces should come into power, may continue to dream of doing away with Pakistan's nuclear arsenal. On the other hand, one might argue that the opportunity for doing that with a preemptive strike would appear to have passed, and the agreement between the two nations on that score would appear to have ratified that "fact." But it is nonetheless a partly plausible scenario, one in which India would use its conventional superiority to drive deep into Pakistan in order to capture and destroy its nuclear weapons and infrastructure, in a situation where the relatively larger Indian nuclear arsenal would be assumed to counterdeter any Pakistani threat to use nuclear weapons. But then, finally, a large-scale conventional war could possibly occur on the basis of escalation from a merely minor crisis, and India might decide, the pressures of outside powers notwithstanding, to go much further than merely teaching its neighbor a "lesson."

India can be described as one of the few modernizing countries that, despite its modest level of per capita GDP (gross domestic product), has made extensive strides in the direction of acquiring, absorbing, and even producing some of the new military technologies that comprise the MTR-RMA. It has, of course, designed and built a number of surface-to-surface missiles, ranging from battlefield systems all the way to medium-range ballistic missiles, and a space launch missile that may be reconfigurable at intercontinental ranges. India has put satellites into space, and one recalls Robert Butterworth's comments in a previous chapter to the effect that there is "no MTR without space." At present India has deployed some reconnaissance satellites (it has launched five since 1991), albeit perhaps only with general area coverage with remote sensing cameras on the Landsat model, and without serious capabilities for delivering real-time and useful tactical intelligence.[9] But it is moving ahead on that front, as well as toward some capability for communications and signals intelligence satellites.

India's defense planners have, of course, absorbed the lessons of the U.S. tour de force in the Gulf War, noting in particular the fast tempo at which that war was conducted.[10] But they are also aware that they cannot match the United States when it comes to MTR technologies, which is, at any rate, a somewhat irrelevant point because they do not anticipate having to fight the United States. Mostly they are interested in developing some degree of MTR capability with Pakistan and China in mind, but also in some asymmetric technological developments that might be useful if, unexpectedly, the United States were on the other side of the line.

But the lessons of the Gulf War apparently did initiate a debate among Indian defense specialists, perhaps similar to the earlier debate in the United States over AirLand Battle which, in retrospect, can be seen as having been largely over the efficacy of the weapons and tactics it later used in the Gulf War. The debates in India over the lessons of that war have been all the more serious because, on the one hand, the Indian military had long been modeled on Soviet as well as British doctrine, and, on the other, the Iraqi military had also relied on Soviet doctrine as well as weapons. Indian analysts saw the Iraqis as having failed miserably in the use of armored forces, and in looking at the success of U.S. forces, they saw the MTR weapons as having given advantages to the offensive side and to maneuver over mass. The value of a volunteer army over one based on conscripts was also noted. But at the same time many of those analysts saw India as not well suited to compete in a technological-qualitative arms race, with the important exceptions of a few "strategic enclaves" such as ballistic missiles and nuclear weapons. Indeed, before the Gulf War, India had seen itself as a technological peer of Egypt and Iraq, and then saw the war as having demonstrated the vast gap between the first and second tiers of military powers.[11]

Indian analysts were perhaps being too modest on this score. Numerous outside observers had long noted that India was the only country other than Israel among the third world nations that had demonstrated the capability to conduct a sustained combined arms offensive, as it did in East Bengal in 1971. There was an awareness that India's British-trained senior officer corps was indeed a big cut above its peers in countries such as Egypt and Iraq, and that that was important even if India's Soviet-supplied weapons arsenal was very similar to theirs. Neither Iraq nor Egypt has come close to India's achievements in space and in nuclear technology, and the latter's huge reservoir of computer programmers and technicians, now a major factor driving relocation of some Western computer firms, is a further indicator of its ability to compete in the new world of the MTR.

Further, Indian analysts emphasized not only the advantage given to the offense over the defense by MTR technologies, but also the role of technological surprise. One analyst claimed that "technological surprise may evolve as the preeminent factor influencing the outcome of a conflict."[12] Particularly as regards reconnaissance-strike complexes and C^3I (command, control, communications, and intelligence) technologies, it was further averred that these technologies "may be expected to tilt the balance of military power heavily in favor of offence, thus placing an increasing premium on mobility, firepower, and offensive action."[13] On a more specific level, Indian Gulf War analysts were impressed with the use of stealth aircraft, with the increasing vulnerability and lack of utility of static defenses (long favored by Indian doctrine as borrowed from the USSR and with Pakistan and China in mind), with the superiority of U.S. tanks (particularly infrared equipment and tank armor) with the invidious comparison of the U.S. Abrams with the Soviet T-72 that is the backbone of the Indian armored forces, and with the importance of light armor and self-propelled artillery.[14] Above all there was the obvious lesson of the importance of combined arms, particularly regarding close air support in destroying armor.

With these lessons in mind the Indian military is apparently working hard with regard to new acquisitions and tactical applications and on new communications and intelligence capabilities. Particularly it is trying to improve its electronic warfare capabilities, having noted how the U.S. forces had used ELINT (electronic intelligence) and countermeasures aircraft to jam and confuse Iraqi radars and to render their fire control systems naked. With war against a superpower in mind and requiring an asymmetric strategy of the weak, India is apparently concentrating on ways to combat superior electronic warfare capabilities and stealth aircraft. But above all the Indians were impressed with the extent to which airpower had become the dominant factor on the battlefield, noting the critical role of U.S. air standoff systems and the ability to conduct round-the-clock operations provided by newer systems such as the U.S. LANTIRN (low-altitude navigation and targeting infrared for night). Hence according to one analyst attributing the views of Indian military planners, "precision weapons like cruise missiles and laser bombs, aided by satellite surveillance which can give exact information of strategic targets, have a telling effect on the military power of the adversary."[15]

Pakistan will have drawn many of the same conclusions from the Gulf War, and its military planners are also trying to figure out how to acquire some of the emerging MTR capabilities.[16] It is, of course, somewhat hampered in that respect by the embargo on U.S. military technology, now being loosened, by an overreliance on a technologically backward China, and by the absence of the kind of technological and scientific base that has allowed India to develop and produce a range of its own weapons systems, albeit many of which rely heavily on imported technology.

What then might be the impact of all of this on respective Indian and Pakistani military developments and strategies and on various possible scenarios for future combat between the two countries? What then is the connection to the obvious facts of the geography of the conflict, involving such issues as strategic depth, the frontal width of the frontiers, terrain, and weather?

As we have noted India has planned on relatively limited wars presumed to be fought out where most of the fighting took place in the previous two wars in the Punjab, in front of Lahore, and south of the bordering mountains to the north. In the past airpower was used primarily to deal with the other side's air forces, on the ground and in the air, but with both sides making only limited use of ground-support air operations. Similarly, neither side was able to do very much in the way of large-scale degradation of the other's industrial or communications infrastructure. The wars were fought at a level of sophistication in between what the United States and Israel were able to do in 1991 and 1982, respectively, and what Iran and Iraq were able to do to each other. But in the future in South Asia, that could change.

Several major questions are posed here. One is whether India can move forward with the development of its air force (both in absolute terms and in relative terms vis-∞-vis Pakistan) so as to achieve total or near-total dominance of the air, in imitation of what Israel and the United States have done in their most recent wars. India, however, would have to do that with a still largely Russian-supplied air force, supplemented by its own designs and some French aircraft. Whether India can add enough effective technology of its own to these systems so as to provide for a dominant capability first in control of the air and then with respect to air-to-ground operations is a crucial question. Pakistan, even if the U.S. F-16 pipeline were reopened and with it access to some other critical systems that could be placed aboard these platforms, would appear to be fated to second place in this race. But could Pakistan prevent India from controlling the air completely, and hence the ground war as well? A related question involves the possibly increased use of surface-to-surface missiles as terror weapons against cities, in imitation of the Iran-Iraq equation. So far, neither side appears to have acquired a significant quantity of such weapons.

A second major question would appear to be whether India, banking on its larger forces, and relying on its interpretation of the lessons of the Gulf War, particularly those pertaining to mobility and the advantage of the offensive, will begin to think more in terms of using the whole length of its frontier with Pakistan. Rather than fighting meat-grinder battles again in the area between the border and Lahore against layered Pakistani defenses and relatively equal-sized forces, could it succeed in mounting major offensives further south in the Great Indian Desert or out of the Rann of Kutch marshes so as to spread the less numerous Pakistani forces thin? This, of course, might run against the apparent Indian doctrine that stresses a short and limited war, but actually that doctrine runs against the grain of everything now being learned from the Gulf War that has to do with factors such as mobility, round-the-clock warfare, and preparation of the battlefield from the air for a massive ground offensive. India here appears to have a fundamental dilemma involving the relationship between its basic strategy and presumptive war aims, on the one hand, and the realities of the emerging battlefield, on the other. On a more specific and narrower level, it might also raise the question of whether India (or Pakistan) might be inclined to take into account the role of the seasons and weather, reminiscent of the previous Indian offensives in December at a time the winter weather makes the mountains impassable, desert warfare in the south relatively more manageable, and the monsoons are merely looming. Whether these considerations will be affected by acquisitions on both sides of MTR technologies raises interesting questions about who is advantaged by what and when.

The possibility that India might attempt to make use of a broader front for an offensive raises some interesting questions with regard to satellite reconnaissance. We noted earlier the big advantage that Iraq had (one could also point to U.S. satellite reconnaissance assistance to Britain during the Falklands War) because of its access to U.S. satellite photos, which apparently enabled it to move mobile strategic reserves in response to impending Iranian offensives that were shifted around to different parts of the long front. There were rumors that the United States provided similar assistance to Israel in 1973, which may have enabled it to find the gap in the Egyptian defenses that provided for the crossing of the Suez Canal after the battle of the Chinese Farm. India will have some satellite capabilities of its own in the near future, but perhaps only later will they be good enough to provide this kind of information. Pakistan is unlikely to have that kind of satellite capability on its own any time soon. But during a war, would it receive assistance of this sort from China? Might Russia provide that sort of assistance to India, its long-time arms client, even if Pakistan is now also purchasing some Russian military equipment and the cold war alignments are more or less voided? Would the United States provide such assistance to either side, say Pakistan, under pressure from Pakistan's friends in Saudi Arabia? Such information might be critical, let us say, if the Indian army was stalled in front of Lahore and decided to attempt a large-scale offensive further south in the desert area, the success of which might depend on achieving strategic surprise.

There might also be some interesting geographic aspects to arms resupply, if such a war were to last long enough to bring that factor into play. Russia, if it were to mount an airlift on behalf of India, would have to overfly some of the countries in Central Asia. So might China if it were to attempt an airlift to Pakistan, despite the small corridor connecting the two countries. A sealift by China on behalf of Pakistan would, presumably, run into the problem of Indian sea control off Karachi, virtually a given in a future war. Whether Pakistan could obtain supplies routed through Iran in the face of the Indian navy is another question, as is the issue of whether Pakistan could be resupplied by air from friendly countries such as Saudi Arabia and Egypt if India should achieve near total control of the air space over Pakistan. Overall the prospects for Indian control of the air and of the sea offshore from Pakistan would appear to create a difficult problem for the latter with regard to arms resupply in the case of a more protracted conflict than those of 1965 and 1971. Presumably the Pakistani military is aware of this and has planned for a longer war with respect to its readiness stocks and spare parts.

If India were to succeed in doing to Pakistan what the United States did to Iraq in Desert Storm in a future conflict, and particularly if it could succeed not only in destroying extensive elements of Pakistan's infrastructure but also in moving its ground forces forward to invest Lahore and Karachi, and perhaps beyond to threaten Islamabad (and the Kahuta nuclear center), Pakistan would be faced with a strategic dilemma not totally dissimilar to Israel's hypothesized last-resort scenario. It could, of course, attempt to bring outside pressures to bear to halt the conflict, banking on India's presumed disinclination to control still more millions of Muslims. But what if India had dismemberment of its rival in mind, or the destruction of the heart of its nuclear capability? Would Pakistan's threat to go nuclear be credible in the light of the much larger Indian nuclear forces and the near impossibility of Pakistan's achieving any kind of counterforce nuclear strikes, all the less likely if its air force had been badly attrited in the preceding part of the war? One major question is whether Pakistan might consider CW or the use of tactical nuclear weapons as intermediate rungs on the ladder of escalation. Indian and Pakistani CW developments have not received as much attention as those in Iraq, Iran, and Syria, but they are presumably there, although in what kind of military sub-balance is not known here. Threat of tactical nuclear use by Pakistan might require a large nuclear arsenal, so as to involve a backup at a more strategic level, and here, too, India's forces would presumably have some level of escalation dominance. It is a strategic no-man's land, raising anew the issues addressed by van Creveld that we noted earlier, about whether such a nuclear specter would, a priori, lower the probabilities of such a conflict or the pursuit of it beyond a limited stage merely in repeat of the 1965 and 1971 wars.[17] The roles of China and the United States in such scenarios, working either together or at cross-purposes, would be critical—in the realm of the unknown.

Notes

Introduction

1. See, for instance, Stanislau Lunev, "Future Changes in Russian Military Policy," *Prism,* February 9, 1996.

Chapter 1

1. Walter Russell Mead, "On the Road to Ruin," *Harper's,* vol. 280, no. 1678 (March 1990), pp. 59–64; Jeffrey Garten, *A Cold Peace* (Times Books, 1992); Edward Luttwak, *The Endangered American Dream* (Simon and Schuster, 1993); and Lester Thurow, *Head to Head: The Coming Economic Battle among Japan, Europe and America* (William Morrow, 1992).

2. Max Singer and Aaron Wildavsky, *The Real World Order* (Chatham, N.J.: Chatham House, 1993).

3. Richard Rosecrance, "A New Concert of Powers," *Foreign Affairs,* vol. 71, no. 2 (Spring 1992), pp. 64–82.

4. "The New World Order: Back to the Future," *The Economist,* January 8, 1994, pp. 21–23.

5. Paul Kennedy, *The Rise and Fall of the Great Powers: Economic Change and Military Conflict from 1500 to 2000* (Random House, 1987).

6. See, in particular, the various writings of George Modelski, especially his *Long Cycles in World Politics* (University of Washington Press, 1985), and also Joshua Goldstein, "Kondratieff Waves as War Cycles," *International Studies Quarterly,* vol. 29 (December 1985), pp. 411–44, and W. R. Thompson, ed., *Contending Approaches to World System Analysis* (Beverly Hills, Calif.: Sage, 1983). A good

review of the long cycle literature is in Richard Rosecrance, "Long Cycle Theory and International Relations," *International Organization,* vol. 41, no. 2 (Spring 1987), pp. 283–301.

7. Oyvind Osterud, "The Uses and Abuses of Geopolitics," *Journal of Peace Research,* vol. 25, no. 2 (1988), pp. 191–99.

8. Ibid., p. 193.

9. David Wilkinson, "Spykman and Geopolitics," in Ciro Zoppo and Charles Zorgbibe, eds., *On Geopolitics: Classical and Nuclear* (Dordrecht, Netherlands: Martinus Nijhoff, 1985). More recently one article posited the importance of pivot states (as contrasted with pivot regions) to evolving U.S. security interests in the 1990s, in which Algeria, Egypt, Turkey, India, and Pakistan were represented in the greater Middle East region, along with Mexico, Brazil, and Indonesia. See Robert S. Chase, Emily B. Hill, and Paul Kennedy, "The Pivotal States," *Foreign Affairs,* vol. 75, no. 1 (January–February 1996), pp. 38–51.

10. See Saul Cohen, *Geography and Politics in a World Divided* (Random House, 1963), esp. pp. 83–87.

11. James Fairgrieve, *Geography and World Power,* 8th ed. (University of London Press, 1941).

12. Cohen, *Geography and Politics in a World Divided,* p. 233.

13. Samuel Huntington, "The Clash of Civilizations," *Foreign Affairs,* vol. 72, no. 3 (Summer 1993), pp. 22–49.

14. Ibid., p. 22.

15. Ibid., p. 23.

16. Ibid., p. 31.

17. Ibid., pp. 31–32.

18. Ibid., p. 39.

19. Fouad Ajami, "The Summoning," *Foreign Affairs,* vol. 72, no. 4 (September–October 1993), pp. 2–9.

20. John L. Esposito, *The Islamic Threat: Myth or Reality?* (Oxford University Press, 1992), especially pp. 168–212.

21. Samuel Huntington, "Response: If Not Civilizations, What?" *Foreign Affairs,* vol. 72, no. 4 (November–December 1993), pp. 187–94.

22. Cited by Colin S. Gray in his important article "The Continued Primacy of Geography," *Orbis,* vol. 40, no. 2 (Spring 1996), pp. 247–59.

23. These issues are covered in more detail in chapter 3.

24. Bernard Lewis, *The Shaping of the Modern Middle East* (Oxford University Press, 1994), pp. 3–23.

25. *Websters New Geographical Dictionary* defines the Middle East as: An extensive region comprising the countries of Southwest Asia and Northeast Africa; term formerly also included Afghanistan, Pakistan, India, and Burma [see *Websters New Geographical Dictionary* (G & C Merriam Company, 1972) p. 760]. The *New Encyclopaedia Britannica* provides a comprehensive definition: Middle East, the lands around the southern and eastern shores of the Mediterranean Sea, extending from Morocco to the Arabian Peninsula and Iran and sometimes beyond. The central part of this general area was formerly called the Near East, a name given to it by some of the first modern Western geographers and historians, who tended to divide the Orient into three regions. Near East applied to the region nearest Europe, extending from the Mediterranean Sea to the Persian Gulf; Middle East, from the Gulf to Southeast Asia; and Far East, those regions facing the Pacific Ocean [see

Micropaedia, vol. 8, 15th ed. (Encyclopaedia Britannica, 1992), p. 108]. According to *The Oxford Companion to Politics of the World,* the name Middle East has a recent origin. For many centuries European geographers used the term Near East. During World War II, the British designated their military headquarters in the region as the "Middle East Command." Thereafter, the term Middle East came into general use, and today this Eurocentric term remains the name of choice for both scientific and popular discourse, not only for the West but also for the region itself. The Middle East is not a concept rooted in physical geography. Natural boundaries do not mark it off unequivocally from other areas of habitation such as Africa and Central Asia. Rather the Middle East takes meaning from human history, geopolitics, culture, and political economy. Though alternative regional conceptions exist for some of the same territories, among them the Arab world, the Islamic world, and the Mediterranean Basin, the Middle East has proved in recent times to be the most useful and enduring analytical framework [see *The Oxford Companion to Politics of the World* (Oxford University Press, 1993), p. 585].

26. For a more theoretical definition of how regions are demarcated see B. M. Russett, *International Regions and the International System* (Rand McNally, 1967); L. J. Cantori and S. L. Spiegel, eds., *The International Politics of Regions* (Prentice-Hall, 1970); R. A. Falk and S. H. Mendlovitz, *Regional Politics and World Order* (San Francisco: Freeman, 1973); George Liska, *The Ways of Power* (Cambridge, Mass.: Basil Blackwell, 1990), pp. 365–69; W. Howard Wriggins with F. Gregory Gause, III, Terence P. Lyons, and Evelyn Colbert, *Dynamics of Regional Politics: Four Systems on the Indian Ocean Rim* (Columbia University Press, 1992); David J. Myers, ed., *Regional Hegemons: Threat Perception and Strategic Response* (Boulder, Colo.: Westview Press, 1991); Barry Buzan, "Third World Security in Structural and Historical Perspective," in Brian L. Job, ed., *The Insecurity Dilemma: National Security of Third World States* (Boulder, Colo.: Lynne Rienner, 1992), pp. 167–89; Barry Buzan, "A Framework for Regional Security Analysis," in Barry Buzan and Gowher Rizvi *South Asian Security and the Great Powers* (St. Martin's Press, 1986), pp. 3–33; Barry Buzan, *People, States & Fear,* 2d ed. (Boulder, Colo.: Lynne Rienner, 1991).

27. For instance, a recent edition of the Institute for National Strategic Studies, *Strategic Assessment 1995: U.S. Security Challenges in Transition,* uses the phrase Greater Middle East to cover North Africa, the Levant, the Persian Gulf, Central and South Asia: the Rand Corporation now has a Greater Middle East Program.

28. Mahnaz Z. Ispahani, *Roads and Rivals: The Political Uses of Access in the Borderlands of Asia* (Cornell University Press, 1989), chapter 1, pp. 1–30.

29. *The New Columbia Encyclopedia, Webster's New Geographical Dictionary,* and the *Lippincott Gazetteer,* place the Caspian Gates (a defile, long, narrow pass) by Derbent. According to Webster's, the name Caspian Gates refers to a specific wall that the Persians built in the sixth century A.D. According to the *New Columbia Encyclopedia,* another name for the Caspian Gates is the Iron Gates.

30. George Rawlinson, *The Sixth Great Oriental Monarchy or the Geography, History, and Antiquities of Parthia* (London: Longmans, Green, 1873), p. 66.

31. Ibid., pp. 65–66.

32. Ispahani, *Roads and Rivals,* pp. 87–88.

33. During World War I, the "Straits," including the Dardanelles, the Sea of Marmara, and the Bosporus, were the only link from the Black Sea to the Mediterranean. When Turkey sided with Germany and Austria-Hungary, Britain and

France realized the strategic importance of wresting control of the straits from Turkey. Subsequent efforts to secure control of the Gallipoli Peninsula resulted in some of the bloodiest fighting of the war.

34. This material comes from Geoffrey Kemp, *The Control of the Middle East Arms Race* (Washington, D.C.: Carnegie Endowment for International Peace, 1991), pp. 210–12.

35. Mordechai Gichon, "The West Bank: The Geostrategic and Historical Aspects," historical supplement to Aryeh Shalev, *The West Bank Line of Defense* (Praeger, 1985), pp. 180–81.

36. Richard F. Nyrop, ed., *Syria: A Country Study* (Washington, D.C.: Government Printing Office, 1979), p. 196.

37. Mark A. Heller, *A Palestinian State: The Implications for Israel* (Harvard University Press, 1983), pp. 17–18.

Chapter 2

1. The early parts of this chapter draw upon material published by one of the author's elsewhere, though the text has been significantly edited. See Geoffrey Kemp, "Maritime Access and Maritime Power: The Past, the Persian Gulf, and the Future," in Alvin J. Cottrell, Robert J. Hanks, Geoffrey Kemp, and Thomas H. Moorer, *Sea Power and Strategy in the Indian Ocean* (Beverly Hills, Calif.: Sage, 1981), and Geoffrey Kemp and John Maurer, "The Logistics of Pax Britannica: Lessons for America," in Uri Ra'anan, Robert L. Pfaltzgraff, Jr., and Geoffrey Kemp, eds., *Projection of Power: Perspectives, Perceptions and Problems* (Hamden, U.K.: Archon Books, The Shoe String Press, 1982).

2. Bruce Swanson, *Eighth Voyage of the Dragon: A History of China's Quest for Seapower* (Annapolis, Md.: Naval Institute Press, 1982), pp. 36–41. During his voyage, Zheng He, who was a Muslim, was accompanied by other Chinese Muslims who spoke Persian and Arabic. His expedition in 1405 got under way with more than 27,000 men aboard 317 ships.

3. F. H. Parry, in E. E. Rich and C. H. Wilson, eds., *The Cambridge Economic History of Europe,* vol. 4 (Cambridge University Press, 1967), pp. 163–65.

4. According to Samuel Morison, the price of cloves went from two to ten ducats per hundredweight over a five-year period in the 1530s. Samuel Eliot Morison, *The Great Explorers: The European Discovery of America* (Oxford University Press, 1978), p. 655. In addition to spices, trade with the East also included Chinese and Persian silks, Indian cotton, Chinese emeralds, and rubies and sapphires from India, Burma, and Ceylon.

5. Horatio Brown, *The Venetian Republic* (London: Bedford Street Press, 1902), p. 65.

6. P. B. Holt, A. Lambton, and B. Lewis, eds., *Cambridge History of Islam,* vol. 1A (Cambridge University Press, 1970), p. 305.

7. Frederic C. Lane, Eric F. Goldman, and Erling M. Hunt, *The World's History* (New York: Harcourt, Brace and Company, 1947), p. 299.

8. Bernard Brodie, *Sea Power in the Machine Age* (Princeton University Press, 1941), pp. 115–16.

9. Ibid., p. 115. For example, before the outbreak of hostilities with Spain in 1898, Admiral George Dewey used the British base of Hong Kong as if it were American. Had the British government chosen to, it could have greatly hindered the movement of Dewey's squadron to the Philippines by refusing him access to their coaling station. Similarly, the famous trip of the American battle fleet around the world in 1907–09, the cruise of the Great White Fleet, depended on British bases and supplies of coal for much of its journey (John D. Alden, *The American Steel Navy* [Annapolis, Md.: Naval Institute Press, 1972], pp. 333–47).

The best example of how Britain's command of the maritime coal trade effectively controlled the movements of the fleet of another great power occurred during the Russo-Japanese War of 1904–05. In order to mass enough naval forces in the Far East to defeat the Japanese navy, Czar Nicholas II ordered the Russian Baltic fleet to travel 18,000 miles to link up with the Russian Pacific fleet based in Port Arthur. On average each Russian warship burned from three to ten tons of coal for each day of travel; by the time the fleet reached the Far East, it would consume approximately half a million tons. Russia would have to obtain coal from foreign sources, since the Czarist merchant marine did not possess nearly enough colliers to keep the fleet supplied as it moved from the Baltic into the Atlantic. Furthermore, Russia possessed no colonial coaling stations at which the fleet could stop en route. Thus Russia would have to depend on the other three great powers, Germany, France, and Britain, for coal to move its fleet. Britain was the one best able to help but for a series of reasons chose not to. Russia then turned to her continental ally, France. However, France's ability to aid the Russian fleet was limited, since she did not have sufficient colliers to supply the Baltic fleet to the Far East. In order to find the colliers it needed Russia signed an agreement with Germany's Hamburg-American Line.

In addition to his commercial and strategic motives, the Kaiser intended to use the supply of coal to the Russian fleet as a coercive diplomatic weapon to break the Franco-Russian encirclement of Germany. Even with German help, the voyage of the Russian fleet around Africa to the Pacific presented formidable logistical problems; after weathering many hardships, the Russian fleet was annihilated by Admiral Heihachiro Togo in the famous Battle of Tsushima on May 27, 1905. (Lamar J. R. Cecil, "Coal for the Fleet that Had to Die," *The American Historical Review,* vol. 69, no. 4 (1964), pp. 991, 993). Cited in Kemp and Maurer, "The Logistics of Pax Britannica," p. 33.

10. See Paul M. Kennedy, "Imperial Cable Communications and Strategy, 1870–1914," *English Historical Review,* vol. 86, no. 141 (1971), p. 730, cited in Kemp and Maurer, "The Logistics of Pax Britannica," p. 35.

11. Kemp and Maurer, "The Logistics of Pax Britannica," p. 35.

12. Mahnaz Z. Ispahani, *Roads and Rivals: The Political Areas of Access in the Borderlands of Asia* (Cornell University Press, 1989), pp. 40–41.

13. Firuz Kazemzadeh, "Russia and the Middle East," in Ivo J. Lederer, ed., *Russian Foreign Policy: Essays in Historical Perspective* (Yale University Press, 1962), p. 490.

14. Peter Hopkirk, *The Great Game: The Struggle for Empire in Central Asia,* (New York: America Inc., 1992), p. 2.

15. Ibid., pp. 3, 5.

16. Ibid., p. 439.

17. Ibid., p. 445.

18. Ispahani, *Roads and Rivals,* pp. 94–95.

19. See Robert K. Massie, *Dreadnought: Britain, Germany and the Coming of the Great War* (Random House, 1991), p. 784.

20. Winston S. Churchill, *The World Crisis* (New York: Scribner, 1923), p. 137.

21. C. Gareth Jones, "The British Government and the Oil Companies 1912–1924: The Search for an Oil Policy," *The Historical Journal,* vol. 20, no. 3 (1977), p. 648. Also see M. Kent, *Oil and Empire: British Policy and Mesopotamian Oil* (London: Macmillan, 1976).

22. Jones, "The British Government and the Oil Companies," pp. 649, 652.

23. Ibid., p. 271.

24. Ibid., p. 661.

25. Ibid., p. 661.

26. Hugh Thomas, *The Suez Affair* (Harmondsworth, U.K.: Penguin, 1970), p. 39, as quoted in Jones, "The British Government and the Oil Companies," p. 661.

27. Jones, "The British Government and the Oil Companies," p. 664.

28. Paul Foley, "Petroleum Problems of the World War," *United States Naval Institute Proceedings,* vol. 50, no. 261 (November 1924), pp. 1802–32.

29. See V. H. Rothwell, "Mesopotamia in British War Aims, 1914–1918," *The Historical Journal,* vol. 13, no. 2 (1970), pp. 289–91. Also see Kent, *Oil and Empire,* pp. 125–26.

30. Jones, "The British Government and the Oil Companies," p. 672.

31. A. P. Herbert, *Leave My Old Morale Alone* (Doubleday, 1948), pp. 35–36.

32. Daniel Yergin, *The Prize: The Epic Quest for Oil, Money and Power* (New York: Touchstone, 1993), p. 395.

33. David A. Rosenberg, *The U.S. Navy and the Problem of Oil in a Future War: The Outline of a Strategic Dilemma 1945–1950,* Naval War College Review, vol. 29 (Summer 1976), p. 55.

34. Ibid., p. 56.

35. Anthony Cave Brown, ed., *Dropshot: The American Plan for World War III Against Russia in 1957* (New York: Dial Press, 1978), p. 156.

36. Yergin, *The Prize,* p. 409.

37. Geoffrey Kemp, *Forever Enemies: American Policy and the Islamic Republic of Iran* (Washington, D.C.: Carnegie Endowment for International Peace, 1994), p. 20.

38. The analysis and data in this section are drawn from Robert E. Harkavy, *Great Powers Competition for Overseas Bases: The Geopolitics of Access Diplomacy* (Pergamon, 1982); and *Bases Abroad: The Global Foreign Military Presence* (Oxford University Press, 1989).

39. George Kirk, *The Middle East: 1945–1950* (Oxford University Press, 1954), pp. 56–83.

40. Yergin, *The Prize,* pp. 630, 635, and 653.

41. Ibid., pp. 603–65.

42. Ibid., pp. 688–706.

43. Ibid., pp. 710–18.

44. The following data are based on International Energy Agency, *Global Energy, The Changing Outlook* (Paris, OECD/IEA, 1991), p. 41.

45. Had Saddam Hussein been able to hold on to Kuwait, he would have controlled 20 percent of OPEC's production, and 25 percent of the world reserves.

The Gulf War led to a short-term loss of more than 4 million barrels a day of oil previously supplied by Iraq and Kuwait.

46. Gulf oil exports have grown slowly since 1991 and hardly at all since 1993. The reason is that the large exporters continue to observe OPEC quota discipline (which most in OPEC do not), and their share of total exports is shrinking. Depending on how fast demand grows beyond the year 2000 and how rapidly new oil sources now on the horizon are developed, and how much longer the GCC states continue to follow an oil strategy that does not serve their long-term interest, Gulf exports could again expand significantly beyond 2000.

Chapter 3

1. For more on Turkey's new role, see Graham Fuller and Ian Lesser with Paul B. Henze and J.F. Brown, *Turkey's New Geopolitics: From the Balkans to Western China* (Boulder, Colo.: Westview Press, 1993).

2. The New International Policy: A Statement by Finance Minister Manmohan Singh, July 24, 1991.

3. See Ross Munro, "The Loser: India in the Nineties," *The National Interest,* Summer 1993, p. 62.

4. The material in this section draws upon Geoffrey Kemp and Janice Gross Stein, "Enduring Sources of Conflict in the Persian Gulf: Predicting Shocks to the System," in Geoffrey Kemp and Janice Gross Stein, eds., *Powderkeg in the Middle East: The Struggle for Gulf Security* (Lanham, Md.: Rowman and Littlefield, 1995).

5. In considering the impact of population dynamics on regional conflict, it is important to distinguish the impact of the three primary population variables: population size, population distribution, and population composition. A population variable can "change" in basically three ways: it can increase, remain constant, or it can decrease. Population *size* refers to the overall size of the population unit, be it an empire, a nation-state, a state, country, or other definable entity. Population *distribution* refers to the location and movement of population. Thus changes in population distribution could refer to migratory patterns within national frontiers or between national frontiers. The voluntary or forced migration of people from rural to urban areas would be an example of the first pattern; the legal or illegal, voluntary or forced emigration or immigration of people from one country to another would be an example of the second. Changes in population distribution therefore are not necessarily synonymous with absolute increases in the size of particular units. The *composition* of a population unit can be broken down in an almost infinite number of ways. However, it would include the basic characteristics of *population structure* (age and sex); *ethnic characteristics* (such as race, color, and language); *social and economic characteristics* (such as education, religion, family status, income, and occupation); *physiological characteristics* (such as height, weight, and life expectancy); and *psychological characteristics* (such as attitudes and motivation). Thus it can be seen that changes in the composition of a population unit will usually require that the particular subunit be specified since there are so many of them. Changes in population size, composition, and distribution can be relative or absolute; they can reflect increases, decreases, or zero

change; absolute population size can change but relative composition and distribution remain constant or reflect both increases and decreases. The intervening variables can range from zero (simple relationships) to many (complex relationships); the specified intervening variables can refer to general relationships among groups or to specific named groups, such as ethnic groups or age groups.

6. All population figures are from the Population Reference Bureau, *World Population Data Sheet* (Washington, D.C., 1995).

7. Abdel R. Omran and Farzaneh Roudi, "The Middle East Population Puzzle," *Population Bulletin,* vol. 48, no. 1 (July 1993), p. 34.

8. See, for instance, Neil W. Chamberlain, *Beyond Malthus: Population and Power* (New York: Basic Books, 1970), pp. 49–54.

9. Philippe Fargues, "From Demographic Explosion to Social Rapture," *Middle East Report,* vol. 24, no. 5 (September–October 1994), p. 7.

10. This does not, of course, include the Gaza Strip, where population-related issues are at the top of the agenda.

11. Philip Robins, "The Overlord State: Turkish Policy and the Kurdish Issue," *International Affairs,* vol. 69, no. 4 (1993), pp. 657–76.

12. Geoffrey Kemp, *Forever Enemies? American Policy and the Islamic Republic of Iran* (Washington, D.C.: Carnegie Endowment for International Peace, 1994), pp. 39–40.

13. Some say the figure is closer to 14.3 million (1990). Ted Robert Gurr, *Minorities at Risk* (Washington, D.C.: USIP, 1993), p. 332. The CIA claims there are 15.2 million Azeris in Iran. *The World Factbook 1993* (Washington, D.C.: CIA, 1994), p. 185.

14. Andrew Whitley, "Minorities and the Stateless in Persian Gulf Politics," *Survival,* vol. 35, no. 4 (Winter 1993–94), p. 35.

15. The CIA claims there are close to 1.3 million Baluch in Iran. *The World Factbook 1993,* p. 185.

16. "Martial Law Declared in Sistan Va Baluchestan," *Middle East Intelligence Report,* February 10, 1994.

17. Gurr, *Minorities at Risk.*

18. For information on Islamic friendship and building stronger ties, see PTV Television Network (Pakistan), August 28, 1993, in FBIS-NES, August 31, 1993, pp. 56–57; IRNA, March 22, 1993, from Middle East News Network, March 25, 1993; and an article on Iranian President Rafsanjani's visit to Pakistan in *Deutsche Presse-Agentur,* September 8, 1992. For an article on the Economic Cooperation Organization, see *Jahan-e Eslam,* February 14, 1993, from Middle East News Network, February 26, 1993. For articles on Iran and Pakistan in Afghanistan, see IRIB Television, March 6, 1993, from Middle East News Network, March 9, 1993; *Ettela'at,* February 9, 1993, from Middle East News Network, February 26, 1993; and, for a review of strains caused by Afghanistan, *Interpress,* August 1, 1992. For Indian fears, see Raju Gopalakrishnan, "Pak-Iran Unholy Alliance Against India," *News India,* April 16, 1993, p. 12. Rumors that Iran agreed to underwrite the Pakistani defense budget in exchange for nuclear technology are contained in *The Nation,* April 15, 1993, and Middle East News Network, April 20, 1993.

19. Richard F. Nyrop, ed., *India: A Country Study* (Washington, D.C.: Department of the Army, 1985), p. 167.

20. Ibid., pp. 166–67.

21. Peter Waldman and Miriam Jordan, "Indian Party Gains Loose Grip on Power," *Wall Street Journal,* May 16, 1996, p. A10.

22. John F. Burns, "Hindu Party Head Chosen for Post of India's Premier," *New York Times,* May 16, 1996, pp. A1, A8.

23. *The Europa World Yearbook 1994 Volume 1* (Kent, U.K.: Europa Publications Ltd., 1994), pp. 1426–27.

24. Nyrop, *India: A Country Study,* pp. 407–08.

25. For Iraqi and Saudi Shiites, see Whitley, "Minorities and the Stateless in Persian Gulf Politics," pp. 36–39, 41–42.

26. Lebanon, Walden Country Reports, NEXIS, 1993.

27. See *Strategic Survey 1995/96* (London: International Institute for Strategic Studies, 1996), p. 219.

28. *World Almanac and Book of Facts,* 1994; and "Somalia's Faction Leaders Agree to Form an Interim Government," *New York Times,* March 22, 1994, p. A8.

29. "Algerian Militant Base Uncovered in France," *New York Times,* April 9, 1994, p. A10.

30. Helen Chapin Metz, ed., *Algeria: A Country Study,* 5th ed. (Washington, D.C.: Federal Research Division, Library of Congress, 1994), pp. 89–90.

31. "Algeria to Introduce Berber Language in Schools," Reuters North America Wire, May 28, 1995.

32. Metz, *Algeria: A Country Study,* p. xxxii.

33. Roula Khalaf, "Vigilantes Join Hunt for Algerian Militants," *Financial Times,* March 13, 1995, p. 6.

34. The West Bank aquifer is composed of the eastern aquifer, the northeast aquifer, and the western aquifer.

35. Julian Ozanne and David Gardner, "Middle East Peace Would Be a Mirage Without Water Deal," *Financial Times,* August 8, 1995, p. 3.

36. Syrian Arab Republic Radio, May 28, 1994, as quoted in Daniel Pipes, "Understanding Assad," *Middle East Quarterly,* vol. 1, no. 4 (December 1994), p. 56.

37. Aryeh Shalev, *Israel and Syria: Peace and Security on the Golan,* JCSS Study no. 24 (Tel Aviv: Jaffee Center for Strategic Studies, 1994), p. 111; and M. Zuhair Diab, "Syrian Security Requirements in a Peace Settlement with Israel," *Israel Affairs,* Summer 1995, p. 72.

38. Ozanne and Gardner, "Middle East Peace Would Be a Mirage Without Water Deal," p. 3.

39. Geoffrey Kemp, *The Control of the Middle East Arms Race* (Washington, D.C.: Carnegie Endowment for International Peace, 1991), p. 38.

40. Ali Mohammed Khalifa, *The United Arab Emirates: Unity in Fragmentation* (Boulder, Colo.: Westview Press, 1979), p. 155.

41. Edward J. Perkins and George F. Ward, Jr., "Iraq's Non-Compliance With UN Security Council Resolutions," *Department of State Dispatch,* August 3, 1992, p. 604.

42. Miriam Amie, "UNIKOM Expects Long-term Kuwait Mission," United Press International, April 18, 1994.

43. Ewan Anderson and Jasem Karam, in their article, "The Iraqi-Kuwaiti Boundary," *Jane's Intelligence Review* (March 1995), pp. 120–21, clarify that the Ratga oil field is south of any previously recognized border, placing it in Kuwait proper. They also explain that it differs in structure, oil type, and production

conditions from Rumaila, from which it is separated by a major fault. On the Umm Qasr question, the authors are somewhat vague in their assessment of Iraqi losses. They evasively argue that the land boundary did not place Iraq's naval base in Kuwait while admitting that "the entire *commercial* port, with all the major installations, together with the town of Umm Qasr, remains in Iraq. The loss amounts to two small wooden jetties . . . together with some military accommodation" (emphasis added).

44. Perkins and Ward, "Iraq's Non-Compliance With UN Security Council Resolutions," pp. 604–05.

45. Alasdair Drysdale and Gerald H. Blake, *The Middle East and North Africa: A Political Geography* (Oxford University Press, 1985), p. 126.

46. Ibid., p. 126.

47. Richard Schofield, ed., *Territorial Foundations of the Gulf States* (St. Martin's Press, 1994), p. 39.

48. Shahram Chubin, *Iran's National Security Policy: Capabilities, Intentions and Impact* (Washington, D.C.: Carnegie Endowment for International Peace, 1994), pp. 101–02.

49. Schofield, *Territorial Foundations of the Gulf States,* p. 72fn.

50. Kemp, *Forever Enemies?,* p. 120.

51. John B. Allcock et al., eds., *Border and Territorial Disputes,* 3d ed. (Essex, U.K.: Longman Group, 1992), p. 382.

52. Kemp, *Forever Enemies?,* pp. 47–48.

53. Ahmad Mardini, "Gulf: Regional States Determined to Resolve Border Disputes," Inter Press Service, July 21, 1995.

54. "Commentary Approves Qatar's Approach to Regional Problem Solving," British Broadcasting Corporation, July 12, 1995. Source: Voice of the Islamic Republic of Iran Network 1, Teheran, July 10, 1995.

55. "Saudi Arabia, Qatar Discuss Improvement of Ties," Xinhua General Overseas News Service, May 19, 1993.

56. "Saudi Defense Minister, Qatari Emir Hold Talks," *Reuters Library Report,* May 18, 1993, NEXIS.

57. *Mideast Mirror,* June 27, 1995.

58. *Times Newspapers Ltd.,* June 28, 1995.

59. Andrew Rathmell, "Threats to the Gulf, Part 1," *Jane's Intelligence Review,* vol. 7, no. 3 (March 1995), p. 132.

60. *Mideast Mirror,* June 27, 1995.

61. "Emir of Qatar Comments on Border Disputes, Relations With Iraq and Israel," British Broadcasting Corporation, July 11, 1995, from QBS Radio (Doha), July 9, 1995.

62. Youssef M. Ibrahim, "Saudis Using Oil in Yemeni Dispute," *New York Times,* June 7, 1992, p. 7.

63. "Yemeni Minister Arrives in Riyadh with Message for King Fahd," Agence France Presse, April 30, 1994; "Yemen, Saudi Arabia Postpone New Border Talks," Reuters World Service, April 26, 1994; and James Wyllie, "Perpetual Tensions, Saudi Arabia and Yemen," *Jane's Intelligence Review,* March 1995, p. 7.

64. Agence France Presse, July 8, 1995.

65. Allcock et al., *Border and Territorial Disputes,* pp. 399–400.

66. Nora Boustany, "North Yemeni Troops Seize Oil Field Center," *Washington Post,* May 25, 1994, p. A27.

67. George Joffe, "Yemen: The Reasons for the Conflict," *Jane's Intelligence Review,* August 1994, p. 368.

68. Allcock, et al., *Border and Territorial Disputes,* p. 397.

69. Ibid., p. 397.

70. "Kuwait, Saudi Arabia Discuss Drawing Border," Agence France Presse, July 18, 1995.

71. "GCC Affairs, Saudi-Kuwait Border Talks," *Arab Press Service Diplomat Recorder,* July 22, 1995.

72. Allcock et al., *Border and Territorial Disputes,* pp. 190–91.

73. Drysdale and Blake, *The Middle East and North Africa,* p. 91.

74. "Libyan Claims on Chad Dismissed by World Court," *Facts on File,* February 17, 1994, p. 95.

75. Allcock et al., *Border and Territorial Disputes,* p. 230.

76. Peter H. Gleick, "Water and Conflict: Fresh Water Resources and International Security," *International Security,* vol. 18, no. 1 (1993), pp. 85, 87.

77. John K. Cooley, "Middle East Water: Power for Peace," *Middle East Policy,* vol. 1, no. 2 (1992), p. 8.

78. Natasha Beschorner, "Water and Instability in the Middle East," *Adelphi Paper* 273 (Winter 1992–93), p. 21.

79. Gleick, "Water and Conflict," p. 88.

80. Ibid., pp. 87, 89.

81. The destruction of Iraq's electric power grid by the coalition in the first days of the Gulf War is a precursor of what could happen in future Middle East wars (see chapter 9 for a more detailed discussion on this issue).

82. Cooley, "Middle East Water," p. 2.

83. Ayman al-Safadi and Nirmin Murad, "Negotiator Remarks on Water Agreement," *Jordan Times,* October 19, 1994, p. 1.

84. Natasha Bukhari, "Syria Said 'Diverting' Water from Country's Share," *Jordan Times,* December 3, 1994, p. 12.

85. Ozanne and Gardner, "Middle East Peace Would Be a Mirage Without Water Deal," p. 3. Statistics vary from author to author. John Kolars of the University of Michigan has written that the "absolute limit on West Bank water use per year is about 615 mcm. Of this, Israel and Israeli settlers use 83 percent," in his article "The Course of Water in the Arab Middle East," *American-Arab Affairs,* no. 33 (Summer 1990), p. 66.

86. "Future Water Shortages Threaten Middle East Peace," *The Jerusalem Post,* May 25, 1990, p. 8.

87. Joyce R. Starr, "Water Wars," *Foreign Policy,* no. 82 (Spring 1991), p. 24.

88. Ozanne and Gardner, "Middle East Peace Would Be a Mirage Without Water Deal," p. 3.

89. Giles Trendle, "Whose Water Is it?" *The Middle East,* January 1992, p. 19.

90. Jonathan C. Randal, "A Dwindling Natural Resource: Israel Relies Heavily on Water from Arab Territories," *Washington Post,* May 13, 1995, p. A25.

91. Cooley, "Middle East Water," p. 15.

92. Beschorner, "Water and Instability in the Middle East," pp. 27–29.

93. Hugh Pope, "The Looming Crisis Over the Tigris-Euphrates Waters," *Middle East International,* June 9, 1995, p. 17.

94. "Paper Criticizes Turkey's 'Dangerous' Water Policy," FBIS-NES-92-146, July 29, 1992, p. 19. Source: Cairo MENA in Arabic, July 26, 1992. See

also "Turkey Sets Arab Alarm Bells Ringing," *The Middle East Today,* July 27, 1992.

95. Ibid.

96. Pope, "The Looming Crisis Over the Tigris-Euphrates Waters." The 1987 protocol suggested that the Euphrates would continue to flow into Syria at a rate of 500 cm per second, or 15.8 cu km per year, approximately half the river's flow. The lack of any official guarantees to that effect has left Syria somewhat nervous about the security of its water supply.

97. Beschorner, "Water and Instability in the Middle East," p. 37.

98. Jonathan C. Randal, "Euphrates Dam Aids Turkish Rebels," *Washington Post,* May 15, 1992, p. A27.

99. Beschorner, "Water and Instability in the Middle East," pp. 43–44.

100. Gleick, "Water and Conflict," p. 86.

101. Starr, "Water Wars," p. 21.

102. World Resources Institute in collaboration with the United Nations Environmental Programme and the United Nations Development Programme, *World Resources 1994-5: A Guide to the Global Environment* (Oxford University Press, 1994), p. 345.

103. Starr, "Water Wars," p. 19.

104. World Resources Institute, p. 183.

105. Beschorner, "Water and Instability in the Middle East," p. 45.

106. Cooley, "Middle East Water," p. 6.

107. Jamil Al Alawi and Mohammed Abdulrazzak, "Water in the Arabian Peninsula: Problems and Perspectives," in Peter Rogers and Peter Lydon, eds., *Water in the Arab World: Perspectives and Prognoses* (Harvard University Press, 1994), pp. 178–80, 63.

108. Ibid., p. 63.

109. Leon Awerbuch, Manager for Power and Desalination, the Bechtel Corporation, cited in Starr, "Water Wars," pp. 20–21.

Chapter 4

1. These are figures prepared by the International Energy Agency in its *World Energy Outlook 1996.*

2. Ibid., p. 3.

3. Throughout this text there are frequent references made to future world demand and supply for oil and, to a lesser extent, natural gas. There are numerous sources to turn to but many provide different estimates based upon assumptions concerning economic growth, oil prices, capacity constraints, and the political environment. The figures used in this text have been taken from official publications, which have been cited in specific endnotes. However, the overall trends are alluded to are based on extensive discussions by the authors with specialists in the field, including James Placke of Cambridge Energy Associates, William Martin, chairman of the Washington Policy and Analyses Inc., and Anthony Cordesman of the Center for Strategic and International Studies. The study has benefited greatly from a draft report to the Trilateral Commission prepared by William Martin, Ryukichi Imai, and Helga Steeg, *Maintaining Energy Security in a Global Context*

(Vancouver, Canada, April 1996). Also consulted was the Global Energy Modeling Working Session prepared by Washington Policy and Analyses/Los Alamos National Laboratory, June 13, 1996, Washington, D.C.; the annual publications of the International Energy Agency, including the *World Energy Outlooks* for 1994, 1995, and 1996; and the *International Energy Agency Middle East Oil and Gas* (OECD, Paris, 1995). Extremely detailed energy maps have been prepared by the *Petroleum Economist* in London, including the energy map of the Gulf (1993) and the energy map of the Caspian/Black Sea Basin (1995). We also had consultations with Dr. Kong Wu of the East-West Center, Honolulu, Hawaii, and benefited from the new book by Kent E. Calder, *Pacific Defense: Arms, Energy, and America's Future in Asia* (William Morrow, 1996). Other sources are cited in the various tables.

4. "Worldwide Look at Reserves and Production," *Oil and Gas Journal,* vol. 92, no. 52 (December 26, 1994). The *International Petroleum Encyclopedia* also offers a figure of 66% as cited by the Department of Defense, Office of International Security Affairs, "United States Security Strategy for the Middle East" (Washington, D.C., 1995). However, *World Oil* cites a figure approaching 54%, "Estimated Proven World Reserves, 1994 Versus 1993," *World Oil,* August 1995, p. 30.

5. Estimates of Caspian Basin oil and gas reserves vary dramatically, with the numbers ranging from 30 billion barrels to 200 billion. According to a recent study, the figure of 90 billion is often used by industry analysts [Rosemarie Forsythe, *The Politics of Oil in the Caucasus and Central Asia,* Adelphi Paper 350 (London: The International Institute for Strategic Studies, 1996), p. 6].

6. In making its estimates for future oil supply and demand the IEA uses two cases, the capacity constraints case and the energy savings case. The first assumes that improvements in energy efficiency will be based on historical trends and that energy prices will increase. The second assumes high efficiency in energy use and flat energy prices. The first case assumes that oil production cannot expand quickly enough to meet rising demand, hence higher prices and a subsequent moderation in energy demand. The second case assumes that much greater energy savings can be made by consuming countries and this can hold prices flat. The difference between the two sets of numbers generated by these assumptions vary from region to region but would be slightly greater for the more advanced OECD countries than the ROW or FSU. For more details, see *World Energy Outlook 1996,* pp. 15–17.

7. Ibid., pp. 1–2.

8. *World Energy Outlook 1994,* p. 38.

9. Ibid., p. 34.

10. *World Energy Outlook 1996,* p. 234.

11. *Annual Energy Outlook 1996* (Washington, D.C.: Energy Information Administration, 1995), p. 48.

12. Ibid., p. 48.

13. Phebe Marr and William Lewis, eds., *Riding the Tiger: The Middle East Challenge After the Cold War* (Boulder, Colo.: Westview Press, 1993), p. 47.

14. British Petroleum, *Statistical Review of World Energy,* London 1993, p. 16, in " The Economic Relations of the Middle East: Toward Europe or within the Region", by Rodney Wilson in *The Middle East Journal,* p. 269.

15. Mudassar Imran and Philip Barnes, "Energy Demand in the Developing Countries: Prospects for the Future," Staff Commodity Working Paper 23 (Wash-

ington, D.C.: World Bank, 1990), pp. 1–3, cited in Edward R. Fried and Philip H. Trezise, *Oil Security: Retrospect and Prospect* (Brookings, 1993), pp. 18–19.

16. Ibid., p. 19.

17. "Power to the People: A Survey of Energy," *Economist,* June 18–24, 1994, p. 11.

18. *Middle East Economic Digest,* May 30, 1994 (Headline: Middle East: Asian Oil Boom For Middle East?).

19. Fereidun Fesharaki, "The Outlook for Oil and Gas Supply and Demand in Asia," unpublished paper presented at a conference on Energy and National Security in the 21st Century, National Defense University, Washington D.C., November 10, 1994.

20. *The Oil Daily,* April 12, 1994, p. 7.

21. Energy Information Administration, *International Energy Outlook 1996* (Washington, D.C., May 1996).

22. Julia Nanay, "The Outlook for Iran's Natural Gas Sector," private paper prepared for the Nixon Center for Peace and Freedom, Washington, D.C., January 1997, p. 12.

23. Ibid., pp. 22–26.

24. International Energy Agency, *Oil, Gas and Coal Supply Outlook* (Paris, 1995), pp. 53–58.

25. Thomas Stauffer, "Trends in Oil Production Costs in the Middle East, Elsewhere" *Oil and Gas Journal,* vol. 92, no. 12 (March 21, 1994), p. 106.

26. Ibid.

27. Ibid.

28. Eliyahu Kanovsky, *The Economic Consequences of the Persian Gulf War* (Washington Institute for Near East Policy, Policy Paper no. 30, 1994), p. 11. Earlier, in a 1987 paper, Kanovksy argued that the trend toward lower prices would continue through the 1990s based on two factors: the revenue needs of oil-exporting countries and the competition for markets. Studying the economies of the Gulf producers, he concluded that with a few major exceptions, OPEC countries face serious financial difficulties and will continue to need an uninterrupted flow of oil revenues. Kanovsky also noted that since diversification programs in these economies have not produced revenue and thus continue to depend on state subsidies, oil will remain the countries' most important source of income in the 1990s. Kanovsky, *Another Oil Shock in the 1990s? A Dissenting View* (Washington Institute for Near East Policy, Policy Paper no. 6, 1987). He asserts that the Gulf War had an aggravating effect on the Gulf economies, increasing their dependence on oil revenues, since payments for Gulf War expenditures by Saudi Arabia and the UAE will require increased oil revenues in the future. After the end of the war, according to Kanovsky, the Saudis, uninhibited by Iraq, were left with only one option: "increase the volume of production and exports" (see Kanovsky, Policy Paper no. 30, p. 18). Kuwait's need for reconstruction funds will trigger a second expansion of oil production capacity. Kanovsky extends this same conclusion to Iraq; once the embargo is lifted, he says, Iraq will expand its production as well, since all the plans for reconstruction and development require increased oil revenues.

29. "Power to the People," p. 14.

30. Ibid., p. 14.

31. *World Energy Outlook 1994,* p. 67.

32. Congress of the United States, Office of Technical Assessment, *U.S. Oil Import and Vulnerability: The Technical Replacement Capability,* OTA-E-503 (Government Printing Office, October 1991), p. 18.

33. Ibid., p. 13.

34. Ibid., p. 10.

35. Ibid., p. 8.

36. The volatility of the situation is demonstrated by the fact that the most serious recent internal security threat to the United Arab Emirates occurred when Hindu and Muslim expatriates began rioting following the December 1992 destruction of a mosque at Ayodhya in central India by Hindu extremists. Citizens of the UAE had never heard of Ayodhya, but were terrified by visions of foreigners rampaging in downtown Dubai. With foreigners making up over 80% of the population and half of these coming from South Asia, the UAE and other Gulf countries cannot remain insulated from the long-standing conflict between India and Pakistan.

37. Private conversation with a U.S. official who has had extensive contacts with the Chinese military establishment.

38. China's decision to lay the foundations for a blue water navy capable of patrolling the world's oceans is based on very traditional geopolitical concerns. These include encroachment of its vast coastal waters by adversaries; exploitation of the sea for minerals, including oil; protection of its coastal economy, which accounts for 70 percent of its gross national product; protection of its huge and growing seaborne trade; protection of its ocean-going fishing industry. According to Chinese scholars, China "is embarking on a massive modernization program and transition to a blue water power. Its objective is to become a new world Pacific power in the twenty-first century." [See You Ji and You Xu, "In Search of Blue Water Power: The PLA Navy's Maritime Strategy in the 1990s," *The Pacific Review,* vol. 4, no. 2 (1991), pp. 137–49.]

39. Mamdouh G. Salameh, "China, Oil, and the Risks of Regional Conflict," *Survival,* Winter 1995–96, p. 141. See also Michael Leifer, "Chinese Economic Reform and Security Policy: The South China Sea Connection," *Survival,* Summer 1995, pp. 45–59.

40. In describing the emerging military capabilities of the Asian powers in the twenty-first century, Paul Dibb speculates: "The Chinese navy will probably have an aircraft carrier capability by 2010 and the variant force assumes two or three aircraft carriers of some 50,000 tons each in operational service by that date or later (2015). . . . A Chinese aircraft carrier force would also prompt a more capable Indian submarine force" [Paul Dibb, *Towards a New Balance of Power in Asia,* Adelphi Paper 295 (London: International Institute for Strategic Studies, 1995), pp. 89–90].

41. Forsythe, *The Politics of Oil,* p. 6.

42. Clive Schofield and Martin Pratt, "Claims to the Caspian Sea," *Jane's Intelligence Review,* February 1996, p. 78.

43. Ibid.

44. Forsythe, *The Politics of Oil,* p. 31.

45. Schofield and Pratt, "Claims to the Caspian Sea," p. 76.

46. Umit Enginsoy, "Russia Urged Not to Scare Investment Away from the Caspian Sea," *Turkish Daily News,* July 19, 1995, pp. 1, A8, in FBIS-WEU-95-141, July 24, 1995, pp. 42–43.

47. "Hot Water," *Russian Petroleum Investor,* December 1996–January 1997, p. 46.

48. Michael Binyon, "Russia Challenges Turkey Over Caspian Sea Pipeline," *The Times,* September 15, 1995.

49. *World Energy Outlook 1994,* pp. 72–73. The IEA cautions that such predictions are highly speculative, however, and that every small change in the growth rate of FSU economies could have a significant impact on exports.

50. Eugene M. Khartukov, Olga V. Vinogradova, and Dmitry A. Surovtsev, "Former Soviet Union: At the Threshold," *World Oil,* August 1995, p. 96.

51. John Greenwald, "Black Gold Rush," *Time,* July 4, 1994, p. 54.

52. Carl Goldstein, "Wells of Hope: Kazakstan Signs Massive Oil Deal," *Far Eastern Economic Review,* April 23, 1993, p. 77.

53. Agis Salpukas, "Caspian Sea Oil Search to Honor Bird and Beast," *New York Times,* January 13, 1994, p. D2.

54. Greenwald, "Black Gold Rush," p. 54.

55. Khartukov, Vinogradova, and Surovtsev, "Former Soviet Union: At the Threshold," p. 96.

56. Ibid., pp. 96–97.

57. Ibid., p. 94.

58. Ibid.

59. "Mitsubishi Studies Feasibility of Gas Pipeline," *Moscow Interfax,* June 16, 1995, in FBIS-SOV-95-117, pp. 104–05.

60. Khartukov, Vinogradova, and Surovtsev, "Former Soviet Union: At the Threshold," p. 100.

61. "Natural Gas Reserves, Development Prospects Viewed," *Pravda,* January 18, 1995, p. 1; from FBIS-SOV, January 23, 1995, p. 60.

62. Steve Levine and Robert Corzine, "Turkmenistan: A Catalogue of Promises Unfulfilled," *Financial Times,* August 22, 1995, p. 4.

63. Erik Nurshin, "Oil Industry Suffers from Caspian Sea Flooding," *Ekspress,* June 20, 1995, in FBIS-SOV-95-122, pp. 84–85.

64. Salpukas, "Caspian Search for Oil to Honor Bird and Beast."

65. Ibid.

66. Transportation of Russian oil and gas is also problematic. All primary gas pipelines through which Gazprom exports 93 percent of its gas to Europe cross Ukrainian territory and are badly maintained. At the beginning of 1995, 60 percent of the gas turbines had outlived their natural life and 14.5 percent of the gas pipelines had exhausted their amortization term and required repair and renovation. It has been argued that 39 of 120 gas pipeline emergency situations occurred because the anticorrosive protective coating of the pipelines had aged (see Svitlana Korotych, "Corrosion Destroys Gas Pipelines and the Primary Trump Card in a Gambling Game With Russia," Kiyevskiye Vedomosti, June 30, 1995, in FBIS-SOV-95-128, "Low Technical State of Gas Pipelines Detailed," p. 61).

67. Kenneth Howe, "Chevron Struggling in Tengiz: Plenty of Oil in Kazakhstan —But It's Hard to Get Out," *San Francisco Chronicle,* September 25, 1995, p. D1.

68. "Shevardnadze Wants Two Oil Pipelines From Azerbaijan," *Interfax,* August 1, 1995, in FBIS-SOV-95-148, p. 85.

69. Sheila Marnie, RFE/RL News Brief, December 10–11, 1992.

70. Benjamin A. Holden and Andy Pasztor, "Chevron Slashes Outlays for Kazakhstan Oil Field," *Wall Street Journal,* February 13, 1995, p. A3. Chevron

reportedly cut capital spending to zero (see Anne Reifenberg, "Mobil, Kazakhstan Discuss Sale of Part of Tengiz Oil Field," *Wall Street Journal,* October 16, 1995, p. A10).

71. Anne Reifenberg, "Caspian Pact May Bolster Chevron Effort," *Wall Street Journal,* March 11, 1996, p. A3.

72. Reifenberg, "Mobil, Kazakhstan Discuss Sale of Part of Tengiz Oil Field."

73. Richard H. Matzke, "Challenges of Tengiz Oil Field and other FSU Joint Ventures," *Oil and Gas Journal,* July 4, 1994, p. 62.

74. "More Question Marks over Caspian-Novorossisk Pipleline Plan," *Monitor,* vol. 2, no. 47 (March 7, 1996).

75. Pipeline Politics II, April 1994. A Petro Finance Special Report.

76. "Azerbaijani Officials on Pipeline, U.S. Visit," *Turkish Daily News,* June 15, 1995, in FBIS-WEU-95-119, pp. 50–51.

77. "Caspian Oil Deal May Isolate Russia," FBIS-SOV-94-203, pp. 35–36.

78. Izzet Dagistanli, "Gasanov Statement on Oil Route, Russian Policy," *Anatolia,* August 26, 1995, in FBIS-SOV-95-166, p. 83.

79. "Protocol on Export of Oil Through Russia Signed," *Turan,* June 24, 1995, in FBIS-SOV-95-122, pp. 92–93.

80. Steve Levine and Bruce Clark, "Compromise Deal Today Over Caspian Oil Route," *Financial Times,* October 10, 1995, p. 2.

81. Steve Levine and Bruce Clark, "Turkish Port to Gain From Oil Deal," *Financial Times,* October 10, 1995, p. 4.

82. Enginsoy, "Russia Urged Not to Scare Investment Away from the Caspian Sea."

83. Bhushan Bahree and Anne Reifenberg, "Russia Wins First Round in Battle to Control Oil Flowing from Caspian," *Wall Street Journal,* October 9, 1995, p. A8.

84. Anne Reifenberg and Bhushan Bahree, "Consortium Sets Initial Route for Caspian Oil," *Wall Street Journal,* October 10, 1995, p. A16. Elaine Holoboff cited an estimated total cost of $250 million for constructing a pipeline from Baku to Tbilisi to replace the existing rail line, upgrading the Tbilisi-Batumi section, building a new section to Supsa if this is the destination, and carrying out either the refurbishment or building of port facilities ("Russia and Oil Politics in the Caspian," *Jane's Intelligence Review,* February 1996, p. 83).

85. "National Assembly Chairman Comments," FBIS-SOV-94-221.

86. See Mikhail Konstantinov and Boris Filippov, "Genie Awaits His Hour," *Rossiyskaya Gazeta,* August 17, 1995, p. 7, in FBIS-SOV-95-162, August 22, 1995, p. 70.

87. "Mitsubishi Studies Feasibility of Gas Pipeline." The cost of the project is also estimated at $12 billion, according to other sources, including Steve Levine, "Way Sought to Pass Russia with Oil Line," *New York Times,* September 9, 1995, pp. A1, A3.

88. "Project Discussed With Turkmen Officials," *Voice of the Islamic Republic of Iran First Program Network,* June 10, 1995, FBIS-NES-95-114, pp. 63–64.

89. Vladimir Socor, "Turkey Asserts Role in Ex-Soviet Orbit," *Prism,* August 4, 1995, p. 3.

90. "Niyazov Signs Joint Investment Contracts With Iran," *Segodnya,* July 6, 1995, FBIS-SOV-95–130, p. 57.

91. Igor Rotar, "Will Natural Gas Save Turkmenistan?" *Prism,* September 8, 1995, p. 11.

Chapter 5

1. This is discussed in, among others, Edward Luttwak, *Strategy: The Logic of War and Peace* (Cambridge, Mass.: Belknap Press, 1987), chapters 8 and 9; B. H. Liddell-Hart, *Strategy: The Indirect Approach* (Praeger, 1954), esp. part 4; and G. D. Sheffield, "Blitzkrieg and Attrition: Land Operations in Europe 1914–45," in Colin McInnes and G. D. Sheffield, eds., *Warfare in the Twentieth Century* (London: Unwin Hyman, 1988), chapter 3; Hew Strachan, *European Armies and the Conduct of War* (London: George Allen and Unwin, 1983), esp. chapters 8–10; and Chris Bellamy, *The Future of Land Warfare* (St. Martin's Press, 1987), esp. pp. 296–99. The last attempts to dispel (p. 136) the simplistic view that there is a straightforward divide between "maneuver" and "attrition." See also Martin van Creveld, *Military Lessons of the Yom Kippur War: Historical Perspectives* (Beverly Hills, Calif.: Sage, 1975), esp. chapter 1. Related to the dichotomy of attrition and maneuver warfare, van Creveld discusses the three main "elements" of warfighting—striking power, mobility, protection—in the context of historical alternation.

2. On grand strategy, see Edward Luttwak, *Strategy,* part 3; and John M. Collins, *Grand Strategy: Principles and Practices* (Annapolis, Md.: Naval Institute Press, 1973).

3. Among the numerous works on the military aspects of the Iran-Iraq war, see in particular, Anthony H. Cordesman and Abraham R. Wagner, *The Lessons of Modern War: The Iran-Iraq War,* vol. 2 (Boulder, Colo.: Westview Press, 1970); Efraim Karsh, *The Iran-Iraq War: A Military Analysis* (London: International Institute for Strategic Studies, 1987), Adelphi Paper No. 220; and Edgar O'Ballance, *The Gulf War* (London: Brassey's, 1988).

4. Regarding the 1971 war, see, among others, D. K. Palit, *The Lightning Campaign: The Indo-Pakistan War, 1971* (Salisbury: Compton Press, 1972). For the 1965 war, see in particular Russell Brines, *The Indo-Pakistan Conflict* (London: Pall Mall Press, 1968); and Hari Ram Gupta, *The India-Pakistan War,* vols. 1 and 2 (Delhi: Prakashan, 1967).

5. Cordesman and Wagner, *The Lessons of Modern War,* vol. 1, esp. pp. 37–44.

6. Randolph S. Churchill and Winston S. Churchill, *The Six Day War* (London: Heinemann, 1971).

7. The basic geography of the 1973 war is covered in Chaim Herzog, *The War of Atonement, October 1973* (Little, Brown, 1975).

8. This is discussed in Shirin Tahir-Kheli, "Defense Planning in Pakistan," in Stephanie Neuman, ed., *Defense Planning in Less-Industrialized States* (Lexington, Mass.: D. C. Heath, 1984), chapter 7. According to her (p. 212), quoting a Pakistani officer, "Pakistan feels exposed because its lines of communication and the highly developed canal system that irrigates the fertile areas of Pakistan that are critical to its economic survival run close to the Indo-Pakistani border." It was further stated that India's capture of just twenty-five miles would wipe out Pakistan because its communications, irrigation, industry, and population are "all together within that depth." Palit, *The Lightning Campaign,* p. 77, claims that "the Pakistani High Command has for a number of years nursed a pipedream about launching a massive, surprise offensive deep into Indian territory spearheaded by armored formations, à la Moshe Dayan."

9. See, among others, Chaim Herzog, *The War of Atonement, October 1973,* esp. chapter 1 under "The New Strategic Concept," and chapter 18, "Lessons and Implications."

10. Patrick O'Sullivan, "The Geography of Wars in the Third World," in Stephanie Neuman and Robert E. Harkavy, eds., *The Lessons of Recent Wars in the Third World: Comparative Dimensions* (Lexington, Mass.: D. C. Heath, 1987), pp. 39–40.

11. Palit, *The Lightning Campaign,* chapter 7 and appendix B; "Indian Army Seeks to Tighten Ring Around Isolated Pakistani Forces in East," *New York Times,* December 7, 1971, p. A1; and "Indians Cross Wide River and Drive on Dacca," *New York Times,* December 11, 1971, p. A1.

12. There is a parallel here with Germany's dilemma in 1914, embodied in the Schlieffen Plan. See Robert E. Harkavy, *Preemption and Two-Front Conventional Warfare: A Comparison of 1967 Israeli Strategy with the Pre-World War I German Schlieffen Plan,* Leonard Davis Institute, Hebrew University, Jerusalem Papers on Peace Problems, no. 23 (1978). The basics of the Schlieffen Plan are rendered in Barbara Tuchman, *The Guns of August* (Dell, 1962), esp. chapters 2–7; and Gerhardt Ritter, *The Schlieffen Plan* (London: Oswald Wolff, 1958). See also Jack Snyder, *The Ideology of the Offensive* (Cornell University Press, 1984).

13. An excellent description of Israel's desperate battle to hold on to the Golan at the outset of the 1973 war is in Avigdor Kahalani, *The Heights of Courage: A Tank Leader's War on the Golan* (Westport, Conn.: Greenwood Press, 1984).

14. Cordesman and Wagner, *The Lessons of Modern War,* vol. 2, *The Iran-Iraq War,* pp. 72 and 97.

15. Nadav Safran, *From War to War* (New York: Pegasus, 1969), appendix C.

16. Palit, *The Lightning Campaign,* chapters 6 and 7.

17. Cordesman and Wagner, *The Lessons of Modern War,* vol. 2, p. 224.

18. Ibid., p. 252.

19. Cordesman and Wagner, *The Lessons of Modern War,* vol. 3, pp. 62 and 174.

20. van Creveld, *Military Lessons of the Yom Kippur War,* and for a later interpretation, Edward Luttwak and Dan Horowitz, *The Israeli Army* (Harper and Row, 1975), chapter 10.

21. Iraq's ability to contribute to an Arab war with Israel is discussed in Michael Widlanski, ed., *Can Israel Survive a Palestinian State?* (Jerusalem: Institute for Advanced Strategic and Political Studies, 1990), pp. 16–18, under "The Iraqi Expeditionary Force."

22. Cordesman and Wagner, *The Lessons of Modern War,* vol. 2, pp. 479–84, under "Close Air Support," wherein it is asserted that "Both Iran and Iraq generally proved unable to use fixed-wing airpower effectively to provide close air support of their military forces."

23. Cordesman and Wagner, *The Lessons of Modern War,* vol. 3, pp. 124–40.

24. Ibid., p. 136.

25. Peter Cary, "The Lessons of the Desert Fox," *U.S. News and World Report,* January 14, 1991, p. 31.

26. Cordesman and Wagner, *The Lessons of Modern War,* vol. 3, p. 52.

27. Cordesman and Wagner, *The Lessons of Modern War,* vol. 1, p. 169.

28. Cordesman and Wagner, *The Lessons of Modern War,* vol. 2, pp. 142 and 423–35.

29. Ibid., pp. 449–50.

30. This is detailed in the Insight Team of the *London Sunday Times, The Yom Kippur War* (Doubleday, 1974), part 2, chapter 1, under "The Y-Day Onslaught;" and Peter Allen, *The Yom Kippur War* (Charles Scribner's Sons, 1982), chapter 4.

31. Cordesman and Wagner, *The Lessons of Modern War,* vol. 1, pp. 185–203.

32. The subsequent analysis is drawn from Palit, *The Lightning Campaign,* chapter 9, under "The War in the Air and at Sea," and chapter 5, under "Pakistan Attacks in the West."

33. Ibid., p. 143.

34. Ibid., pp. 142–43.

35. Cordesman and Wagner, *The Lessons of Modern War,* vol. 2, p. 456.

36. Ibid., p. 457.

37. Ibid., p. 485.

38. Ibid., pp. 492–94 (493).

39. Cordesman and Wagner, *The Lessons of Modern War,* vol. 3, pp. 169–219.

40. Ibid., pp. 180–91.

41. Ibid., p. 180.

42. Ibid., pp. 192–205.

43. Ibid., p. 203.

44. This list was suggested in a private communication from Michael Eisenstadt.

45. John M. Collins, unpublished but forthcoming textbook on "military and naval geography," in a chapter on "oceans and seashores."

46. William Burrows and Robert Windrem, *Critical Mass* (New York: Simon and Schuster, 1994), pp. 311–12.

47. Palit, *The Lightning Campaign,* chapter 9.

48. Ashley Tellis, "The Naval Balance in the Indian Subcontinent," *Asian Survey,* vol. 25, no. 12 (1985), pp. 1186–213 (1206). See also Devin T. Hagerty, "India's Regional Security Doctrine," *Asian Survey,* vol. 31, no. 4 (1991), pp. 351–63.

49. Yedidia Ya'ari, "The Littoral Arena: A Word of Caution," *Naval War College Review,* vol. 48, no. 2 (1995), pp. 7–21 (17).

50. Palit, *The Lightning Campaign,* p. 149.

51. Cordesman and Wagner, *The Lessons of Modern War,* vol. 2, p. 558.

52. Ibid., pp. 558–59.

53. See Jerrold F. Elkin and Major W. Andrew Ritezel, "New Delhi's Indian Ocean Policy," *Naval War College Review,* vol. 40, no. 4 (1987), pp. 50–63.

54. Peter Wallensteen, "Conflicts and Conflict Resolution in 1988," in Peter Wallensteen, ed., *States in Armed Conflict 1988* (University of Uppsala, 1988), pp. 1–17. See also Wallensteen et al., "Major Armed Conflicts," *SIPRI Yearbook 1989* (Oxford University Press, 1989), pp. 339–55.

55. The wars are generally fought with less sophisticated weapons (more on the insurgency side), featuring those associated with light infantry, and with a much lesser role for mechanized forces, airpower, and seapower. The wars are fought by less-sophisticated forces, reflecting socioeconomic development levels and as measured by per capita gross national product. The wars, if basically guerilla insurgencies against an incumbent regime, are asymmetric, involving the wholly different opposing strategies and tactics of insurgency and counterinsurgency, respectively. The wars do not involve an identifiable moving "front" (analogous to a line of scrimmage in football), but rather a much more fluid and chaotic pattern

of conflict not easily represented on maps (which may or may not involve extensive regional areas of control by insurgents).

56. These and other factors related to the military geography of LIC are discussed in Patrick O'Sullivan, "A Geographical Analysis of Guerrilla Warfare," *Political Geography Quarterly,* vol. 2, no. 2 (1983), pp. 139–50.

Chapter 6

1. Among numerous works are James F. Dunnigan and Austin Bay, *From Shield to Storm* (William Morrow, 1992); Michael J. Mazaar, Don M. Snider, and James A. Blackwell, *Desert Storm* (Boulder, Colo.: Westview Press, 1993); Eliot A. Cohen, "A Revolution in Warfare," *Foreign Affairs,* vol. 75, no. 2 (March-April 1996), pp. 37–54; Lawrence Freedman and Efraim Karsh, *The Gulf Conflict 1990–1991* (Princeton University Press, 1993); Rick Atkinson, *Crusade: The Untold Story of the Gulf War* (Houghton Mifflin, 1993); and Bernard Trainor and Michael Gordon, *The Generals' War: The Inside Story of the Conflict in the Gulf* (Little, Brown, 1995).

2. Department of Defense, *Conduct of the Persian Gulf War,* Final Report to Congress, Pursuant to Title V of the Persian Gulf Conflict Supplemental Authorization and Personnel Benefits Act of 1991 (Public Law 102-25), Washington, D.C., April 1992 (hereinafter, The Pentagon Report); and U.S. House of Representatives, Committee on Armed Services, *Defense for a New Era* (Government Printing Office, March 30, 1992).

3. These papers, unpublished and edited by Patrick J. Garrity, were written under the auspices of the Center for National Security Studies, Los Alamos National Laboratory, New Mexico, circa 1993. A summary of the conclusions of these studies is Patrick J. Garrity, "Why the Gulf War Still Matters: Foreign Perspectives on the War and the Future of International Security," Los Alamos National Laboratory, Center for National Security Studies, Report no. 16, July 1993.

4. Joseph Alpher, ed., *War in the Gulf: Implications for Israel* (Boulder, Colo.: Westview Press, 1992), Report of a Jaffee Center for Strategic Studies Study Group.

5. John O'Loughlin, Tom Mayer, and Edward S. Greenberg, eds., *War and Its Consequences: Lessons from the Persian Gulf Conflict* (HarperCollins, 1994).

6. "Warfare 2020," *U.S. News and World Report,* August 5, 1996, pp. 34–42 (34–35).

7. The Pentagon Report, pp. xvii and xxii–xxiii.

8. Ibid., p. xxi.

9. Edward Luttwak, "The Gulf War in Its Purely Military Dimension," in O'Loughlin, Mayer, and Greenberg, eds., *War and Its Consequences,* pp. 33–50 (48).

10. Dunnigan and Bay, *From Shield to Storm,* p. 214.

11. Ibid., p. 194.

12. The Pentagon Report, p. 296. See also pp. 243–55 for further analyses.

13. Dunnigan and Bay, *From Shield to Storm,* p. 224.

14. The Pentagon Report, p. 138.

15. Dunnigan and Bay, *From Shield to Storm,* pp. 192–93.

16. Ibid., p. 233.

17. Ibid., p. 231.

18. The Pentagon Report, pp. 4 and 282.

19. Dunnigan and Bay, *From Shield to Storm,* p. 235.

20. The Pentagon Report, p. xxiv and 354–56, and Dunnigan and Bay, *From Shield to Storm,* p. 445.

21. Dunnigan and Bay, *From Shield to Storm,* p. 306.

22. This almost paradoxical point is made in Dunnigan and Bay, *From Shield to Storm,* p. 235.

23. See James E. Wilson, "The Fourth Dimension of Terrain," *Military Review,* vol. 26, no. 6 (September 1946), pp. 49–55.

24. The Pentagon Report, pp. 31–32.

25. Ibid., p. 69.

26. Ibid., p. 530. See also Dunnigan and Bay, *From Shield to Storm,* pp. 236–73, wherein the roles of Bedouin guides, Special Forces ground patrols, satellites, and GPS are all discussed in relation to trafficability. The Pentagon Report, p. 250, says that one Special Forces team used low-light cameras and probing equipment to determine if the terrain north of the border would support armored vehicles.

27. The Pentagon Report, pp. xiv, 750, 245.

28. Ibid., pp. 757–65.

29. Ibid., p. 256.

30. Ibid., p. 124.

31. Molly Moore, "Desert Defies Machine Age: Saudi Sun, Sand Hobble Weaponry," *Washington Post,* September 3, 1990, p. A1. See also "In the Table-Flat Land, Sand, Wind, and Flies," *New York Times,* February 6, 1991, p. A7.

32. See Dunnigan and Bay, *From Shield to Storm,* pp. 192–93, 392, and also p. 372 with regard to Special Operations Forces.

33. Molly Moore, "On the Job Training in the Saudi Desert," *Washington Post National Weekly Edition,* December 3–9, 1990, p. 15.

34. The wet weather may adversely have affected air operations, but at the same time may have aided the movement of armored forces. Hence according to Dunnigan and Bay, *From Shield to Storm,* p. 288, "some commanders reported that the two days of rainy weather during the offensive may have given them a bit of a break since that kept the sand down on the ground and out of their vehicles' engine filters." The broader problem of wet weather is also discussed on p. 276. See also Atkinson, *Crusade,* chapter 14.

35. The Pentagon Report, p. 68.

36. From Incirlik in Turkey missions were flown by F-16s, F-15Cs, F-111Es, and F-4Gs. From air bases in Saudi Arabia, Tabuk, King 'Abd Al-'Aziz, King Fahd, Al Kharj, and Taif, missions were flown by various combinations of F-15Cs, AV-8Bs, A-10s, AC-130s, F-15Es, F-16As, F/A-16As, EF-111s, and F-111Fs, as well as by F-117A stealth aircraft. Still other bases were available in Bahrain (Shaikh Isa), the United Arab Emirates (Al Minhad and Al-Dhafra), and Qatar (Doha). Within the theater, aircraft based at the above were supplemented by those (F-14, F/A-18, A-6E, AE-6B, A-7E) based on six aircraft carriers, three of which (USS *Midway, Ranger,* and *Roosevelt*) were on station in the Persian Gulf, and the other three (USS *America, Kennedy,* and *Saratoga*) in the Red Sea just south of the Sinai peninsula. But some strike missions were mounted from much further afield,

with, of course, the aid of tanker refueling. In the opening hours of the air campaign, B-52s were flown all the way from Barksdale AFB in Louisiana, traveling more than eleven hours to launch ALCMs (air-launched cruise missiles) against Iraqi military communications sites and power generation and transmission facilities. Throughout the air assault, other B-52 raids were mounted from Morón Air Base in Spain and RAF Fairford in Britain, both long-utilized U.S. bases in the context of the cold war (see The Pentagon Report, pp. 106, 108–09).

37. Ibid., pp. 402, 108.

38. Ibid., p. 403.

39. *Gulf War Airpower Survey (GWAPS)* (Government Printing Office, 1993), six volumes, survey directed by Eliot A. Cohen. The GAO report is discussed in "'Smart' Weapons Were Overrated, Study Concludes," *New York Times,* July 9, 1996, p. A1.

40. *GWAPS,* vol. 4, pp. 78, 273, 274.

41. Ibid., pp. 267, 268.

42. Ibid., pp. 222, 356.

43. Ibid., p. 353.

44. Ibid., p. 266.

45. *GWAPS,* vol. 2, p. 183.

46. Ibid., pp. 98, 99, 152.

47. Ibid., p. 110.

48. Ibid., pp. 124, 117.

49. For these adumbrations, see Robert S. Dudney, "Lebanon, Falklands: Tests in High-Tech War," *U.S. News and World Report,* August 16, 1982, pp. 24–25.

50. James R. FitzSimonds and Jan M. Van Tol, "Revolutions in Military Affairs," *Joint Forces Quarterly,* Spring 1994, pp. 24–31.

51. Eliot A. Cohen, "A Revolution in Warfare," *Foreign Affairs,* vol. 75, no. 2 (March-April 1996), pp. 37–54 (38).

52. "Get Smarter on Smart Weapons," *New York Times,* July 11, 1996, p. A22.

53. William A. Owens, "Introduction," in Stuart E. Johnson and Martin C. Libicki, eds., *Dominant Battlespace Knowledge* (Washington, D.C.: National Defense University, 1996), pp. 1–14.

54. FitzSimonds and Van Tol, "Revolutions in Military Affairs," pp. 24–31 (26).

55. Andrew Krepinevich, "Cavalry to Computer," *The National Interest,* Fall 1994, pp. 30–42. Krepinevich claims there appear to have been as many as ten military revolutions since the fourteenth century, when the Hundred Years' War (1337–1453) spawned two of them. Prior to the current MTR were the nuclear revolution after World War II, and another during the interwar period that featured improvements in internal combustion engines, aircraft design, and the exploitation of radio and radar that made possible the *blitzkrieg,* carrier aviation, modern amphibious warfare, and strategic aerial bombardment. He sees military revolutions as comprising four elements: technological change, development, operational innovation, and organizational adaptation.

56. Ibid.

57. Michael J. Mazaar, Jeffrey Shaffer, and Benjamin Ederington, *The Military Technical Revolution: A Structural Framework* (Washington, D.C.: Center for Strategic and International Studies, March 1993); and Alvin Toffler and Heidi Toffler, *War and Anti-War* (Little, Brown, 1993). See also Winn Schwartau, *Infor-*

mation Warfare (New York: Thunder's Mouth Press, 1994); and Martin C. Libicki, "What Is Information Warfare?" *Strategic Forum,* no. 28 (Washington, D.C.: Institute for National Strategic Studies, National Defense University, May 1995).

58. William J. Perry, "Desert Storm and Deterrence," *Foreign Affairs,* vol. 70, no. 4 (Fall 1991), pp. 65–82.

59. Owens, "Introduction," in *Dominant Battlespace Knowledge,* p. 3.

60. Gordon R. Sullivan and James M. Dubik, "Land Warfare in the 21st Century," *Military Review,* vol. 73, no. 9 (September 1993), pp. 13–32.

61. Anticipating our subsequent discussion of the relationship of the RMA to the geography of warfare, some analysts write of the advent of a "nonlinear battlefield." Hence, according to Kendall: The use of the term "nonlinear" implies operations that are not constrained to a linear front line. This concept can certainly include operations in depth and laterally using precision strike systems as one type of "nonlinear" operation. It also envisions a more open or disjointed form of maneuver warfare in which forces are widely separated and the front line is not clearly defined or even a valid concept. The first interpretation emphasizes the multidimensional battlefield as the area of operations in which enemy "centers of gravity" are addressed by fires in depth and laterally as well as by maneuver against linear positions. In a sense, this concept is similar to the transition from traditional chess to the three-dimensional version. (There are obviously other aspects of nonlinear operations as well, such as vertical envelopment and air and space operations in general.) [See Frank Kendall, "Exploiting the Military Technical Revolution: A Concept for Joint Warfare," *Strategic Review,* vol. 20, no. 2 (Spring 1992), pp. 23–30 (26).] See also Michael J. Mazaar, *The Revolution in Military Affairs: A Framework for Defense Planning* (Carlisle, Penn.: U.S. Army War College, Strategic Studies Institute), p. 19, who states that "rather than large units moving solidly in a single line of advance, future warfare might therefore see a more confused patchwork of dispositions, with U.S. and allied units in front of, among, and behind enemy forces." The advent of the nonlinear battlefield during the Gulf War is remarked upon in Michael Mazaar, Don Snider, and James Blackwell, *Desert Storm: The Gulf War and What We Learned* (Washington, D.C.: Center for Strategic and International Studies, 1993), pp. 142–43.

Still other writers have conjured up more colorful images of a future battlefield. Martin Libicki and Alvin Bernstein foresee the future of a "pop-up battlefield" and "fire ant warfare" (see Alvin H. Bernstein, "Conflict and Technology: The Next Generation," in Werner Kaltefleiter and Ulrike Schumacher, *Conflicts, Options, Strategies in a Threatened World* (Christian-Albrechts-University Press, 1995), pp. 145–57 (151); and Martin C. Libicki, *The Mesh and the Net: Speculations on Armed Conflict in a Time of Free Silicon* (Washington, D.C.: National Defense University, March 1994), McNair Paper no. 28. Specifically regarding the future of Middle Eastern warfare, and preceding the Gulf War, an earlier analysis of emerging MTR weapons technologies is Hirsh Goodman and W. Seth Carus, *The Future Battlefield and the Arab-Israeli Conflict* (New Brunswick, N.J.: Transaction Publishers, 1990), esp. the appendix and chapter 8 under "The Operational Art of War." Libicki refers to the overall architecture of future battlefield systems as a "mesh," conveying the possibility of incredible sensor systems whose information can be rapidly fused (data compilation and analysis, communications to firing systems), in real time, to allow for truly massive, simultaneous strikes against enemy forces (see Libicki, *The Mesh and the Net,* pp. 24–28).

Hence, according to Bernstein: The pop-up battlefield of the future will be flooded with systems whose sole purpose will be to acquire the signatures of the enemy's weapons platforms. The battlefield will be covered with *drones* (unmanned aircraft), *loitering missiles* (missiles that hover and collect information), *autonomous land crawlers* (small, unmanned vehicles that move over the terrain), *submersibles* (tiny, unmanned submarines), and small satellites. In the future they will cost far less than they do today, they will sense much more, and collectively they will be able to receive signals from every part of the *electromagnetic spectrum* (a force field that is made up of electric and magnetic components given off by all kinds of machinery). These technologies will completely transform the way we control and command our forces in the future and will also call into question the future effectiveness of such current weapons platforms as tanks, manned aircraft, surface ships, and submarines (see Bernstein, *Conflict and Technology,* p. 151).

62. FitzSimonds and Van Tol, "Revolutions in Military Affairs," p. 27.

63. See Owens, "Introduction," in *Dominant Battlespace Knowledge,* pp. 1–14.

64. "Warfare 2020," pp. 34–42 (39).

65. Ibid., p. 36.

66. Douglas A. MacGregor, "Future Battle: The Merging Levels of War," *Parameters,* vol. 22, no. 4 (Winter 1992–93), pp. 33–47 (44).

67. See Kendall, "Exploiting the Military Technical Revolution," p. 29, who states that the emerging new type of warfare "raises the potential benefit of a decisive preemptive strike."

68. For a good review of these principles, with an emphasis on military geography, see David G. Chandler, *Atlas of Military Strategy* (New York: The Free Press, 1980), esp. pp. 9–10. See also U.S. Department of the Army, *FM100-5 Operations,* Washington, D.C., May 1986, esp. chapters 6–10.

69. Some of these prospective changes are reviewed in Sullivan and Dubik, "Land Warfare in the 21st Century," pp. 13–32. See also Libicki, *The Mesh and the Net,* pp. 19–30.

70. See Johnson and Libicki, eds., *Dominant Battlespace Knowledge* (Washington, D.C.: National Defense University, 1996).

71. Thomas J. Welch, "Some Perspectives on the Revolution in Military Affairs," unpublished paper, Office of Net Assessment, Office of the Secretary of Defense, Washington, D.C., circa 1996, pp. 9, 10.

72. Paul Bracken, "The Significance of DBK," in Johnson and Libicki, eds., *Dominant Battlespace Knowledge,* pp. 51–65 (52).

73. While most of the RMA literature seems to project a coming advantage to the offense, Bernstein, "Conflict and Technology," appears to demur on behalf of the defense. So, indeed, does Libicki, *The Mesh and the Net,* p. 45, who says that "mesh warfare favors defense" but notes that in the future it will be "possible to destroy an opponent's above-the-ground civilization without being able to occupy its territory."

74. See, for instance, Harvey M. Sapolsky, "Non-Lethal Warfare Technologies: Opportunities and Problems." A report based on a conference held June 2–3, 1993, Lexington, Mass.

75. Still others made the following list recently published by *Harpers: Antitraction technology* (or "slickum"): Using airborne delivery systems or human agents, we can spread or spray Teflon-type, environmentally neutral lubricants on railroad tracks, ramps, runways, even on stairs and equipment, potentially denying

their use for a substantial period. *Roach motels* (or "stickum"): Polymer adhesives, delivered by air or on the ground, can "glue" equipment in place and keep it from operating. *Window breaking via sonic boom:* Not a totally nonlethal process for people on the streets below high-rise buildings, this could be accomplished by low-altitude supersonic flight. *Supercaustics:* Supercaustics can be millions of times more caustic than hydrofluoric acid. A round of jellied superacids could destroy the optics of heavily armored vehicles, penetrate glass, or silently destroy key weapons systems. *The grime from Hell:* A layer of paint weighing less than a gram per square meter can totally block light through a windshield, viewing window, or sensor lens. Alternatively, encountering very fine dust at the speed of a fast aircraft can microcrater (sandblast) a windshield. *Foam, sticky or hard:* To immobilize people or to render them less effective. *Anesthetics:* "Gas" could be used to put people to sleep. Tiny darts loaded with an appropriate anesthetic (fentanyl, for instance) could do this more selectively and with a wide safety margin. *Infrasound:* Very low-frequency sound generators could be tuned to incapacitate humans by causing disorientation, nausea, vomiting, or bowel spasms. The effect ceases as soon as the generator is turned off, with no lingering physical or environmental damage. (See "Secret Weapons for the CNN Era," *Harper's Magazine,* October 1994, pp. 17–18.)

76. Mazaar, Shafter, and Ederington, *The Military Technical Revolution,* p. 48. Also stressed here is the potential role of special operations forces, whose agility, stealth, and precision are said to make them ideal tools for irregular combat operations.

77. The diffusion of MTR to "peer powers" and to second-tier nations is currently under investigation at CSIS by a team of researchers headed by Daniel Gouré, pending forthcoming publication. It is also discussed in Krepinevich, "Cavalry to Computer," who notes that an initial lead in an MTR is often quickly dissipated, and that the initial leader sometimes cannot hold a lead, as was the case for the French navy in the nineteenth century.

78. Keith Krause, *Arms and the State: Patterns of Military Production and Trade* (Cambridge University Press, 1992).

79. See "The Revolution in Military Affairs," *Strategic Forum,* no. 11 (Washington, D.C.: Institute for National Strategic Studies, November 1994), which provides a taxonomy of future threats to the United States in classifying potential MTR competitors as "peer," "niche," and "regional."

80. Robert L. Butterworth, "Economic Constraints on the Revolution in Military Affairs: There Are No Competitors without Space," Arlington, Va., Aries Analytics, Inc., April 18, 1994, p. 5, prepared for the National Defense University, Institute for National Strategic Studies. See also Butterworth, "Space Systems and the Military Geography of Future Regional Conflicts," Center for National Security Studies, Los Alamos National Laboratory, Report no. 14, January 1992; and Libicki, *The Mesh and the Net,* pp. 40–43.

81. Anthony Cordesman, "Weapons of Mass Destruction in the Middle East" (Washington, D.C.: Center for Strategic and International Studies, January 22, 1996), pp. 8 and 6.

82. Robert L. Butterworth, "Economic Constraints on the Revolution in Military Affairs."

83. Robert L. Butterworth, private communication.

84. Discussions of possible asymmetric strategies on the part of various third world countries are scattered about in several of the unpublished papers emerging from the Los Alamos National Laboratory, on the lessons of the Gulf War. See, in particular, James P. Thomas, "Indian Military Lessons Learned from the Gulf War," pp. 14–16, under "Asymmetric Strategies." In particular, see Patrick J. Garrity, "Why the Gulf War Still Matters," pp. 87–90, under "The Outlines of an Asymmetrical Strategy."

85. There are some indications that newer satellite imaging systems are increasingly able not only to see through clouds, but to penetrate underground, so to speak. Hence Dunnigan and Bay, *From Shield to Storm,* p. 192, reporting on the use of the U.S. Lacrosse satellite in the Gulf War, claimed that not only can it "see through clouds and other atmospheric obstructions," but that "this satellite can detect items buried up to ten feet underground to pinpoint missiles and other equipment hidden in trenches and bunkers."

86. David E. Jeremiah, "What's Ahead for the Armed Forces," *Joint Force Quarterly,* no. 1 (Summer 1993), pp. 33–34.

87. See "For U.S. Pilots, Chief Worry in Bosnia is Weather," *New York Times,* February 16, 1994, p. A6. Libicki, *The Mesh and the Net,* pp. 84–90, has speculated on the effect of MTR developments on unconventional warfare, in rural and urban settings. He sees initially a minimal impact on rural conflict, saying that "in such realms warfare is light on platforms and heavy on cover, physical (e.g., jungle canopies) and virtual (e.g., peasant by day, fighter by night)." He says the changes in relative advantage between irregular and state forces are not easily predicted, "state forces are already easier to track; they tend to move in larger units on well-known paths," and "jungle feet need far more sensors to detect than do road trucks." Hence "until sensors become absolutely ubiquitous information technology may, if anything, increase the vulnerability of state forces." A paradox! Likewise, he says that information technology can help (and thus hurt) both sides in urban conflict.

88. Libicki, *The Mesh and the Net,* p. 44.

Chapter 7

1. "Strategic Mobility, Forward Presence, and the Defense of American Interests," final report of a conference organized by The School of International Affairs, Georgia Tech University and the U.S. Army War College, Atlanta, Georgia, September 6–7, 1991, prepared by Daniel S. Papp. See also James F. Dunnigan and Austin Bay, *From Shield to Storm* (William Morrow, 1992), pp. 238–41.

2. Albert Wohlstetter, "Illusions of Distance," *Foreign Affairs,* vol. 46, no. 2 (January 1968), pp. 242–55, who actually downplayed the distance factor.

3. Amid a fairly prolific literature in the 1980s on competitive U.S.-Soviet power projection into the Persian Gulf area, see Thomas McNaugher, *Arms and Oil* (Brookings, 1985), esp. part 1.

4. See, in particular, Jacqueline K. Davis, *Forward Presence and U.S. Security Policy: Implications for Force Posture, Service Roles and Joint Planning* (Cambridge, Mass.: Institute for Foreign Policy Analysis, 1995), National Security Paper no. 16; and "Forward Presence and U.S. Security Planning," workshop summary

prepared by National Security Planning Associates, Washington, D.C., September 29, 1994.

5. "China's Li Peng scoffs at reports of access to Burma's ocean ports," *Boston Globe,* December 29, 1994, p. 11; and William Branigan, "Asian Nations Express Concern about China's New Weapons and Bases," *Washington Post,* March 31, 1993, p. 7.

6. Michael Brzoska and Frederic Pearson, *Arms and Warfare: Escalation, De-Escalation, and Negotiation* (University of South Carolina Press, 1994); and Robert E. Harkavy, "Arms Resupply During Conflict: Framework for Analysis," *The Jerusalem Journal of International Relations,* vol. 7, no. 3 (1985), pp. 5–41.

7. One other point bears mentioning here by comparison with the recent past, and that has to do with the recent dramatically changed nature of the arms supplier markets. After a long period that saw the United States and the Soviet Union dominate those markets, up to about 65–70 percent between them, the United States has come more completely to dominate them, in addition to (see later discussion) having a huge advantage with regard to the requisite logistics and basing access. (See "Going Up, Up in Arms," *Time,* December 12, 1994, pp. 46–57, wherein it is indicated that by 1994, the United States had come to account for about 70 percent of global arms sales.) See also National Defense University, Institute for National Strategic Studies, *Strategic Assessment 1995,* Washington, D.C., 1995, chapter 11 under "Arms Transfers and Export Controls." Still, as was made clear throughout the Iran-Iraq war, other major and midlevel suppliers, France, Britain, China, both Koreas, Brazil, and Israel, do have the capacity to conduct significant intrawar supplier operations, particularly as pertains to small arms, artillery ammunition, and spare parts.

8. The tortuous and confusing historical record of U.S. policies regarding arms supplies to combatants in war, particularly with regard to embargoes, is reviewed in Robert E. Harkavy, *The Arms Trade and International Systems* (Cambridge, Mass.: Ballinger, 1975), chapter 7.

9. This relationship between arms transfers (and arms resupply) and basing access is discussed in Harkavy, *Bases Abroad,* pp. 324–46; and Harkavy, "The New Geopolitics: Arms Transfers and the Major Powers' Competition for Overseas Bases," in Stephanie Neuman and Robert Harkavy, eds., *Arms Transfers in the Modern World* (Praeger, 1979), pp. 131–51.

10. Richard Millett, "The State Department's Navy: A History of the Special Service Squadron, 1920–1940," *The American Neptune, vol. 35* (1975), pp. 118–38; and Stephen S. Roberts, "The Decline of the Overseas Station Fleets: The United States Asiatic Fleet and the Shanghai Crisis, 1932," Professional Paper no. 208 (Arlington, Va.: Center for Naval Analyses, November 1977).

11. James Cable, *Gunboat Diplomacy: 1919–1991,* 3d ed. (St. Martin's Press, 1994), esp. chapter 2.

12. For general, long-range analyses, see Cable, *Gunbook Diplomacy,* and Barry Blechman and Stephen S. Kaplan, *Force Without War* (Brookings, 1978).

13. Bruce Watson, *Red Navy at Sea: Soviet Naval Operations on the High Seas, 1956–1980* (Boulder, Colo.: Westview Press, 1982).

14. U.S. Navy, Deputy Chief of Naval Operations; Resources, Warfare Requirements, and Assessments (N8), "Force 2001: A Program Guide to the U.S. Navy, Edition '95," Washington, D.C., 1995, p. 4, and Paul Huth, *Extended Deterrence*

and the Prevention of War (Yale University Press, 1988), chapter 2 under "Conceptualizing Deterrence."

15. Generally, regarding interwar bases, see Robert E. Harkavy, *Great Power Competition for Overseas Bases: The Geopolitics of Access Diplomacy* (Pergamon Press, 1982), chapter 3.

16. Regarding the development of the British imperial basing system, see among others, D. H. Cole, *Imperial Military Geography,* 12th ed. (London: Sifton Praed, 1956).

17. Harkavy, *Great Power Competition for Overseas Bases,* pp. 75 and 78.

18. "Spain reportedly urges U.S. to quit air base near Madrid," *International Herald Tribune,* February 25, 1975, which claims that U.S. tankers based at Torrejon were used to assist the airlift to Israel in 1973.

19. These issues are discussed in Duncan L. Clarke and Daniel O'Connor, "U.S. Base-Rights Payments After the Cold War," *Orbis,* vol. 37, no. 3 (Summer 1993), pp. 441–57.

20. "United States military transport aircraft traveling to and from Persian Gulf will stop refueling in India," *New York Times,* February 20, 1991, p. A13

21. "African and Mideast Bases Aid Somalia Airlift," *New York Times,* December 18, 1992, p. A9.

22. Charles Krauthammer, "'Cheap hawks' are threatening to shoot down the B-2 bomber," *Pittsburgh Post Gazette,* July 17, 1995, p. A10.

23. James Kitfield, "The New Way of Logistics in Europe," *Air Force Magazine,* August 1994, pp. 60–63.

24. These are all listed as continuing to be operational in International Institute for Strategic Studies, *The Military Balance: 1994–1995* (London: Brassey's for the IISS, 1994).

25. "U.S. Doubling Air Power in Persian Gulf," *Washington Post,* October 28, 1994, p. A1. Therein it was reported that the United States would be able to preposition equipment for one armored brigade in each of Qatar and Kuwait, each consisting of 108 Bradley armored fighting vehicles and 58 M1A1 Abrams tanks. The A-10s would be based at the Al Jaber base in Kuwait. Earlier, regarding U.S. basing access in the mid-1980s in connection with the contingency of a Soviet thrust through Iran toward the Persian Gulf, see McNaugher, *Arms and Oil,* esp. pp. 53–64, with maps and aircraft range radii depicting U.S. strike capabilities from eastern Turkey, Dhahran, Incirlik, Abadan, and Israel, using F-4E, A-10, F-16, A-7D/E, F-111 D/E, A-6E, and F-15E aircraft.

26. "With Thai Rebuff, U.S. Defers Plan for Navy Depot in Asia," *New York Times,* November 12, 1994, p. 6. Even in Japan, U.S. bases might be in jeopardy, apart from the possible connection to trade issues, as see "Some Leaders in Japan Begin to Question U.S. Bases," *New York Times,* August 28, 1994, p. A1.

27. "Vietnamese Hint the U.S. Could Use Port Again," *New York Times,* November 24, 1994, p. A12. Also, the United States had acquired enhanced access to Singapore and Brunei in the aftermath of the Gulf War, as reported in "U.S. Seeks Access to Bases in Asia," *New York Times,* July 27, 1991, p. A1.

28. John Mintz, "A Floating Arsenal of Democracy for the 21st Century," *Washington Post National Weekly Edition,* July 1–7, 1996, p. 29.

29. A critique of the U.S. Navy's plans for its arsenal ships, also containing a plea for the retention of the Iowa-class battleships for similar purposes, is in

William L. Stearman, "The Navy's Plans for the Arsenal Ship Won't Fly," *Washington Post National Weekly Edition,* July 22–26, 1996, p. 23.

30. Regarding underwater cables prior to World War I, see Paul M. Kennedy, "Imperial cable communications and strategy, 1870–1914," *English Historical Review,* vol. 86, no. 141 (1971), pp. 728–52.

31. There is an enormous range of communications facilities, running along a frequency spectrum from extra-low frequency to ultrahigh frequency, with each part of the frequency band corresponding to, or being preferable for, certain kinds of military activity. Under the general heading of intelligence facilities are a variety of ground links for reconnaissance, SIGINT, early warning and other military satellites, satellite control stations, early warning ground radars, nuclear detection arrays, telescopes for detecting others' satellites, underwater sonar arrays for detecting submarines, navigational aids, solar flare detectors, and others. These matters are discussed in Jeffrey Richelson and Desmond Ball, *The Ties That Bind* (Boston, Mass.: Allen and Unwin, 1985); William Arkin and Richard Fieldhouse, *Nuclear Battlefields: Global Links in the Arms Race* (Cambridge, Mass.: Ballinger, 1985); Richelson, *The U.S. Intelligence Community* (Cambridge, Mass.: Ballinger, 1985); and Robert E. Harkavy, *Bases Abroad: The Global Foreign Military Presence.*

32. "Scientists Fight Navy Plan to Shut Far-Flung Underseas Spy System," *New York Times,* June 12, 1994, p. 1. The scope of the cold war era U.S. SOSUS global network is revealed in Jeffrey Richelson, *The Ties That Bind,* pp. 200–02. But a recent U.S. Navy publication, *Force 2001: A Program Guide to the U.S. Navy* (Washington, D.C.: Deputy Chief of Naval Operations, 1995), p. 97, discusses an upgrade program for SOSUS. These are all listed as continuing to be operational in IISS, *The Military Balance: 1994–1995.* See also "Pentagon Adds 83 Bases to Europe Cutbacks," *Washington Post,* January 31, 1992, p. A6, which reports that by that date, the United States had closed 463 of the 1,402 military sites that had existed during the cold war.

33. John Pomfret and David B. Ottoway, "From Pariah to Pal," *Washington Post,* National Weekly Edition, November 27–December 3, 1995, p. 16.

34. An additional earlier source in this area is J. R. Blaker, S. J. Tsagronis, and K. T. Walter, *U.S. Global Basing: U.S. Basing Options,* Report for the U.S. Department of Defense, HI-3916-RR (Alexandria, Va.: Hudson Institute, October 1989).

35. In addition to the projections outlined in the U.S. Navy's *Force 2001,* see Donald C. F. Daniel and Bradd C. Hayes, "The Future of U.S. Sea Power," Carlisle, Penn.: Strategic Studies Institute, U.S. Army War College, May 19, 1993, pp. 13–14, wherein are outlined, from a 1993 perspective, the Bush Administration's "Base Force," and then Secretary of Defense Aspin's Force C and Force B options. More generally, regarding future budgetary and force structure projections, including the U.S. Navy, see Harlan K. Ullman, *In Irons: U.S. Military Might in the New Century* (Washington, D.C.: National Defense University, 1995).

36. IISS, *The Military Balance: 1994–1995,* p. 31.

37. Daniel and Hayes, "The future of U.S. Sea Power," p. 24.

38. Overflights are discussed in R. E. Harkavy, *Bases Abroad,* pp. 95–100, drawing on a previous work by P. M. Dadant, "Shrinking International Airspace as a Problem for Future Air Movements, A Briefing," Report R-2178-AF (Santa Monica, Calif.: RAND Corporation, 1978).

39. These and immediately subsequent data are drawn from IISS, *The Military Balance: 1994–1995.*

40. Ibid., pp. 171–72.

41. In addition to sources previously cited, see also Ross H. Munro, "China's Waxing Spheres of Influence," *Orbis,* vol. 38, no. 4 (Fall 1994), pp. 585–605.

42. IISS, *The Military Balance: 1994–1995,* p. 176.

43. The development of the Indian navy and its strategic implications are covered in Matthew Gurgel, "The Indian Navy: Striving for Protection or Domination?" *Harvard International Review,* vol. 14, no. 2 (Winter 1991–92), pp. 54–59.

44. IISS, *The Military Balance: 1994–1995,* pp. 154–55.

45. Ibid., pp. 128.

46. For a discussion of some of these matters, see "The Paladin of Jihad," *Time,* May 6, 1996, pp. 51–52.

47. Johnnie E. Wilson, "Power Projection Logistics Now . . . and in the 21st Century," *Army,* October 1994, pp. 137–43.

48. IISS, *The Military Balance: 1994–1995,* p. 30.

49. Hans Mark, "C-17 brings new era in airlift," *Aerospace America,* March 1994, pp. 26–30; and David F. Bond, "MAC Faces Widening Gap in Peacetime, Crisis Needs," *Aviation Week and Space Technology,* September 9, 1991, pp. 48–49.

50. Information in unpublished information paper provided the authors by the Logistics Management Institute, McLean, Va.; under the heading of "Strategic Airlift Acquisition."

51. IISS, *The Military Balance: 1994–1995,* p. 113.

52. Ibid., p. 46.

53. Ibid., p. 26.

54. Dan Morgan, "Military Shipbuilders Procure a Windfall: The Sealift Program Gets a $1 Billion Lifeline from the House," *Washington Post National Weekly Edition,* July 8–14, 1996, p. 29.

55. Information provided in unpublished information paper provided the authors in "Strategic Sealift Status." See also the U.S. Navy's *Force 2001,* pp. 37–38, under "Strategic Sealift," which indicates that the DOD Mobility Requirements Study has concluded that "the Navy should acquire additional roll-on–roll-off (ro-ro) ships to provide required surge lift for two army heavy divisions."

56. IISS, *The Military Balance: 1994–1995,* p. 26.

57. Ibid., pp. 31–33.

58. Les Aspin, Secretary of Defense, "Report on the Bottom-Up Review," Washington, D.C., Department of Defense, October 1993.

59. IISS, *The Military Balance: 1994–1995,* p. 32.

60. Anthony Cordesman, "U.S. Forces in the Gulf: Resources and Capabilities," Center for Strategic and International Studies, Washington, D.C., February 1996, pp. 48–49 and 51, 53.

61. Ibid.

62. Ibid., p. 50.

63. *Force 2001,* p. 125.

64. Ibid., pp. 7–9.

65. *U.S. News and World Report,* "Warfare 2020," August 5, 1996, pp. 34–42.

66. The material on the key choke points, Suez, Bab el Mandeb, and Hormuz, is taken from Geoffrey Kemp with R. Adm. Harold Bernsen (Ret.), *Challenges in Strategic Waters: Suez Canal, Bab el Mandeb, Strait of Hormuz,* CRM 93-235 (Alexandria, Va.: Center for Naval Analysis, January 1994).

Chapter 8

1. This is discussed in Anthony Cordesman, *After the Storm: The Changing Military Balance in the Middle East* (Boulder, Colo.: Westview Press, 1993), pp. 484–89.

2. "Nuclear Trafficking in Europe," U.S. House of Representatives, House Republican Research Committee, Washington, D.C., November 30, 1992; and "Soviet Breakup Creates a Bazaar in Nuclear Goods," *Washington Post,* May 15, 1993, p. A1.

3. Whether merely anecdotal or not, this is mentioned in Alvin Bernstein, "Conflict and Technology: The Next Generation," in W. Kaltefleiter and U. Schumacher, eds., *Conflicts, Options, Strategies in a Threatened World* (Christian-Albrechts-University, 1995), p. 147. Elsewhere, this has been attributed to Indian general K. Sundarji, in George Quester and Victor Utgoff, "No-First-Use and Nonproliferation: Redefining Extended Deterrence," *Washington Quarterly,* vol. 17, no. 2 (Spring 1994), pp. 103–14 (107).

4. Martin van Creveld, *Nuclear Proliferation and the Future of Conflict* (New York: Free Press, 1983). See also Kenneth Waltz and Scott Sagan, *The Spread of Nuclear Weapons: A Debate* (Norton, 1995).

5. Kenneth Waltz's thesis was propounded earlier in "The Spread of Nuclear Weapons: More May Be Better," Adelphi Paper no. 171 (London: Brassey's for the IISS, 1987). For specific application to the Arab-Israeli equation, an argument on behalf of nuclear stability was made earlier by Shai Feldman, *Israeli Nuclear Deterrence: A Strategy for the 1980s* (Columbia University Press, 1982); and in Steven J. Rosen, "Nuclearization and Stability in the Middle East," in Onkar Marwah and Ann Schulz, eds., *Nuclear Proliferation and the Near-Nuclear Countries* (Cambridge, Mass.: Ballinger, 1975). In a related vein the impending advent of a "new mini-MAD strategy for the Mideast" is heralded in Dan Raviv and Yossi Melman, "The Mideast Goes MAD," *Washington Post,* July 15, 1990, p. B12. See also Gerald M. Steinberg, *Deterrence, Defense or Arms Control? Israeli Perception and Response for the 1990s* (Santa Monica, Calif.: California Seminar on International Security and Foreign Policy, 1990). For an approach more closely tied to various aspects of international relations theory, particularly involving the impact of systems structure, see Zachary S. Davis and Benjamin Frankel, eds., *The Proliferation Puzzle* (London: Frank Cass, 1993). Among the earlier relevant works, see in particular Leonard Spector and Jacqueline R. Smith, *Nuclear Ambitions: The Spread of Nuclear Weapons, 1989–1990* (Boulder, Colo.: Westview Press, 1990), and Leonard Spector, *The Undeclared Bomb* (Cambridge, Mass.: Ballinger, 1988). Specifically, with regard to the Middle East, see Anthony Cordesman, *Weapons of Mass Destruction in the Middle East* (London: Brassey's, 1991); Shlomo Aronson, *The Politics and Strategy of Nuclear Weapons in the Middle East* (State University of New York Press, 1992); and Mitchell Reiss and Robert S. Litwak, eds., *Nuclear*

Proliferation After the Cold War (Washington, D.C.: The Woodrow Wilson Center, 1994).

6. "China Raises Nuclear Stakes on the Subcontinent," *New York Times,* August 27, 1996, p. A6.

7. Cordesman, *After the Storm,* p. 484.

8. Ibid., p. 485. Regarding missile ranges, see also "Missile and Launch Capabilities of Selected Countries," *The Nonproliferation Review,* vol. 2, no. 3 (Spring-Summer 1995), pp. 203–06.

9. Cordesman, *After the Storm,* p. 485.

10. Ibid., pp. 493, 488, 491, 492.

11. Ibid., pp. 493–94.

12. Ibid.

13. Anthony H. Cordesman, *Weapons of Mass Destruction,* CSIS, Washington, D.C., January 22, 1996, pp. 11–13.

14. Cordesman, *After the Storm,* p. 417.

15. Ibid., pp. 418, 419.

16. Soviet anxieties over the range implications of this testing are discussed in "Battle of Jericho: Moscow's Minuet Over a Missile," *Time,* August 10, 1987, p. 22; and "Soviet Cautions Israel Against a New Missile," *New York Times,* July 29, 1987.

17. Cordesman, *Weapons of Mass Destruction,* p. 7.

18. Marc Dean Millot, Roger Molander, and Peter A. Wilson, "The Day After . . . Study—Nuclear Proliferation in the Post–Cold War World," vol. 2, Santa Monica, Calif.: RAND Corporation, 1993, p. 43.

19. This was discussed earlier in Robert E. Harkavy, *Specter of a Middle Eastern Holocaust: The Strategic and Diplomatic Implications of the Israeli Nuclear Weapons Program,* Monograph Series in World Affairs (University of Denver, 1977), pp. 33–41.

20. R. Jeffrey Smith, "The Arming of Pakistan," *Washington Post Weekly Edition,* September 2–8, 1996, p. 16.

21. "Nuclear Ambitions," *U.S. News and World Report,* pp. 42–43.

22. Angelo Codevilla, "Defenseless America," *Commentary,* vol. 102, no. 3 (September 1996), pp. 51–52.

23. "'Special Ops': The Top-Secret War," *Newsweek,* March 18, 1991, p. 32.

24. William E. Burrows and Robert Windrem, *Critical Mass* (New York: Simon and Schuster, 1994), p. 286, who refer to "bunkers cut into the Judean hills just west of the town of Zekharyeh"; and Seymour Hersh, *The Samson Option* (New York: Random House, 1991), pp. 215–16, who refers to "three or more missile launchers in place and operational at Hirbat Zachariah."

25. Seth Carus, *Cruise Missile Proliferation in the 1990s* (Westport, Conn.: Praeger, 1992), The Washington Papers, no. 159. For general background on cruise missiles, see also Ronald Huisken, *The Origins of the Strategic Cruise Missile* (Praeger, 1981).

26. Carus, *Cruise Missile Proliferation in the 1990s,* pp. 17–27.

27. Ibid., pp. 40, 54–57, 59, 60, 66, 68.

28. This is emphasized in Spector, *The Undeclared Bomb,* pp. 47–57, who discusses the technical requirements for aircraft delivery of nuclear weapons.

29. Gordon M. Burck and Charles C. Floweree, *International Handbook on Chemical Weapons Production* (New York: Greenwood Press, 1991), p. 508.

30. Burrows and Windrem, pp. 349–52, discuss such crises in 1984, 1990, and 1992, claiming also that in 1984 both sides were threatening to attack each other's nuclear facilities. Seymour Hersh thinks that Pakistan and India have come close to a nuclear war. See his "On the Nuclear Edge," *New Yorker,* March 29, 1993, pp. 56–73. His claims are called exaggerated in Lewis Dunn, "Rethinking the Nuclear Equation," *Washington Quarterly,* vol. 17, no. 1 (Winter 1994), footnote 4 on p. 24.

31. Burrows and Windrem, p. 286, refer to nuclear-armed "black squadrons" based at Tel Nof.

32. Eric H. Arnett, *Gunboat Diplomacy and the Bomb: Nuclear Proliferation and the U.S. Navy* (Praeger, 1989), pp. 55–56, 67–69.

33. Ibid., p. 106, even speculates that SSTs (submarines carrying torpedoes) "would also provide India with an interesting new threat to China," though it is deemed a low priority.

34. See Robert E. Harkavy, *Preemption and Two-Front Conventional Warfare,* Jerusalem Papers on Peace Problems (The Hebrew University of Jerusalem, 1977).

35. This is discussed in Harkavy, *Specter of a Middle Eastern Holocaust,* p. 91.

36. But in an escalating technological race, the United States is developing various "counter-proliferation" technologies to deal with the problems of dispersion and deep underground facilities. See "Critical Mass," *U.S. News and World Report,* April 17, 1995, esp. pp. 40–41 under "Battlefield 2000."

37. The following analysis and data are drawn from Office of Technology Assessment, *Proliferation of Weapons of Mass Destruction: Assessing the Risks,* U.S. Congress, OTA-ISC-559, Washington, August 1993; and Cordesman, *Weapons of Mass Destruction.*

38. Cordesman, *Weapons of Mass Destruction in the Middle East,* p. 24.

39. Ibid., p. 20.

40. Ibid.

41. Steve Coll, "Nuclear Goods Traded In Post-Soviet Bazaar," *Washington Post,* May 15, 1993, p. A1.

42. Ibid.

43. Ibid. See also "The Russia Connection," *Time,* July 8, 1996, pp. 32–36.

44. "Nuclear Trafficking in Europe," p. 1.

45. Ibid., p. 5. See also John Pomfret and David B. Ottoway, "Keeping the Pipeline Well-Armed," *Washington Post National Weekly* Edition, May 20–26, 1996, p. 14, wherein arms smuggling routes to Bosnia and Croatia from Eastern Europe and from Turkey and Iran (by air and sea) are mapped and discussed.

46. Ibid., p. 12.

47. "The Russia Connection," *Time,* July 8, 1996, pp. 32–36.

Chapter 9

1. *Conduct of the Persian Gulf War.* Pursuant to Title 5 of the Persian Gulf Conflict Supplemental Authorization and Personnel Benefit Factors of 1991 (Public Law 102-25), (Washington, D.C.: Department of Defense, April 1992).

2. Ibid., p. 148.

3. Ibid., pp. 150, 151, 158.

4. Ibid., p. 159.

5. In the heyday of nuclear strategy the concept of "assured destruction" was introduced into the lexicon to provide defense planners with guidelines as to the requirements for survivable nuclear forces needed to assure various levels of destruction against an adversary's population and industrial base. It was U.S. secretary of defense Robert McNamara who quantified the concept. His analysis of the requirements for deterrence can be reduced to several distinct but related components: (1) Nuclear deterrence requires that the United States maintain the capability to deter deliberate attack upon both itself and its allies. (2) This can be achieved by "maintaining a highly reliable ability to inflict unacceptable damage upon any single aggressor or combination of aggressors at any time during the course of a strategic nuclear exchange, even after absorbing a surprise first attack. This is called the "assured destruction" capability. (3) When calculating the nuclear forces required to maintain this posture, two criteria are necessary: (a) all estimates of both a potential aggressor's capability and his intentions must be conservative and assume a "worst plausible case"; and (b) it is necessary to determine the level of potential destruction that would have to be achieved to maintain the deterrence posture. In his book, *The Essence of Security,* published in 1968, McNamara stated the requirement as follows: "In the case of the Soviet Union, I would judge that a capability on our part to destroy, say, one-fifth to one-fourth of her population and one-half of her industrial capacity would serve as an effective deterrent. (4) By translating these goals into force levels, it can be shown that the required damage would be achieved if between 200 and 300 MTE (megaton-equivalence) was optimally delivered against Soviet industrial and population centers. [Taken from Geoffrey Kemp, *Nuclear Forces for Medium Powers: Part I: Targets and Weapons Sytems,* Adelphi Papers no. 106 (London: International Institute for Strategic Studies, 1974).]

Are there any lessons from this doctrine that apply in the era of advanced conventional munitions? The first point to note is that Mr. McNamara's checklist of targets includes both civilian population and industrial base. It does not refer to an attack only on industrial targets. Yet when examining the potential uses of advanced conventional munitions, the deliberate targeting of population centers is rarely, if ever, mentioned. For good reason: the utility of ACMs is their accuracy and ability to destroy both hard and soft targets with limited amounts of explosive power. Thus while they are ideally suited for military targets, both mobile and static, hard and soft, as well as specific industrial facilities and infrastructure, their utility against general population centers would be little better than traditional conventional munitions. In this regard the history of counter-population attacks with conventional weapons suggests that unless used on a massive scale similar to the World War II strategic bombing campaign against Germany and Japan or the destruction of cities by the artillery of advancing armies, the disparity between nuclear, biological, or chemical weapons and conventional munitions for this role is very great.

6. See David Rodman, "Regime-Targeting: A Strategy for Israel," *Israel Affairs,* vol. 2, no. 1 (Autumn 1995), pp. 153–67.

7. Anthony Cordesman, *After the Storm: The Changing Military Balance in the Middle East* (Boulder, Colo.: Westview Press, 1993), pp. 282–82.

8. Ibid., p. 283. On p. 281, Cordesman speculates that given the accuracy of the SS-21 missiles, "Syria could eventually fire nerve agents successfully at Israeli air bases, C^3I sites, Dimona, and mobilization centers—and seriously degrade Israeli conventional and nuclear capabilities."

9. Ibid., p. 214.

10. Ibid. See also Robert W. Swartz, "Gulf War Lessons Learned: Middle Eastern Perspectives," Los Alamos National Laboratory, 1993, p. 5, unpublished paper. Swartz addresses the problem of escalation dominance in connection with Syria's possible use of chemical weapons. Hence, "Another problem for Syrian reliance on unconventional weapons as a deterrent may be Israeli escalation superiority because of Jerusalem's perceived nuclear weapon capabilities. This escalation superiority might enable Israel to conduct a conventional strategic bombing campaign in a future war with Syria, along the lines of Operation Desert Storm, while Damascus is unable to use its chemically armed SSM's [surface-to-surface missiles] for fear of being attacked with nuclear weapons."

11. Patrick Garrity, *Why the Gulf War Still Matters: Foreign Perspectives on the War and the Future of International Security,* Center for National Security Studies, Los Alamos National Laboratory, Report no. 16, July 1993, p. 81.

12. On this point, to the contrary, p. 80, Garrity states that: "The Gulf War in particular served to reinforce the Israeli military view that territorial depth remains an essential means of hedging against strategic surprise and the possibility of initial defensive failure." See also Swartz, "Gulf War Lessons Learned," p. 9, wherein: "Israeli leaders also paid close attention to the role of territory in the military operations of the second Gulf War, and concluded that, despite earlier claims to the contrary, land still matters."

13. Garrity, *Why the Gulf War Still Matters,* pp. 86–90, under "The Outlines of an Asymmetrical Strategy."

14. John Pomfret, "Some Neighborly Advice: Turkey Strengthens Its Ties to Israel," *Washington Post National Weekly Edition,* June 10–16, 1996, p. 16.

15. This is discussed in Insight Team of the London Sunday *Times, The Yom Kippur War* (Doubleday, 1974), pp. 282–84.

16. A more detailed analysis of this is found in Robert E. Harkavy, "After the Gulf War: The Future of Israeli Nuclear Strategy," *Washington Quarterly,* vol. 14, no. 3 (Summer 1991), pp. 161–79.

17. There are alternative scenarios that would result not only from significant political changes, but also from military force changes, in the latter instance, that probably would entail the coming to fruition of significant and irreversible (via preemption) nuclear programs in Iraq, Iran, and maybe Egypt. On the political side, the possibilities can run in any number of directions. For the United States the most ominous development would be the coming to fruition of a larger and hostile regional coalition, presumably based on Islamic or pan-Arab solidarity or both, which would pose a much greater challenge than the standard scenario, based on the Gulf War, of a conflict with either Iraq or Iran. That expanded coalition could involve some combination, or all, of the following: Iran, Iraq, Syria, Egypt, Pakistan, Turkey, Libya, Jordan, perhaps even some Gulf states. At the extremes that could involve Islamic revolutions in Egypt, Turkey, Jordan, and maybe Saudi Arabia. If such an extreme situation were to develop, it would mean multiple sources of WMD threats (to what extent coordinated would be an interesting question), as well as, obviously, either the complete or near-complete absence of basing access for the United States in the region. It might also pose a greater threat to the United States via asymmetric strategies, specifically in such a case, that of terrorism. One major question in such a scenario would be whether such a large-scale Islamic or pan-Arab coalition would be availed of outside support in the form

of arms resupply during a conflict, perhaps from Russia or (maybe unlikely unless NATO unravels) Western Europe, or maybe China, begging all manner of questions in relation to a priori force structures or orders of battle, logistics, etc.

Whatever the number and identities of actors involved in various combinations, several other general factors will form the basic framework of future conflict scenarios: objectives of the adversary(ies), extent of intraregional, that is, Arab, support for U.S. or coalition objectives, the extent of Western support, and, of course, military factors, particularly the development of RMA or partial RMA capabilities at various future points in time. Objectives will be scenario dependent (defeat of Israel, overrunning of Kuwait, overturning of regimes in GCC states, etc.), but can also be classified along the lines of all-out war and total defeat of enemies versus more limited goals, such as "seize and hold" (as in 1973 and perhaps in 1980 on behalf of Iraq in Khuzestan), punitive operations to teach opponents a lesson, humiliation of adversaries, testing of resolve, and salami tactics. Arab political support will obviously represent degrees all the way from one end to the other of our limiting case scenarios, no doubt highly dependent on the maintenance of the status quo vis-à-vis the trend toward Islamic fundamentalism and anti-Westernism, but also no doubt dependent on regional perceptions about U.S. resolve, staying power, and ability to protect friends and clients. The extent of Western support involves fundamental questions about the future cohesion of NATO, both in general and with respect to coordination of Middle Eastern policies, always involving the possibility of a divorce when it comes to Israel and oil. Broader international support can vary as well and that ramifies into questions of basing access (India, Thailand, Singapore, and Russia were all at issue in 1990–91) as well as the possibilities of U.S. foes receiving arms resupply during wars.

18. See, for instance, "United States military transport aircraft traveling to and from Persian Gulf will stop refueling in India," *New York Times,* February 20, 1991, p. A13; and "Decision to allow United States military aircraft to refuel in Bombay's international airport en route from Pacific to Persian Gulf sets off political outcry that could bring down government of Prime Minister Chandra Shekhar," *New York Times,* January 30, 1991, p. A7.

19. Even these forces were reported suffering from a lack of "readiness," that is, a weakening in terms of supplies, ammunition, weapons of all sorts, and training. There were numerous reports and analyses pointing to the re-emergence of the "hollow" army that had been filled up and strengthened during the fat years of the Reagan era. This theme is pursued in detail in Harlan K. Ullman, *In Irons: U.S. Military Might in the New Century* (Washington, D.C.: Institute for National Strategic Studies, National Defense University, 1995). In particular, there were questions about personnel, relative to the finely honed force that had fought in 1990–91, almost self-consciously as a comeback from the perceived defeats and humiliations of the Vietnam era. For instance, one recent report, commenting on the drawdowns called for by the Bottom-Up Review and follow-on budgetary decisions by the Clinton Administration, claimed they will represent a 27 percent overall cut in military manpower since Desert Storm, a 45 percent cut in active army divisions, a 50 percent cut in army reserve components, a 37 percent cut in combat ships including a 27 percent cut in carriers, a 23 percent cut in active air carrier wings, a 50 percent cut in reserve carrier wings, a 45 percent cut in U.S. Air Force fighter wings, and a 42 percent cut in U.S. Air Force reserve active wings. That will translate, on a more specific level, into cuts of 34, 41, 17, 25, 53, and

31 percent, respectively, of tanks, other armored fighting vehicles, artillery, surface-to-air missile launchers, short-range air defense systems, and tactical wheeled vehicles. The U.S. Air Force will have cut its bomber strength by 44 percent and its active fighter-attack strength by 48 percent. The U.S. Navy will have similar levels of cuts all across the board (Anthony Cordesman, "U.S. Forces in the Gulf: Resources and Capabilities," Center for Strategic and International Studies, Washington, D.C., February 1996, pp. 32–33).

20. The Clinton/Aspin "Bottom-Up Review" called for U.S. plans to meet potential aggressors (Iran, Iraq, North Korea) capable of fielding forces of 400,000–700,000 men, 2,000–4,000 tanks, 3,000–3,500 armored fighting vehicles, 2,000–3,000 artillery pieces, 500–1,000 combat aircraft, 100–200 naval vessels, up to 50 submarines and 100–1,000 Scud-class missiles (maybe armed with nuclear, chemical, and biological warheads). See Cordesman, "U.S. Forces in the Gulf," p. 14.

21. Patrick Garrity, *Why the Gulf War Still Matters,* pp. 87–92.

22. Ibid., p. 89. See also James Thomas, "Indian Military Lessons Learned from the Gulf War," Los Alamos National Laboratory, Center for National Security Studies, New Mexico, August 1992, unpublished paper, for a specific discussion of asymmetric strategies being studied by the Indian military as part of a "high-low" mix of technological responses, particularly regarding electronic warfare and the countering of stealth technologies by the employment of electro-optical devices.

23. It is recognized that some kinds of military or counterterrorist operations depend largely on surprise of a sort that can happen only once. Israel, for instance, will presumably never again have the favorable conditions that it had for the Entebbe raid, for the simple reason that in a similar situation, the terrorists would have the sense to move the hostages well away from the airport. Similarly, the Osirak raid of 1981 can never again be repeated, because any aspiring nuclear power will now have the sense to move its nuclear facilities underground and to disperse them. But whether an Iran or an Iraq, knowing the history and the technological underpinnings of the "left hook" of the past, will be able to counter it with an improved, presumably more mobile and dispersed strategy, is not easy to say. Capabilities may not so easily follow cognitions. But any strategy that would result in greater casualties for a newer coalition effort might be a success in the sense that to the extent it was anticipated by U.S. intelligence, it might deter a repeat of the previous successful and almost unchallenged operation.

24. The emerging technological basis for preemptive counterproliferation is outlined in "Critical Mass," *U.S. News and World Report,* April 17, 1995, pp. 39–45, esp. pp. 40–41 under "Battlefield 2000."

25. John Donnelly, "Reimer: U.S. Stronger Now Against Mass-Destruction Weapons," *Defense Week,* June 24, 1996, p. 7. See also "The Return of the Jedi?" *U.S. News and World Report,* April 1, 1996, p. 29 regarding new defensive systems against "regional" missile threats.

26. *U.S. News and World Report,* see "Warfare 2020," August 5, 1996, pp. 34–42.

Chapter 10

1. Steven E. Plaut, "The Middle East and the Venice Model of Development," *Mediterranean Quarterly,* Summer 1996, pp. 72–86.

2. *Claiming the Future: Choosing Prosperity in the Middle East and North Africa* (Washington, D.C.: World Bank, 1995), p. 1.

3. Ibid., p. 2.

4. Peter Stalker, *The Work of Strangers: A Survey of International Labour Migration,* International Labor Organization, 1994, p. 241.

5. See Bhikhu Parekh, "The Indian Diaspora: Spreading Wings," *Ethnic Newswatch,* February 28, 1994; "India Airlifts Citizens from Gulf," *Reuter Library Report,* January 14, 1991.

6. Stalker, *The Work of Strangers,* p. 242.

7. The Council on Foreign Relations and the World Economic Forum, "The Casablanca Report: Results of the Middle East/North Africa Economic Summit, October 30–November 1, 1994, Casablanca, Morocco."

8. Sarah Helm, "Mideast Water Wars," *World Press Review,* vol. 42, no. 1 (January 1995), p. 37.

9. Two other canal projects are under consideration within the context of the peace process. One is the Mediterranean-Dead Sea canal beginning at Haifa, which will channel water from the Mediterranean to the north of the Jordan Valley for desalination in reverse osmosis plants and storage in a new lake in the Rift Valley for domestic and agricultural use. The remaining brine would be disposed of through a concrete-lined canal through which it would pass into the Dead Sea. The other is the Dead Sea canal beginning in the Gaza Strip where 3,177 cubic feet of sand-free water would be pumped per second and discharged through a pressure pipeline into the canal and toward an 800-megawatt power-generating station on the Dead Sea. It has been argued that the Red-Dead canal is the only option likely to receive the full backing of the Jordanian government because of its location. (See Toby Ash, "Projects of Peace; Israeli-Jordanian Investment Projects," *MEED Middle East Business Weekly,* November 25, 1994.)

10. "Paying for Jordan's Peace," *Project and Trade Finance,* no. 140 (December 21, 1994), p. 6.

11. Helm, "Mideast Water Wars," p. 37.

12. Joseph R. Gregory, "Liquid Asset," *World Monitor,* vol. 4, no. 11 (November 1991), pp. 32 and 30.

13. Ibid., p. 32.

14. The United States was proposing nuclear desalinization plants in the Sinai in the mid-1960s.

15. Francois Duchene, *Jean Monnet: The First Statesman of Independence* (Norton, 1994) and Shimon Peres with Arye Nqor, *The New Middle East* (New York: Henry Hall, 1993).

16. For a cogent presentation of this point of view, see Eliyahu Kanovsky, "Middle East Economies and Arab-Israeli Peace Agreements," *Israel Affairs,* vol. 1, no. 4 (Summer 1995), pp. 22–39.

17. Ministry of Transport, the Hashemite Kingdom of Jordan, "Infrastructure in Jordan—In the Regional Context," paper prepared for the Middle East/North Africa Economic Summit, October 30–November 1, 1994, Casablanca, Morocco, p. 3.

18. For a wider view of these and other proposals, see Jim Lederman "The Investments that Cement Arab-Israeli Peace," *Middle East Quarterly,* vol. 3, no. 1 (March 1996), pp. 33–42. Lederman argues that the most likely areas for cooperative ventures are roads, electrical grids, telecommunications, and energy pipelines.

19. *Freight Land Transportation in Egypt: Present Status and Future Prospects* (brochure) (American Chamber of Commerce in Egypt, January 1995), p. 27.

20. Because of Turkey's close ties with Germany, the present road from Germany to Turkey via Austria, Hungary, the former Yugoslavia, and Bulgaria is well traveled and, on the whole, quite good. Early in the next century or possibly even before, a major road link between northern Europe and the Middle East will be possible. From Aberdeen to Aqaba, Edinburgh to Eilat, or Rotterdam to Riyadh by divided highway will be possible since much of the road system is already in place. Consider, for instance, the Aberdeen to Aqaba route. Much of this route is already serviced by good roads. One can now drive from Aberdeen to Dover on divided highway, board a special car train through the Channel Tunnel and then drive all the way to Vienna on excellent highways. The road between Vienna and Budapest has some 50 miles of nondivided highway at present; between Budapest and Belgrade there is another 100-mile section that needs to be completed. A 50-mile gap between Belgrade and Sofia and a 65-mile gap between Sofia and Istanbul still exist. But from Istanbul to Ankara there is a divided road. By far the largest stretch of undivided highway is between Ankara and Adana (at least 250 miles). On reaching Adana there is another 60-mile section of undivided highway before Aleppo, at which point there is a good road system all the way from Aleppo through Amman to Aqaba.

In other words, out of the total distance of some 3,300 miles from Aberdeen to Aqaba, only stretches amounting to 575 miles remain to be double paved, that is, over 80 percent of the route is already divided highway. (This is far more impressive than the Pan-American highway stretching from Alaska to deep into South America. The Pan-American highway is undivided though much of Central America and still has uncompleted sections in Panama.) A divided highway is not essential but it would make a huge difference for mass transportation. A completed highway system of the kind described above would open up the entire European Union to the Middle East and vice versa. The possibilities that could follow upon this are overwhelming. There are excellent roads in Saudi Arabia and when Iraq gets back on its feet after the demise of Saddam Hussein, it, too, will be able to resume its very impressive road building activity, which was started in the late 1970s. In short, if the peace process continues, by early in the next century easy driving from northern Europe to the Persian Gulf and Indian Ocean may become a reality. The economic, social, political, and strategic implications of such a development could rival those of the transcontinental railroads across North America in the nineteenth century, an event that catapulted the United States to world power status by the early twentieth century.

A major road network from the European Union into the Middle East would have profound and probably long-lasting consequences. Driving from the north to the Arabian Peninsula (and then on to India, at some point) has always been possible since the invention of the automobile, but the constraints and barriers along the route have by and large proven far too formidable for all but the most intrepid summer students seeking adventure. But if there were a divided highway to India and no political problems crossing borders, commerce would explode. Americans think nothing of driving across the United States, which is about 3,000 miles coast to coast. Why should Germans not make a trip of the same distance to Pakistan?

21. Driving on India's main highways is a nightmare. Accidents occur every few miles, making the single-lane highways subject to appalling delays. The truck traffic is dominated by a cadre of seasoned and very hardened drivers who use amphetamines and other diversions to help them on their travels. No one who has driven on Indian roads forgets the experience, and the demand for road modernization is reaching decibel levels. The problem is money, as the cost of upgrading the road system is far beyond the means of the central government. New roads can only be built if money can be raised somewhere else. The country is so densely populated and fiercely democratic that there are no easy solutions. India is suffering in a way that was not apparent in the nineteenth century when the United States and Russia pioneered their own great transcontinental routes and did not have to take into account legal opposition from local groups. By the end of the nineteenth century, the transcontinental railroads had transformed and revolutionized North America, and the Russian Empire had expanded into Central Asia. If Indians and Pakistanis could travel as their American counterparts do, the societies would change radically. It might not be for the better, but it would represent a revolution on the subcontinent.

22. Gregory Staple and Zachary Schrag, *TeleGeography 1995: Global Telecommunications Traffic Statistics and Commentary* (Washington, D.C.: Telegeography, 1995), p. xi.

23. These facts are taken from *Telecommunications Map of the World* (London and Washington, D.C.: Petroleum Economist Ltd. and Telegeography, 1996).

24. Ibid. For details on the extraordinary technology that will permit this revolution, see also Staple and Schrag, *TeleGeography 1995,* p. 87.

25. Ibid., pp. 66–73.

Appendix 1

1. Claudia Rosett and Allana Sullivan, "Conoco Tests the Tundra for Oil Profits," *Wall Street Journal,* September 1, 1994, p. A6.

2. IEA, *World Energy Outlook 1994,* p. 72.

3. Margaret Shapiro, "Winter Lashes Fuel-Short Former Soviets, School, Business Limp to a Shivering Halt," *Washington Post,* December 7, 1993, p. A1.

4. Estimate by Cambridge Energy Associates.

5. Martha M. Hamilton, "In the Gulf of Mexico, Drill Waters Run Deep: Vast Untapped Reservoirs Have Oil Experts Gushing," *Washington Post,* November 1, 1996, p. 1.

6. IEA, *World Energy Outlook 1995,* p. 98.

7. Congress of the United States, OTA, *U.S. Oil Import and Vulnerability Technical Replacement Capability, OTA-E-503* (Government Printing Office, October 1991), pp. 16 and 18.

8. IEA, *World Energy Outlook 1994,* pp. 42–43.

9. IEA, *World Energy Outlook 1995,* p. 99.

10. Tamsin Carisle, "Natural Gas Drilling Is Booming North of the Border," *Wall Street Journal,* August 4, 1994, p. B4.

11. IEA, *World Energy Outlook 1995,* p. 116.

12. Ibid., pp. 140–41.

13. *Energy Information Administration/International Energy Outlook 1995,* p. 30.

14. The figure is average for January–May 1994, U.S. Department of Energy, in *Wall Street Journal,* August 2, 1994, p. 1.

15. IEA, *World Energy Outlook 1994,* p. 46.

16. Mamdouh G. Salameh, "China, Oil, and the Risks of Regional Conflict," *Survival,* Winter 1995–96, p. 139.

17. Rigoberto Tiglao, "Troubled Waters, Philippine Offshore Oil Search Roils China," *Far Eastern Economic Review,* June 30, 1994, p. 20.

18. "Brunei: Political Background," *Economist* Intelligence Unit Country Profiles, November 1, 1995.

19. "Far East: Offshore Gas Developments Dominate Activity," *World Oil,* August 1995, p. 131.

20. Central Intelligence Agency, *The World Factbook 1995,* p. 199.

21. "Far East: Offshore Gas Developments Dominate Activity."

22. Michael Di Cicco, "Indonesia Plans Natural Gas Grid," United Press International, July 6, 1995.

23. IEA, *World Energy Outlook 1995,* p. 30.

24. Ibid., p. 223.

25. FBIS-WEU-94–147, August 1, 1994, "Oil Firm To Invest in Libya Despite UN Sanctions," p. 43.

26. IEA, *World Energy Outlook 1996,* p. 23.

27. FDCH, Congressional Testimony before the Committee on Energy and Natural Resources, March 16, 1994.

28. IEA, *World Energy Outlook 1994,* p. 29.

29. Robert T. Enyon, "Will Nuclear Power Be a Large Part of the Energy Future?" Energy Information Administration, presented to the Conference Energy and National Security in the 21st Century, November 10, 1994, Washington D.C.

30. IEA, *World Energy Outlook 1994,* p. 30.

31. Patrick Tyler, "Huge Water Project Would Supply Beijing By 860-Mile Aqueduct," *New York Times,* July 19, 1994, p. A8.

32. "Survey of Energy," *Economist,* June 18–24, 1994, p. 15.

33. International Energy Agency, *Global Energy: The Changing Outlook* (Paris: OECD, 1992), p. 78.

34. Reuters World Service, March 22, 1994

35. International Energy Agency, *World Energy Outlook* (Paris: OECD, 1995), p. 119.

36. Bob Davis, a battery developing engineer for Chrysler, stated that "the timing does not coincide with the mandates," BNA State Environment Daily (c), BNA, Inc.

37. "Solectria Force", a brochure about the Northern Virginia's Electric Vehicle published by Virginia Power.

38. William J. Cook, "Look, Mom, No Gas," *U.S. News and World Report,* September 30, 1996, pp. 52–54.

39. Matthew Wald, "Business Technology; A Battery Built on Wheel Power," *New York Times,* June 29, 1994, p. D1.

40. Matthew Wald, "Flywheel to Power Vehicles," *New York Times,* June 22, 1994, p. D2.

41. Facsimile from National Alternative Fuels Headline for Transportation and Technologies, Arlington, Va., October 10, 1994.

42. Office of Transportation Technologies, Department of Energy, "Technology Assessments of Advanced Energy Storage Systems for Electric and Hybrid Vehicles," April 30, 1993, p. ES-2.

43. Ibid., p. ES-3.

44. Matthew Wald, "Flywheel to Power Vehicles."

45. Matthew Wald, "Chrysler's Electric Race Car Has Turbine and Flywheel," *New York Times,* January 6, 1994, p. D17.

46. S. C. Gwynne, "What's Driving the Rosen Boys?" *Time,* September 23, 1996, pp. 50–52.

47. The forecast is by Dr. Kerry Gravatt, director of the Commerce Department's Office of Technology Commercialization, in *Energy Report,* March 28, 1994.

48. The majority of the 6.5 million AFVs in the world are either the propane-powered vehicles (3.5 million), or ethanol-powered ones (2.5 million). Alternative Energy Network Online Today, July 1, 1994.

49. *Oil and Gas Journal,* July 4, 1994, p. 28.

50. "Pickens: It's make-or-brake time for NGVs," *Gas Daily,* April 22, 1994.

51. "Propane and CNG Expected to Dominate AFV Market by 2010 with 45% Each," Alternative Energy Network, July 1, 1994.

52. "NGV Facts", EDO Corp. Energy Division, 1993.

53. "A Comparison of Alternative Vehicular Fuels with Conventional Gasoline," American Gas Association, Policy and Analysis Issues, August 15, 1991, Table 1.

54. "Gasoline is the 'fuel of the future,' says API president," Alternative Energy Network Online, April 1, 1994; excerpts from a March 28 Houston Chronicle editorial by Charles DiBona, president and CEA of the API.

Appendix 2

1. Patrick O'Sullivan, "A Geographical Analysis of Guerrilla Warfare," *Political Geography Quarterly,* vol. 2, no. 2 (April 1983), pp. 139–50.

2. Patrick O'Sullivan, "The Geography of Wars in the Third World," in Stephanie G. Neuman and Robert E. Harkavy, eds., *The Lessons of Recent Wars in the Third World: Comparative Dimensions* (Lexington, Mass.: D. C. Heath, 1987), p. 49.

Appendix 3

1. J. Mohan Malik, "India's Response to the Gulf Crisis," *Asian Survey,* September 1991, p. 847.

2. See "Indian Workers' Gulf Remittances Exceed $2.2 b," United Press International, April 26, 1994; S. Venkat Narayan, "2,100 Indians Died in Gulf Last Year," *Moneyclips,* May 13, 1994; John Eckhouse, "Migrant Workers' Economic Impact," *San Francisco Chronicle,* July 1, 1991.

3. Randall Palmer, "Arab Ban on Planes, Ships Isolates Indians in Gulf," *Reuters European Business Report,* September 29, 1994; Malik, "India's Response to the Gulf Crisis," p. 847.

4. "More Jockeying as Gulf LNG, Pipeline Talks Gather Pace," *Power Asia,* April 28, 1995.

5. "Qatar Oil Minister Heads for India, China," *Reuters European Business Report,* March 23, 1995.

6. "Bahrain Probing Deal With 'Waking Giant' Says Shirawi; GCC 'Must Bolster Relations With India,'" *Moneyclips,* April 13, 1995.

7. "Gulf-India Ties Improving, Says Report from New Delhi," *Moneyclips,* April 29, 1995.

8. Iftikhar H. Malik, "Pakistan's National Security and Regional Issues," *Asian Survey,* December 1994, p. 1077.

9. "Electric Power Generation Markets in India and Pakistan," U.S. Agency for International Development, 1993.

10. "They Can't Let Go," *Economist,* January 21, 1995, India Survey, p. 20.

11. John Burns, "As India's Economy Fattens, Feeding Is Frenzied," *New York Times,* March 24, 1995, which uses the conservative 150 million figure, while "Hello World," *Economist,* January 21, 1995, is more generous with its estimate of 250 million.

12. Sudeep Chakravarti, "The Middle Class: Hurt But Hopeful," *India Today,* April 15, 1995; John Thor-Dahlburg, "Taking the Pulse of India's Villages," *Los Angeles Times,* May 23, 1994; "The New Indians: An Ambitious, Self-Assured Generation Rewrites the Old Rules," *Asiaweek,* April 14, 1995.

13. "The New Indians."

14. Chakravarti, "The Middle-Class: Hurt But Hopeful," p. 92.

15. Thor-Dahlburg, "Taking the Pulse of India's Villages," p. 1.

16. Chakravarti, "The Middle-Class: Hurt But Hopeful," p. 93.

17. Michael Schuman, "Power Hungry," *Forbes,* April 24, 1995, p. 162.

18. "They Can't Let Go."

19. Sunil Jain, "An Area of Darkness," *India Today,* May 15, 1995, p. 62.

20. Ibid.

21. "They Can't Let Go."

22. At one point Indian politics threatened to force Enron to abandon the Dabhol project. The domestic furor over the Enron project began with the ruling Congress (I) Party's electoral loss in Maharashtra in March 1995. The victorious BJP-Shiv Sena coalition suspected irregularities in the deal and established a four-member committee to review the terms of the project. Critics argued that the deal had imposed all sorts of one-sided obligations on the Indian government without corresponding obligations on Enron's part: Enron was guaranteed one and a half times the fees received by public-sector power plants, as well as an extremely high return on its equity, and guarantees from the state and central governments to backstop any dues that the Maharashtra State Electricity Board (MSEB) might fail to pay. The company in turn argues that these guarantees were necessary, given the woefully inept functioning of the state electricity boards and the high likelihood that they would default on payments. Once in power, the BJP and the Shiv Sena softened their adversarial stances, and there was a fair review of the case. As of August 1996 it was back on track albeit with new arrangements on costs. The Enron

case nonetheless provides the "grittiest test of India's liberalization programme" (see "They Can't Let Go"), symbolizing the classic tension between socialist self-reliance and free-market principles, and serving as a barometer of India's viability to investors around the globe.

23. See "India's Industry: Congestion Floods Car Sector," *Global Financial Markets,* April 19, 1995; "U.S. Auto Parts Firm Allied Signal Announces Joint Venture in India," *Agence France Presse,* May 4, 1995; "Demand for Cars Rides a Boom in India's Economy," *Los Angeles Times,* September 29, 1994; Matthew E. Kahn, "Growing Car Ownership in LDC's: The Impact on the Environment and Trade," *Columbia Journal of World Business,* December 22, 1994, p. 12.

24. D. Patrick Maley, "A Gulf-India Marriage Made in Heaven?" United Press International, November 11, 1994.

25. "Demand for Cars Rides Boom in India's Growing Economy," *Los Angeles Times,* September 29, 1994, part D, p. 5.

26. Ibid.

27. See Mark Nicholson, "India Urged to Speed Rail Reforms," *Financial Times,* April 10, 1995, p. 6; Neelesh Misra, "Indian Industry Urges Fuel Taxes to Pay for Roads," *Reuters Asia-Pacific Business Report,* April 10, 1995.

Appendix 4

1. Seymour Hersh, *The Samson Option: Israel's Nuclear Arsenal and American Foreign Policy* (Random House, 1991); and William E. Burrows and Robert Windrem, *Critical Mass* (Simon and Schuster, 1994). Among other relatively recent works on the Israeli nuclear program, see Shai Feldman, *Israeli Nuclear Deterrence: A Strategy for the 1980s* (Columbia University Press, 1982); Louis Rene Beres, ed., *Security or Armageddon: Israel's Nuclear Strategy* (Lexington, Mass.: D. C. Heath, 1986); Yair Evron, *Israel's Nuclear Dilemma* (Cornell University Press, 1994); and Frank Barnaby, *The Invisible Bomb: The Nuclear Arms Race in the Middle East* (London: I. B. Tauris, 1989). Evron's book takes the position, contrary to Waltz, Rosen, et al., that nuclear weapons in the Middle East are likely to be destabilizing in effect.

2. "Revealed: The Secrets of Israel's Nuclear Arsenal," *Sunday Times* (London), October 5, 1986. Interestingly, Burrows and Windrem have no information on possible Israeli U-235 production facilities. Hersh, p. 201, refers to evidence of a laser isotope separation facility on the basis of the Vanunu revelations.

3. The possible Israeli possession of neutron weapons was bruited by the Vanunu revelations, and it would make sense in the context of possible tactical uses in the Sinai or Golan areas against armored formations. Robert Harkavy was once told by a prominent U.S. physicist, here unnamed, that he was quite sure the "flash over the ocean" was an Israeli neutron weapon test. See also Hersh, *The Samson Option,* pp. 199–200.

4. This was earlier speculated upon in Robert E. Harkavy, *Spectre of a Middle Eastern Holocaust: The Strategic and Diplomatic Implications of the Israeli Nuclear Weapons Program,* Monograph Series in World Affairs (University of Denver, 1977). If Israel actually did have operational nuclear weapons at the outset of the

1967 war, as is apparently believed by Burrows and Windrem and others, that might imply that such weapons were derived from the theft of weapons-grade U-235 from a plant in Apollo, Pennsylvania—a possibility that was extensively investigated by the U.S. government. But Hersh denies the validity of this claim, leaving open the question of whether Israel, contrary to U.S. government estimates, had succeeded in separating Pu-239 at Dimona before the 1967 war or had received weapons-grade material from France.

5. Hersh, *The Samson Option,* p. 222; and Insight Team of the London *Sunday Times, The Yom Kippur War* (Doubleday, 1974), pp. 282–84.

6. Hersh, *The Samson Option,* pp. 198–99. Interestingly, despite the ambiguous and unannounced nature of the Israeli program, the widely used source book, *The Military Balance* (London: Brassey's for IISS, annual) has now taken to assuming nuclear weapons as part of Israel's order of battle. In its 1994–95 edition, it notes that "it is widely believed that Israel has a nuclear capability with up to 100 warheads."

7. "Revealed: The Secrets of Israel's Nuclear Arsenal," *Sunday Times* (London), October 5, 1986.

8. Anthony Cordesman, "Weapons of Mass Destruction in the Middle East," Washington, D.C., Center for Strategic and International Studies, January 22, 1996, pp. 7.

9. See, for instance, Leonard Spector, *The Undeclared Bomb* (Cambridge, Mass.: Ballinger, 1988), pp. 128–48.

10. Burrows and Windrem, *Critical Mass,* chapters 2 and 11.

11. Ibid., pp. 79–80.

12. "Nuclear Ambitions," *U.S. News & World Report,* February 12, 1996, pp. 42–46.

13. Burrows and Windrem, *Critical Mass.* See also Spector, *The Undeclared Bomb,* pp. 80–117, who provides estimates (pp. 92–93) on India's potential annual plutonium production and nuclear weapons potential.

14. Burrows and Windrem, *Critical Mass,* p. 357.

15. "Nuclear Ambitions."

16. Burrows and Windrem, p. 374.

17. "Saddam Spills Secrets," *Time,* September 4, 1995, p. 41. According to this article, "before the Kuwait invasion, the CIA concluded in a secret report that Iraq could cobble together at least one crude nuclear device and detonate it in the western Iraqi desert in a demonstration explosion to shake the teeth of Saddam's Arab neighbors." See also William Safire, "Saddam Wins Again," *New York Times,* October 16, 1995, p. A15; and "Experts Doubt Iraq's Claims on A-Bomb," *New York Times,* August 30, 1995, p. A6.

18. Burrows and Windrem, *Critical Mass,* pp. 275–79; and Amos Perlmutter, *Two Minutes Over Baghdad* (London: Corgi Press, 1982).

19. Burrows and Windrem, *Critical Mass,* p. 26. See also Spector, *The Undeclared Bomb,* pp. 207–18.

20. Burrows and Windrem, *Critical Mass,* p. 45.

21. Cordesman, "Weapons of Mass Destruction in the Middle East," p. 14.

22. In the spring of 1995 there was a flurry of articles about Iran's nuclear program and the U.S. effort at blocking technology transfers from Russia and

elsewhere. See, among others, Steve Coll, "The U.S. Case Against Iranian Nukes," *Washington Post National Weekly Edition,* May 15–21, 1995, p. 23; Thomas W. Lippman, "No More Mr. Nice Guy," *Washington Post National Weekly Edition,* May 15–21, 1995, p. 18; Elaine Sciolino, "Iran's Nuclear Goals Lie in Half-Built Plant," *New York Times,* May 19, 1995, p. A1; "Republicans Warn Russia That Its Deal with Iran Threatens Aid," *New York Times,* May 8, 1995, p. A7; and Kenneth Timmerman, "Iran's Nuclear Menace," *The New Republic,* vol. 212, no. 17 (April 24, 1995), pp. 17–19. See also Burrows and Windrem, *Critical Mass,* pp. 340–43.

23. Burrows and Windrem, *Critical Mass,* p. 342.

24. Cordesman, "Weapons of Mass Destruction in the Middle East," p. 10.

25. Burrows and Windrem, *Critical Mass,* pp. 324–27. Here it is reported that Qaddafi earlier offered India $18 billion for nuclear bomb technology. See also Spector, *The Undeclared Bomb,* for information on Libya's nuclear dealings with India, the Soviet Union, Belgium, Brazil, and Argentina.

26. Burrows and Windrem, *Critical Mass,* pp. 321–24.

27. Gordon M. Burck and Charles C. Flowerree, *International Handbook on Chemical Weapons Proliferation* (New York: Greenwood Press, 1991), p. 221, wherein it is stated that Egypt used CW agents as well as tear gas in the 1963–67 war in Yemen.

28. Ibid., pp. 223–24, in which press reports about Egyptian stocks of Soviet nerve gas in the Sinai are contested. This source says that "any warfare chemicals in the Sinai were probably only smoke and incapacitant" but cites conflicting evidence.

29. Ibid., pp. 341–55. See also Sterling Seagrave, *Yellow Rain* (New York: M. Evans, 1981).

30. Burck and Flowerree, pp. 85–151. See also Burrows and Windrem, *Critical Mass,* pp. 47, 95, 127, 132, and 216; and Cordesman and Wagner, *The Lessons of Modern War,* vol. 2 (Boulder, Colo.: Westview Press, 1970), pp. 506–18.

31. Burck and Flowerree, *International Handbook on Chemical Weapons Proliferation,* pp. 470–72.

32. Ibid., chapter 4. Another useful recent general source is Edward M. Spiers, *Chemical and Biological Weapons: A Study of Proliferation* (St. Martin's Press, 1994).

33. Burck and Flowerree, *International Handbook on Chemical Weapons Proliferation,* p. 189.

34. Burrows and Windrem, *Critical Mass,* pp.46–47.

35. Cordesman, "Weapons of Mass Destruction in the Middle East," p. 12.

36. Burrowes and Windrem, *Critical Mass,* p. 47.

37. Ibid., p. 49.

38. Cordesman, "Weapons of Mass Destruction in the Middle East," p. 13.

39. Burck and Flowerree, *International Handbook on Chemical Weapons Proliferation,* p. 209.

40. Ibid., p. 210.

41. Cordesman, "Weapons of Mass Destruction in the Middle East," p. 9.

42. James Morrison and Martin Sieff, "Israel Plans Attack on Nerve Gas Plant if Syrians Threaten," *Washington Times,* January 11, 1988, p. A1. The Israeli

perspective on Syrian chemical weapons is discussed in Burck and Floweree, pp. 209–11.

43. Ibid., p. 211 and pp. 213–14.

44. Rabta is discussed in Burck and Flowerree, *International Handbook on Chemical Weapons Proliferation*, pp. 274–305; and Burrows and Windrem, *Critical Mass*, pp. 223–30. (Safire's allegation is noted on p. 225.)

45. Burck and Flowerree, p. 275.

46. Ibid., pp. 222–36.

47. Burrows and Windrem, *Critical Mass*, p. 340.

48. Burck and Flowerree, *International Handbook on Chemical Weapons Proliferation*, p. 245.

49. Anthony Cordesman, *After the Storm: The Changing Military Balance in the Middle East* (Boulder, Colo.: Westview Press, 1993), pp. 420–21.

50. Cordesman, "Weapons of Mass Destruction in the Middle East," p. 9.

51. Ibid., p. 10.

52. See "Blast Probers Say Trail Leading to Iraq," *Boston Globe,* February 10, 1995, p. 1.

Appendix 5

1. For assessment of Iran's current military doctrines and capabilities, see Shahram Chubin, *Iran's National Security Policy: Capabilities, Intentions and Impact* (Washington, D.C.: Carnegie Endowment, 1994); and Michael Eisenstadt, *Iran's Military Capabilities and Intentions* (Washington, D.C.: The Washington Institute for Near East Policy, 1996).

2. James P. Thomas, "Iranian Military Lessons Learned from the Gulf War," unpublished report for the Center for National Security Studies, Los Alamos National Laboratory, New Mexico, 1993.

3. Anthony Cordesman, *After the Storm: The Changing Military Balance in the Middle East* (Boulder, Colo.: Westview Press, 1993), pp. 408–10.

4. See Robert Harkavy, *Bases Abroad* (Oxford: Oxford University Press, 1989), p. 96.

5. Ephraim Karsh, "Gulf War Lessons: The Case of Iraq," unpublished report for the Center for National Security Studies, Los Alamos National Laboratory, New Mexico, 1993, p. 13.

6. William Safire, "Saddam Wins Again," *New York Times,* October 16, 1995, p. A15.

7. Karsh, "Gulf War Lessons," p. 21.

8. "IAF Confronts Modernization," *Aviation Week and Space Technology,* July 25, 1994, p. 51.

9. "India Initiates Military Intelligence Overhaul Program," *Space News,* September 4–10, 1995, p. 9.

10. James P. Thomas, "Indian Military Lessons Learned from the Gulf War," unpublished report for the Center for National Security Studies, Los Alamos National Laboratory, New Mexico, 1993, p. 3.

11. Ibid., pp. 4, 5, 6, and 9.

12. Ibid., p. 11.

13. Ibid., pp. 11–12.

14. Ibid., pp. 24–25.

15. Ibid., p. 19.

16. See Graham Kinahan, "Pakistan and Lessons From the Gulf War," unpublished paper for the Center for National Security Studies, Los Alamos National Laboratory, New Mexico, circa 1993.

17. Martin van Creveld, *Nuclear Proliferation and the Future of Conflict* (New York: Free Press, 1993).

Index

Page numbers for entries occurring in figures are suffixed with an f; those for entries occurring in maps, with an m; those for entries occurring in notes, with an n; and those occurring in tables, with a t.

About the Authors

Geoffrey Kemp is the Director of Regional Strategic Programs at the Nixon Center for Peace and Freedom. He served in the White House during the first Reagan administration and was Special Assistant to the President and Senior Director for Near East and South Asian Affairs on the National Security Council staff. Prior to assuming his current position, he was a senior associate at the Carnegie Endowment for International Peace, where he was Director of the Middle East Arms Control Project. In the 1970s, he worked in the Defense Department and for the U.S. Senate Committee on Foreign Relations. He was also on the faculty of the Fletcher School of Law and Diplomacy, Tufts University. Dr. Kemp is the author and editor of several books on U.S. policy in the Middle East and South Asia.

Robert E. Harkavy is a Professor of Political Science at the Pennsylvania State University and is also associated with the University's Institute for Policy Research and Evaluation. He previously served with the U.S. Atomic Energy Commission and the U.S. Arms Control and Disarmament Agency. He also was a visiting research professor at the U.S. Army War College and is a long-time consultant to the Department of Defense. Dr. Harkavy has been a Fulbright scholar in Sweden, an Alexander von Humboldt scholar in Germany, and a visiting lecturer in Japan. He has authored and edited several previous books on the arms trade, overseas bases, nuclear proliferation, third world security policies, and U.S. national security policy.

THE CARNEGIE ENDOWMENT FOR
INTERNATIONAL PEACE

The Carnegie Endowment for International Peace was established in 1910 in Washington, D.C., with a gift from Andrew Carnegie. As a tax-exempt operating (not grant-making) foundation, the Endowment conducts programs of research, discussion, publication, and education in international affairs and U.S. foreign policy. The Endowment publishes the quarterly magazine, *Foreign Policy.*

Carnegie's senior associates—whose backgrounds include government, journalism, law, academia, and public affairs—bring to their work substantial first-hand experience in foreign policy through writing, public and media appearances, study groups, and conferences. Carnegie associates seek to invigorate and extend both expert and public discussion on a wide range of international issues, including worldwide migration, nuclear nonproliferation, regional conflicts, multilateralism, democracy-building, and the use of force. The Endowment also engages in and encourages projects designed to foster innovative contributions in international affairs.

In 1993, the Carnegie Endowment committed its resources to the establishment of a public policy research center in Moscow designed to promote intellectual collaboration among scholars and specialists in the United States, Russia, and other post-Soviet states. Together with the Endowment's associates in Washington, the center's staff of Russian and American specialists conduct programs on a broad range of major policy issues ranging from economic reform to civil-military relations. The Carnegie Moscow Center holds seminars, workshops, and study groups at which international participants from academia, government, journalism, the private sector, and nongovernmental institutions gather to exchange views. It also provides a forum for prominent international figures to present their views to informed Moscow audiences. Associates of the center also host seminars in Kiev on an equally broad set of topics.

The Endowment normally does not take institutional positions on public policy issues. It supports its activities principally from its own resources, supplemented by nongovernmental, philanthropic grants.

Carnegie Endowment for
International Peace
2400 N Street, N.W.
Washington, D.C. 20037
Tel: (202) 862-7900
Fax: (202) 862-2610

Carnegie Moscow Center
Mosenka Plaza
24/27 Sadovaya Samotechnaya
103051 Moscow, Russia
Tel: (7-095) 258-5025
Fax: (7-095) 258-5020